D0078880

FUNDAMENTALS OF METAL FATIGUE ANALYSIS

Julie A. Bannantine, Ph.D.
University of Illinois

Jess J. Comer, Ph.D.
South Dakota School of Mines

James L. Handrock, Ph.D.
University of Illinois
Currently at Sandia National Labs.

Prentice Hall
Englewood Cliffs, New Jersey 07632

Library of Congress Cataloging-in-Publication Data

BANNANTINE, JULIE.
Fundamentals of metal fatigue analysis/Julie Bannantine, Jess Comer, James Handrock.

p. cm.
Includes index.
ISBN 0-13-340191-X

1. Metals—Fatigue. I. Comer, Jess. II. Handrock, James.
III. Title
TA460.B23 1990 88-38653
620.1'63—dc19 CIP

Editorial/production supervision and
interior design: Jennifer Wenzel
Cover design: Ben Santora
Manufacturing buyer: Mary Noonan

A Division of Simon & Schuster
Englewood Cliffs, New Jersey 07632

Printed in the United States of America
10 9 8 7 6 5 4 3

ISBN 0-13-340191-X

Prentice-Hall International (UK) Limited, *London*
Prentice-Hall of Australia Pty. Limited, *Sydney*
Prentice-Hall Canada Inc., *Toronto*
Prentice-Hall Hispanoamericana, S.A., *Mexico*
Prentice-Hall of India Private Limited, *New Delhi*
Prentice-Hall of Japan, Inc., *Tokyo*
Simon & Schuster Asia Pte. Ltd., *Singapore*
Editora Prentice-Hall do Brasil, Ltda., *Rio de Janeiro*

CONTENTS

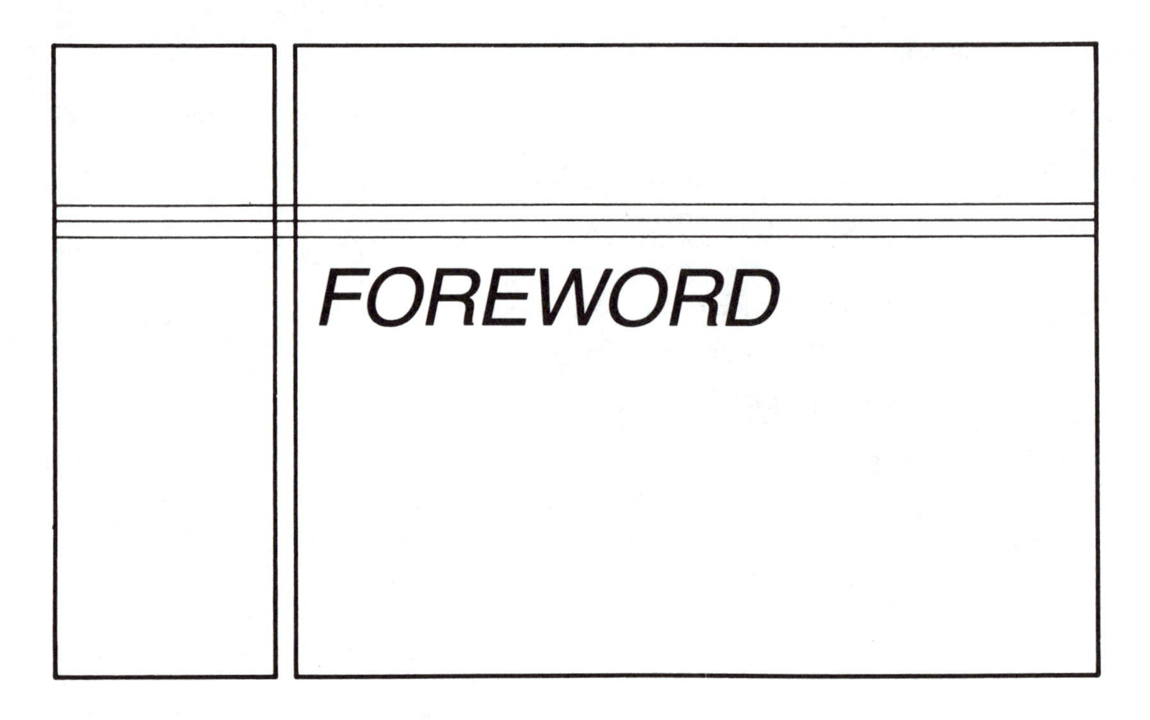

FOREWORD

Fatigue of metals has been studied for over 150 years. August Wöhler, while not the first, is one of the more famous early fatigue researchers. During the period from about 1850 to 1875, experiments were conducted to establish a safe alternating stress below which failure would not occur. Full scale axles as well as smaller laboratory specimens were employed to establish the endurance limit concept for design. Nearly one hundred years of research has been performed to experimentally establish the effects of the many variables that influence the long life fatigue strength of metals.

Bauschinger (Circa 1885) developed a mirror extensometer with the sensitivity to measure one microstrain and for many years studied the relationship between small inelastic strains and the safe stress in fatigue. He believed in a natural elastic limit (measured in cyclic tests) below which fatigue would not occur. Unstrained material exhibited a primitive elastic limit that was not equal to the natural elastic limit. Today we recognize this phenomena as the difference between the monotonic and cyclic yield strength of the material.

In 1903 Ewing and Humphrey, motivated by the work of Wöhler and Bauschinger, published their classic paper entitled "The Fracture of Metals under Repeated Alterations of Stress." Flat fatigue specimens made from high quality Swedish iron were tested in the annealed condition. Optical microscopy was employed to examine the same region of the specimen at various stages of the fatigue life. They stated, "The course of the breakdown was as follows: The first examination, made after a few reversals of stress, showed slip-lines on some of the crystals, the slip-lines were quite similar in appearance to those which are seen when a simple tensile stress exceeding the elastic limit is applied. After many reversals they changed into comparatively wide bands with rather hazily defined edges. As the number of reversals increased this process of broadening continued, and some parts of the surface became almost covered with dark markings. When this stage was reached it was found that some of the crystals had cracked. The cracks occurred along broadened slipbands. In some instances they were first seen on a single crystal, but soon they joined up from crystal to crystal, until finally a long continuous crack was developed across the surface of the specimen." They also observed "Once an incipient crack begins to form across a certain set of crystals, the effect of further reversals is mainly confined to the neighborhood of the crack." Thus, at the turn of the century, to-and-fro slip was established as the cause of fatigue damage. Later work using electron microscopy, x-ray and other powerful tools, has provided further substantiation that the basic cause of fatigue crack nucleation is the result of alternating shear stresses and strains.

Jenkin, in 1923, used what must be the first spring-slider model for simulating the stress strain behavior of metals. By using several parallel elements, Jenkin was able to simulate many of the complex hysteresis loops that were previously reported in the literature by Smith and Wedgewood. In describing this work, Jenkin notes that "about six months ago I wrote a paper . . . and made a model to illustrate a small point in it. It grew too strong for me and took command, and for the last six months I have been its obedient slave—for the model explained the whole of my subject—Fatigue." The importance of cyclic deformation was clearly established in 1923 but largely ignored until forty years later.

At about this time Griffith published his classical paper on fracture. It is not widely appreciated that Griffith was motivated in his studies by the fatigue problem. He acknowledges Professor Jenkin, at whose request the work was commenced. Griffith's work showed that the last cycle of fatigue was nothing more than brittle fracture caused by cyclic growth of a fatigue crack to an unstable length. Little was known or written about the manner in which the nucleated fatigue crack grew to catastrophic proportions. Virtually nothing quantitative was done on this problem of crack propagation until forty years later by Paris.

In 1927, Moore and Kommers published their book, "The Fatigue of Metals." H. F. Moore worked for many years on a number of practical fatigue problems, especially those relating to the railroad industry. This book and the work of Moore and his associates had a large effect on fatigue design, testing, etc., in the United States. Moore was responsible for organizing an ASTM Committee on Fatigue Research which later grew into Committee E-9 on Fatigue. The SAE Committee on Fatigue Design and Evaluation has its origin in a "counter movement" of engineers, who wanted a group that was more practical than the ASTM Committee on Fatigue.

Research in fatigue during the 1930's and 1940's was largely devoted to experimentally establishing the *effects* of the many factors that influence the long-life fatigue strength. Tests were usually conducted in rotating bending and the life range of interest was about 10^6 cycles and greater.

Coffin and Manson began their work during the 1950's and established quantitative relationships between plastic strain and fatigue life. Both were motivated by problems of fatigue of metals at high temperatures where inelastic strain cannot be ignored.

Many significant contributions were made during the 1960's. Irwin and others pioneered the development of fracture mechanics as a practical engineering tool. Paris quantified the relationships for fatigue crack propagation. Paris in "Twenty Years of Reflection on Questions Involving Fatigue Crack Growth" comments on his original work. "The paper was quite promptly rejected by three leading journals, whose reviewers uniformly felt that it is not possible that an elastic parameter such as K can account for the self-evident plasticity effects in correlating fatigue crack growth rates." Smooth specimen simulations of notches and cycle counting methods for variable amplitude loading were developed. Much of this work was made possible by the introduction of closed-loop materials testing systems and the introduction of the digital computer for solving engineering problems.

By the 1970's fatigue analysis became an established engineering tool in many industrial applications. Despite all of this knowledge, unintended fatigue failures continue to occur. More research will not solve most of these problems. More education will. Many of the failures are a result of fatigue technology being in the hands of the "experts" rather than the people who design and build structures and components. This book should help alleviate this problem. It focuses on the various analytical techniques available to engineers to assess the fatigue resistance of metallic structures and components. The book is written on the premise that there is no general fatigue analysis that is best in all design situations. Each technique has its own advantages and limitations and a selection must be based on material, loading history, service environment, component geometry and consequences of component or structural failure. I believe that the authors have considerable insight into the details of fatigue life assessment techniques and have explained them in a manner that is easily understood. This book should be a valuable teaching and reference tool.

Darrell Socie
Professor of Mechanical Engineering
University of Illinois at Urbana-Champaign

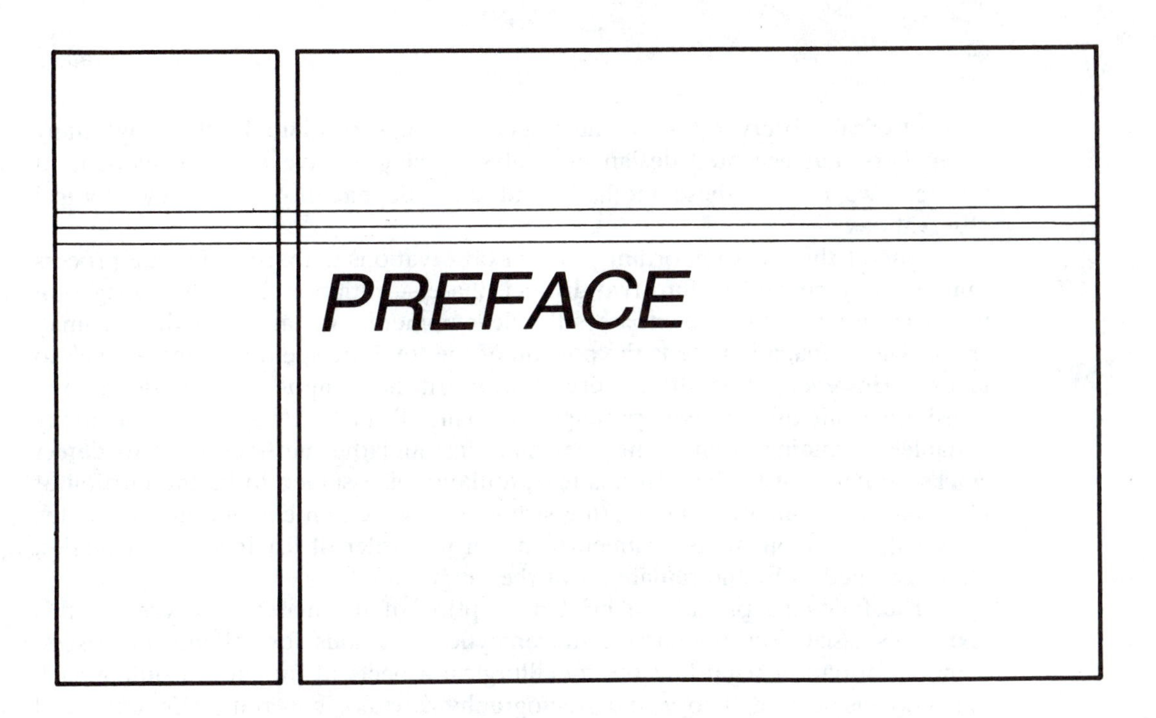

Metal fatigue is a process which causes premature failure or damage of a component subjected to repeated loading. It is a complicated metallurgical process which is difficult to accurately describe and model on the microscopic level. Despite these complexities, fatigue damage assessment for design of components and structures must be made. Consequently, fatigue analysis methods have been developed. It is the intent of this book to cover the primary analytical methods to quantify fatigue damage. It is hoped that this book will provide the student or engineer with information and background that will enable them to become proficient at these analytical methods in an effort to design against fatigue damage.

Three primary fatigue analysis methods are presented in this text. These are the stress-life approach, the strain-life approach, and the fracture mechanics approach. These methods have their own region of application with some degree of overlap between them. The understanding of any one of these methods provides a technique which may be used to perform a fatigue analysis. However, it is the insights gained from the understanding of all three methods which allow the engineer to choose the method or methods that are most appropriate for the given problem.

Historically two over-riding considerations have promoted the development of fatigue analysis methods. The first has been the need to provide designers and engineers methods that are practical, easily implemented, and cost effective. The second consideration has been the need to reconcile these analytical approaches with physical observations. It has been through continued effort by many researchers that accepted design or analysis practices have been developed. In this text we discuss these methods and their dependence upon key physical observations.

One of the most important physical observations is that the fatigue process can generally be broken into two distinct phases—initiation life and propagation life. The initiation life encompasses the development and early growth of a small crack. The propagation life is the portion of the total life spent growing a crack to failure. However, it is often very difficult, if not impossible, to define the transition from initiation to propagation. This distinction depends upon many variables, including component size, material, and the methods used to detect cracks, to name just a few. In this text, initiation is assumed to be the portion of life spent developing an engineering size crack. (Often an engineering size crack for smaller components is assumed to be on the order of 0.1 inch.) Propagation life is assumed to be the remainder of the fatigue life.

The following provides a brief description of the material covered in this text. This discussion emphasizes the analytical methods for fatigue life assessment. Information regarding the metallurgical aspects of fatigue, including such topics as dislocation theory and fractography, are not covered. Also advanced topics such as environmental and high temperature applications are not presented. Instead the reader is referred to texts related to these subjects.

Chapter 1 covers the stress-life approach, which was the first fatigue analysis method to be developed. It is used mainly for long life applications where stresses and strains are elastic. It does not distinguish between initiation and propagation, but deals with total life, or the life to failure of a component.

Chapter 2 covers the strain-life method which was developed in the 1960's. This method is usually considered an initiation approach. It is used when the strain is no longer totally elastic, but has a plastic component. Short (low cycle) fatigue lives generally occur under these conditions.

The fracture mechanics method is presented in Chapter 3. This method is based upon linear elastic fracture mechanics (LEFM) principles which are adapted for cyclic loading. This method is used to predict propagation life from an initial crack or defect. This method is also used in combination with the strain-life approach to predict a total (i.e. initiation and propagation) life.

The fundamental concepts of these three methods are presented in Chapters 1 through 3. Techniques used to apply these methods to the fatigue analysis of notched components are presented in Chapter 4. Chapter 5 presents methods of application for variable amplitude loading. In Chapter 6, a general comparison of these three methods is discussed. Finally, a brief introduction to the area of multiaxial fatigue is given in Chapter 7.

The approaches discussed in this book are the composite effort of many researchers and engineers over a period of years. We have made every effort to cite each individual's contribution in the references. Inadvertently some may have been omitted; this was not intentional. Any comments, suggestions, or corrections regarding this text would be gratefully accepted by the authors.

ACKNOWLEDGMENTS

JoDean Morrow, professor emeritus at the University of Illinois, is gratefully acknowledged. Many of the ideas and techniques presented in Chapters 2, 4, and 5 were developed by him or his students. His friendship and wise advice were greatly appreciated, as well as his generous support, especially allowing us access to his classnotes, homework files, and technical reports. Acknowledgment is also given to Darrell Socie, professor at the University of Illinois, for the initial idea to begin this work, as well as some general thoughts on the structure of the material covered in the book. He is also thanked for his technical advice and suggestions on the scope of the book. Early support for this work was provided by the Fracture Control Program at the University of Illinois. This program is a cooperative effort between research and industry to contribute and disseminate knowledge and procedures for the development and manufacture of safe, reliable, structural or mechanical products.

There are many of our friends and peers, especially those who were with us at the University of Illinois, who aided us with suggestions and reviews, and were supportive and encouraging. To these individuals we extend our sincere thanks.

Finally, and most importantly, our families, especially Sandy and Sarah, deserve our deepest appreciation for their patience, love, and support which allowed this book to be completed.

Julie A. Bannantine
Jess J. Comer
James L. Handrock
January, 1989

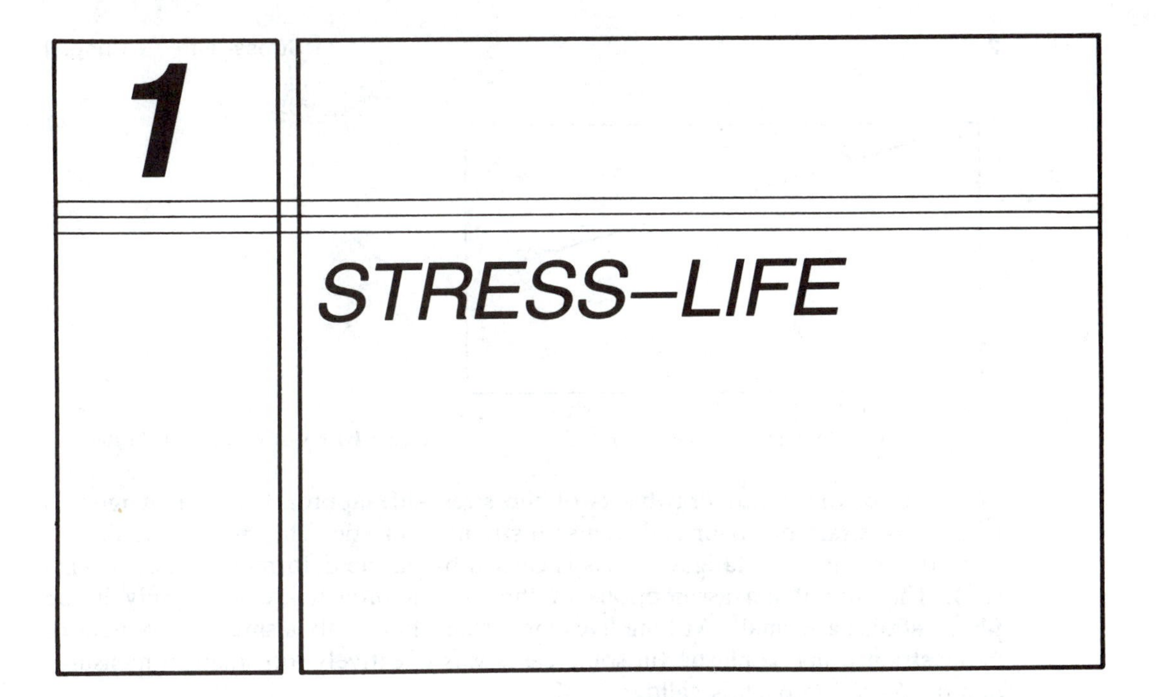

1.1 INTRODUCTION

The stress–life, *S–N*, method was the first approach used in an attempt to understand and quantify metal fatigue. It was the standard fatigue design method for almost 100 years. The *S–N* approach is still widely used in design applications where the applied stress is primarily within the elastic range of the material and the resultant lives (cycles to failure) are long, such as power transmission shafts. The stress–life method does not work well in low-cycle applications, where the applied strains have a significant plastic component. In this range a strain–based approach (Chapter 2) is more appropriate. The dividing line between low and high cycle fatigue depends on the material being considered, but usually falls between 10 and 10^5 cycles.

1.2 *S–N* DIAGRAM

The basis of the stress–life method is the Wöhler or *S–N* diagram, which is a plot of alternating stress, *S*, versus cycles to failure, *N*. The most common procedure for generating the *S–N* data is the rotating bending test. One example is the R. R. Moore test, which uses four-point loading to apply a constant moment to a rotating (1750 rpm) cylindrical hourglass-shaped specimen. This loading produces a fully reversed uniaxial state of stress. The specimen is mirror polished with a typical diameter in the test section of 0.25 to 0.3 in. The stress level at the surface of the specimen is calculated using the elastic beam equation ($S = Mc/I$) even if the resulting value exceeds the yield strength of the material.

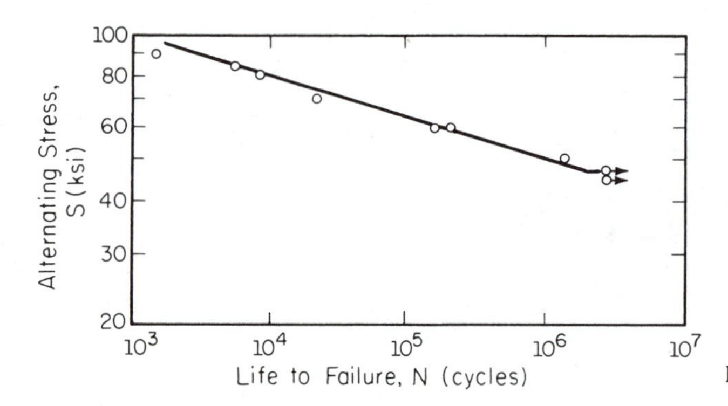

Figure 1.1 *S–N* curve for 1045 steel.

One of the major drawbacks of the stress–life approach is that it ignores true stress–strain behavior and treats all strains as elastic. This may be significant since the initiation of fatigue cracks is caused by plastic deformation (i.e., to–fro slip). The simplifying assumptions of the *S–N* approach are valid only if the plastic strains are small. At long lives most steels have only a small component of cyclic strain which is plastic (in some cases it is effectively too small to measure) and the *S–N* approach is valid.

S–N test data are usually presented on a log–log plot, with the actual *S–N* line representing the mean of the data. Certain materials, primarily body-centered cubic (BCC) steels, have an endurance or fatigue limit, S_e, which is a stress level below which the material has an "infinite" life (see Fig. 1.1). For engineering purposes, this "infinite" life is usually considered to be 1 million cycles. The endurance limit is due to interstitial elements, such as carbon or nitrogen in iron, which pin dislocations. This prevents the slip mechanism that leads to the formation of microcracks. Care must be taken when using the endurance limit since it can disappear due to:

1. Periodic overloads (which unpin dislocations)
2. Corrosive environments (due to fatigue corrosion interaction)
3. High temperatures (which mobilize dislocations)

It should be pointed out that the effect of periodic overloads mentioned above relates to smooth specimens. Notched components may have completely different behavior, due to the residual stresses set up by overloads. This is discussed more fully in Section 1.4.4.

Most nonferrous alloys have no endurance limit and the *S–N* line has a continuous slope (Fig. 1.2). A pseudo-endurance limit or fatigue strength for these materials is taken as the stress value corresponding to a life of 5×10^8 cycles.

There are certain general empirical relationships between the fatigue properties of steel and the less expensively obtained monotonic tension and hardness properties. When the *S–N* curves for several steel alloys are plotted in

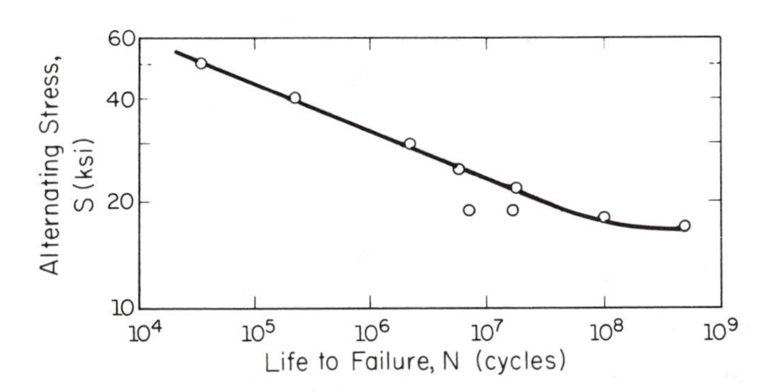

Figure 1.2 S–N curve for 2024-T4 aluminum. (Data from Ref. 1.)

nondimensional form using the ultimate strength, they tend to follow the same curve (Fig. 1.3).

The ratio of endurance limit to ultimate strength for a given material is the fatigue ratio (Fig. 1.4). Most steels with an ultimate strength below 200 ksi have a fatigue ratio of 0.5. It should be noted that this ratio can range from 0.35 to 0.6. Steels with an ultimate strength over 200 ksi often have carbide inclusions formed during the tempering of martensite. These nonmetallic inclusions serve as crack initiation points, which effectively reduce the endurance limit.

Using a rule of thumb relating hardness and ultimate strength [S_u(ksi) $\approx$ 0.5 $\times$ BHN], the following relationships for steel can be given:

Endurance limit related to hardness:

$$S_e(\text{ksi}) \approx 0.25 \times \text{BHN} \quad \text{for BHN} \leq 400 \qquad \text{if BHN} > 400,\ S_e \approx 100 \text{ ksi} \tag{1.1}$$

where BHN is the Brinell hardness number.

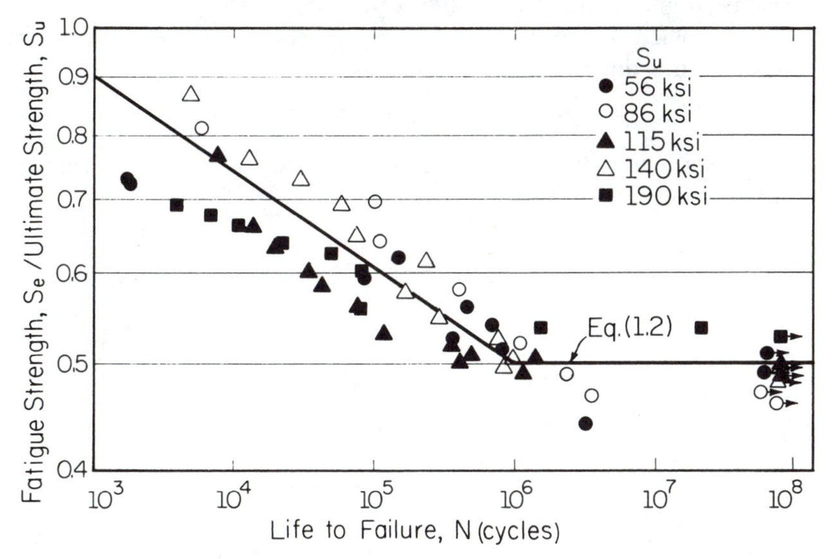

Figure 1.3 S–N curves for several wrought steels, plotted in ratio form (S_e/S_u).

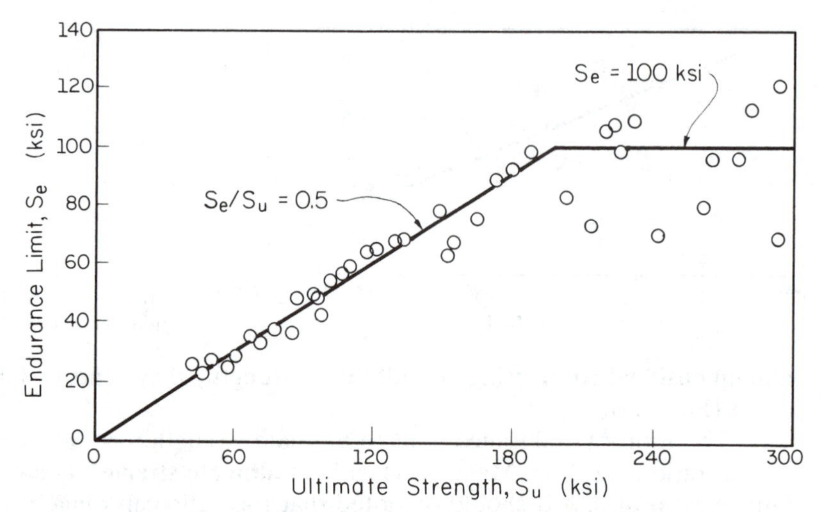

Figure 1.4 Relation between rotating bending endurance limit and tensile strength of wrought steels.

Endurance limit related to ultimate strength:

$$S_e \approx 0.5 \times S_u \quad \text{for } S_u \leq 200\text{ ksi} \qquad \text{if } S_u > 200\text{ ksi}, S_e \approx 100\text{ ksi} \tag{1.2}$$

The alternating stress level corresponding to a life of 1000 cycles, S_{1000}, can be estimated as 0.9 times the ultimate strength. The line connecting this point and the endurance limit is the estimate used for the S–N design line (Fig. 1.5) if no actual fatigue data are available for the material.

In place of the graphical approach shown above, a power relationship can be used to estimate the S–N curve for steel:

$$S = 10^C N^b \quad (\text{for } 10^3 < N < 10^6) \tag{1.3}$$

where the exponents, C and b, of the S–N curve are determined using the two defined points shown in Fig. 1.5:

$$b = -\frac{1}{3}\log_{10}\frac{S_{1000}}{S_e} \qquad C = \log_{10}\frac{(S_{1000})^2}{S_e} \tag{1.4}$$

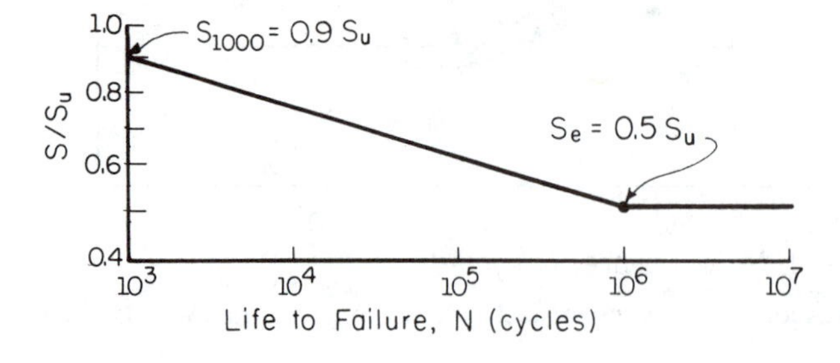

Figure 1.5 Generalized S–N curve for wrought steels on log–log plot.

The equation giving life in terms of an alternating stress is

$$N = 10^{-C/b} S^{1/b} \quad (\text{for } 10^3 < N < 10^6) \tag{1.5}$$

Note that when the estimates for S_{1000} and S_e are made,

$$S_{1000} \approx 0.9 S_u \quad \text{and} \quad S_e \approx 0.5 S_u \tag{1.6}$$

The S–N curve is defined as

$$S = 1.62 S_u N^{-0.085} \tag{1.7}$$

Similar empirical relationships for materials other than steel are not as clearly defined.

Before continuing, certain points about the S–N curve should be emphasized:

1. The empirical relationships outlined in this section are only estimates. Depending on the level of certainty required in the fatigue analysis, actual test data may be necessary.
2. The most useful concept of the S–N method is the endurance limit, which is used in infinite-life or "safe stress" designs.
3. In general, the S–N approach should not be used to estimate lives below 1000 cycles.

Regarding point 3, although there are several methods used to estimate the S–N curve in the range 1 to 1000 cycles they are not recommended. These methods use some percentage of ultimate strength, S_u, or true fracture stress, σ_f, as the estimate for alternating stress at either 1 or $\frac{1}{4}$ cycle. One of the main problems in using this approach is that most materials have an S–N curve which is very flat in the low cycle region. This is due to the large plastic strains caused by high load levels. When doing low cycle analysis a strain-based approach is more appropriate.

1.3 MEAN STRESS EFFECTS

The following relationships and definitions are used when discussing mean and alternating stresses (Fig. 1.6):

$$\Delta\sigma = \sigma_{\max} - \sigma_{\min} = \text{stress range}$$

$$\sigma_a = \frac{\sigma_{\max} - \sigma_{\min}}{2} = \text{stress amplitude}$$

$$\sigma_m = \frac{\sigma_{\max} + \sigma_{\min}}{2} = \text{mean stress}$$

$$R = \frac{\sigma_{\min}}{\sigma_{\max}} = \text{stress ratio} \qquad A = \frac{\sigma_a}{\sigma_m} = \text{amplitude ratio}$$

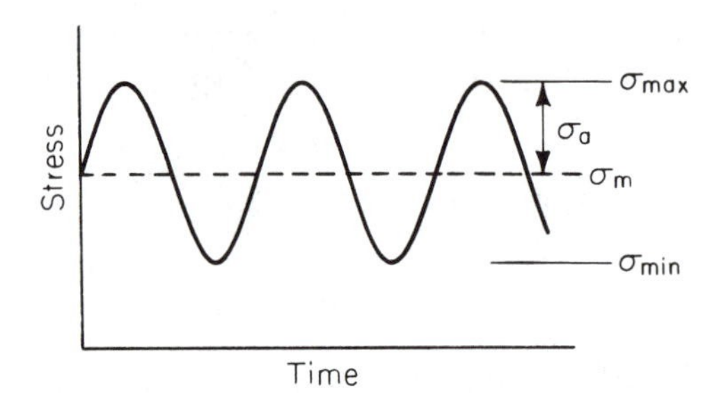

Figure 1.6 Terminology for alternating stress.

The R and A values corresponding to several common loading situations are:

Fully reversed:	$R = -1$	$A = \infty$
Zero to max:	$R = 0$	$A = 1$
Zero to min:	$R = \infty$	$A = -1$

The results of a fatigue test using a nonzero mean stress are plotted on a Haigh diagram (alternating stress versus mean stress) with lines of constant life drawn through the data points (Fig. 1.7). This diagram is sometimes incorrectly called the modified Goodman diagram. The data can also be plotted on a master diagram (Fig. 1.8) which has an extra set of axes for maximum and minimum stress.

Since the tests required to generate a Haigh diagram can be expensive, several empirical relationships have been developed to generate the line defining the infinite-life design region. These methods use various curves to connect the endurance limit on the alternating stress axis to either the yield strength, S_y, ultimate strength, S_u, or true fracture stress, σ_f, on the mean stress axis.

The following relationships are commonly used and are shown on Fig. 1.9:

$$\text{Soderberg (USA, 1930):} \qquad \frac{\sigma_a}{S_e} + \frac{\sigma_m}{S_y} = 1 \tag{1.8}$$

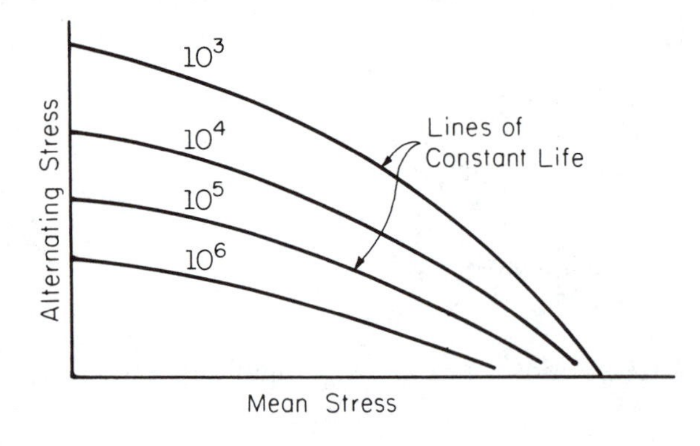

Figure 1.7 Haigh diagram.

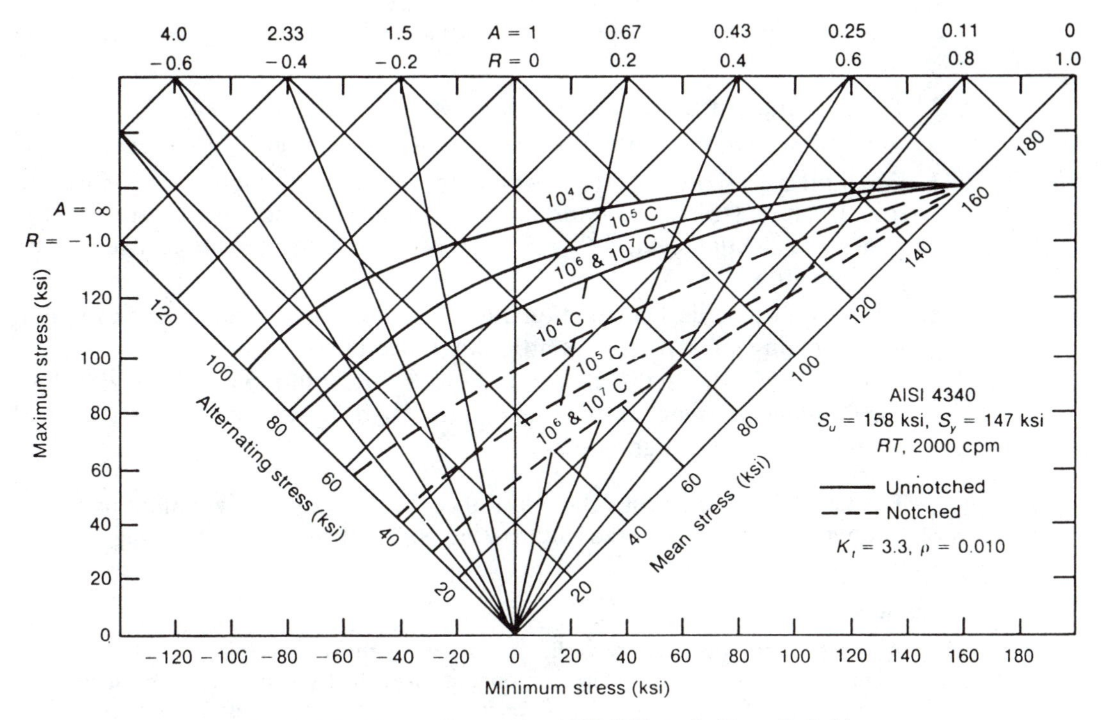

Figure 1.8 Master diagram for AISI 4340 steel. (From Ref. 2.)

Goodman (England, 1899): $$\frac{\sigma_a}{S_e} + \frac{\sigma_m}{S_u} = 1 \qquad (1.9)$$

Gerber (Germany, 1874): $$\frac{\sigma_a}{S_e} + \left(\frac{\sigma_m}{S_u}\right)^2 = 1 \qquad (1.10)$$

Morrow (USA, 1960s): $$\frac{\sigma_a}{S_e} + \frac{\sigma_m}{\sigma_f} = 1 \qquad (1.11)$$

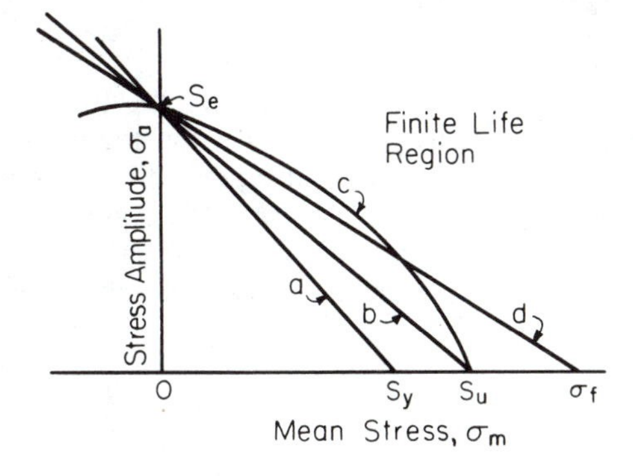

Figure 1.9 Comparison of mean stress equations (a. Soderberg, b. Goodman, c. Gerber, d. Morrow).

The following generalizations can be made when discussing cases of tensile mean stress:

1. The Soderberg method is very conservative and seldom used.
2. Actual test data tend to fall between the Goodman and Gerber curves.
3. For hard steels (i.e., brittle), where the ultimate strength approaches the true fracture stress, the Morrow and Goodman lines are essentially the same. For ductile steels ($\sigma_f > S_u$) the Morrow line predicts less sensitivity to mean stress.
4. For most fatigue design situations, $R < 1$ (i.e., small mean stress in relation to alternating stress), there is little difference in the theories.
5. In the range where the theories show a large difference (i.e., R values approaching 1), there is little experimental data. In this region the yield criterion may set design limits.

For finite-life calculations the endurance limit in any of the equations can be replaced with a fully reversed alternating stress level corresponding to that finite-life value.

Example 1.1

A component undergoes a cyclic stress with a maximum value of 110 ksi and a minimum value of 10 ksi. The component is made from a steel with an ultimate strength, S_u, of 150 ksi, an endurance limit, S_e, of 60 ksi, and a fully reversed stress at 1000 cycles, S_{1000}, of 110 ksi. Using the Goodman relationship, determine the life of the component.

Solution Determine the stress amplitude and mean stress.

$$\sigma_a = \frac{\sigma_{max} - \sigma_{min}}{2} = \frac{110 - 10}{2} = 50 \text{ ksi}$$

$$\sigma_m = \frac{\sigma_{max} + \sigma_{min}}{2} = \frac{110 + 10}{2} = 60 \text{ ksi}$$

Generate a Haigh diagram with constant life lines at 10^6 and 10^3 cycles. These lines are constructed by connecting the endurance limit, S_e, and S_{1000} values on the alternating stress axis to the ultimate strength, S_u, on the mean stress axis (see Fig. E1.1).

When the stress conditions for the component ($\sigma_a = 50$ ksi, $\sigma_m = 60$ ksi) are plotted on the Haigh diagram, the point falls between the 10^3 and 10^6 life lines. This indicates that the component will have a finite life, but the life is greater than 1000 cycles. Next, a line is drawn through the point representing the stress conditions and the ultimate strength, S_u, on the mean stress axis. This represents a constant life line at a life equal to the life of the component. This line intersects the fully reversed alternating stress axis at a value of 83 ksi. Note that this value could also be obtained by solving a modification of Eq. (1.9):

$$\frac{\sigma_a}{S_n} + \frac{\sigma_m}{S_u} = 1$$

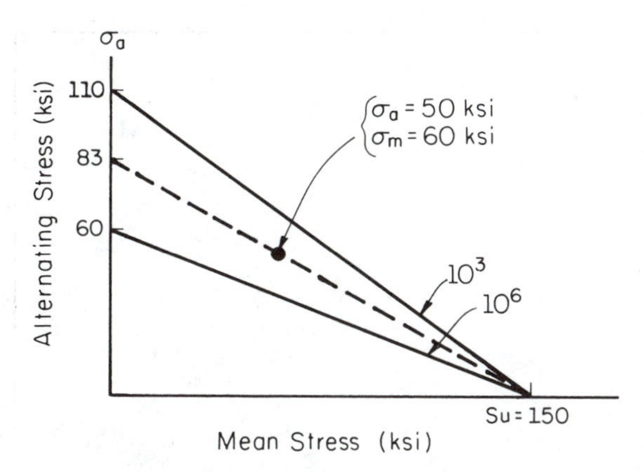

Figure E1.1 Haigh diagram.

where S_n is the fully reversed stress level corresponding to the same life as that obtained with the stress conditions σ_a and σ_m. For this problem,

$$\frac{50}{S_n} + \frac{60}{150} = 1$$

$$S_n = 83 \text{ ksi}$$

The value for S_n can now be entered on the S–N diagram (Fig. E1.2) to determine the life of the component, N_f. (Recall that the S–N diagram represents fully reversed loading). When a value of 83 ksi is entered on the S–N diagram for the material used for the component, the resulting life to failure, N_f, is 2.4×10^4 cycles.

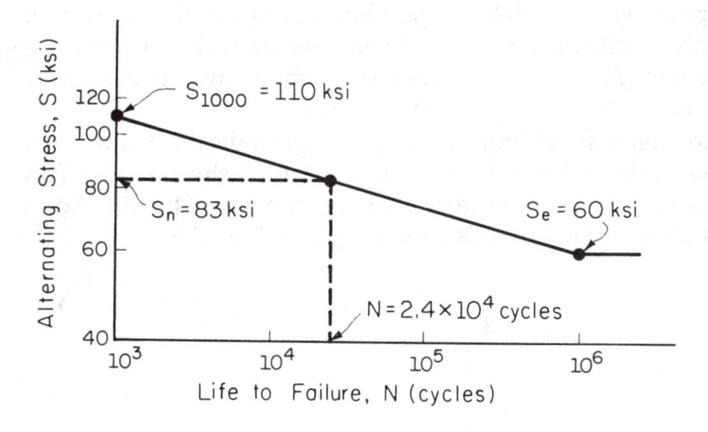

Figure E1.2 *S–N* diagram.

This problem could be redone using the Gerber [Eq. (1.10)], Soderberg [Eq. (1.8)], and Morrow [Eq. (1.11)] mean stress equations. Each technique would provide a slightly different life estimate.

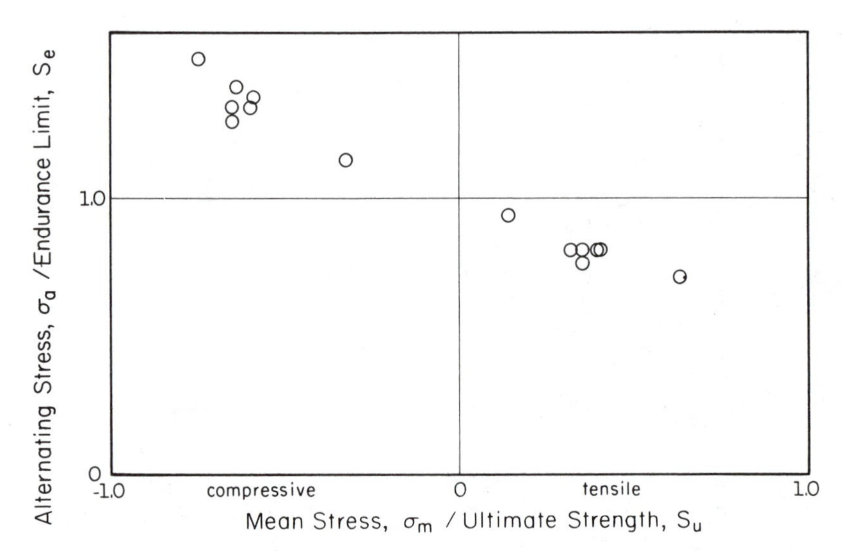

Figure 1.10 Compressive and tensile mean stress effect for smooth specimen. (Data from Ref. 3.)

As seen in Fig. 1.9, the three linear models predict that compressive mean stresses are very beneficial and allow for very large alternating stresses. Experimental results from smooth specimens do indeed indicate that a compressive mean stress is beneficial and increases the life at a given alternating stress (Fig. 1.10). There is difficulty in relating this behavior to notched components. The problems arise when trying to predict the residual stresses generated near the notch root. When extrapolating the Haigh diagram into the range of compressive mean stress, a conservative estimate for notched components is that a compressive mean stress has no effect (Fig. 1.11). At very large compressive mean stresses the design envelope will be set by yield or buckling limits.

Test results from torsion tests of unnotched specimens indicate that a mean shear stress has no effect on life when added to an alternating shear stress. This trend does not appear to hold for notched torsional components. The effect of a mean shear stress on an alternating normal stress is discussed in Chapter 7.

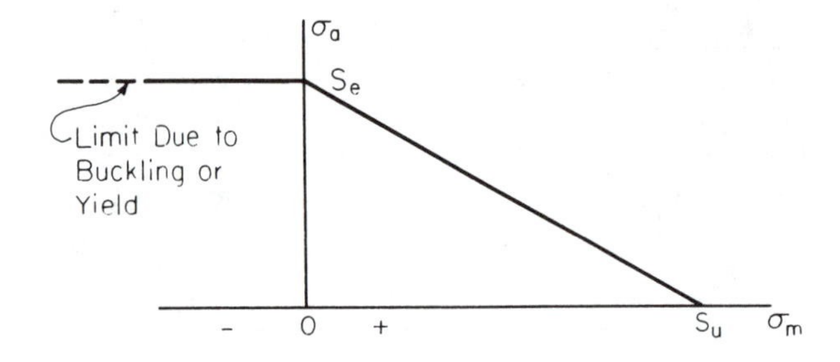

Figure 1.11 Estimate of Haigh diagram for notched components using Goodman line.

1.4 MODIFYING FACTORS

The results of an R. R. Moore test are from the special case of a mirror-polished 0.25 in diameter specimen loaded under fully reversed bending. To denote this, the endurance limit found using the R. R. Moore test is often given a prime, S_e'. The endurance limit needed for design situations, S_e, must take into account differences in size, surface finish, and so on.

For many years the emphasis of most fatigue testing was to gain an empirical understanding of the effects of various factors on the baseline S–N curves for ferrous alloys in the intermediate to long life ranges. The variables investigated include:

1. Size
2. Type of loading
3. Surface finish
4. Surface treatments
5. Temperature
6. Environment

The results of these tests have been quantified as modification factors which are applied to the baseline S–N data.

$$S_e = S_e' C_{\text{size}} C_{\text{load}} C_{\text{surf.fin.}} \cdots \tag{1.12}$$

This modified endurance limit tends to be conservative.

The modification factors are usually specified for the endurance limit, and the correction for the remainder of the S–N curve is not as clearly defined. The general trend is for these modification factors to have less effect at short lives. At the extreme limit of monotonic loading they all approach a value of 1. A conservative estimate is to use the modification factors on the entire S–N curve.

It is very important to remember that these modification factors are empirical models of a phenomenon and may give limited insight into the underlying physical processes. Great care must be taken when extrapolating these empirical modification factors beyond the range of data used to generate them.

1.4.1 Size Effects

The fatigue failure of a material is dependent on the interaction of a large stress with a critical flaw. In essence, fatigue is controlled by the weakest link of the material, with the probability of a weak link increasing with material volume. This differs from bulk material properties such as yield strength and modulus of elasticity. This phenomenon is evident in the fatigue test results of a material using specimens of varying diameters (see Table 1.1). The size effect has been correlated with the thin layer of surface material subjected to 95% or more of the

TABLE 1.1 Influence of Size on Endurance Limit

Diameter (in)	Endurance Limit (ksi)
0.3	33.0
1.5	27.6
6.75	17.3

Source: J. H. Faupel and F. E. Fisher, *Engineering Design,* John Wiley and Sons, New York, 1981. Reprinted with permission.

maximum surface stress. A large component will have a less steep stress gradient and hence a larger volume of material subjected to this high stress (Fig. 1.12). Consequently, there will be a greater probability of initiating a fatigue crack in large components. This concept is backed up by test results which show a less pronounced size effect for axial loading, where there is no gradient, than for bending or torsion. The idea of a highly stressed volume is important when considering stress gradients due to notches (see Chapter 4).

There are many empirical fits to the size effect data. A fairly conservative one is [5], in English units,

$$C_{\text{size}} = \begin{cases} 1.0 & \text{if } d \le 0.3 \text{ in.} \\ 0.869d^{-0.097} & \text{if } 0.3 \text{ in.} \le d \le 10 \text{ in.} \end{cases} \tag{1.13}$$

and in SI units,

$$C_{\text{size}} = \begin{cases} 1.0 & \text{if } d \le 8 \text{ mm} \\ 1.189d^{-0.097} & \text{if } 8 \text{ mm} \le d \le 250 \text{ mm} \end{cases} \tag{1.14}$$

where d is the diameter of the component. Some other points to consider when dealing with the size effect are:

1. The effect is seen mainly at very long lives.
2. The effect is small in diameters up to 2.0 in. even in bending or torsion.
3. Due to the processing problems inherent in large components, there is a greater chance of having residual stresses and various metallurgical variables, which may adversely affect fatigue strength.

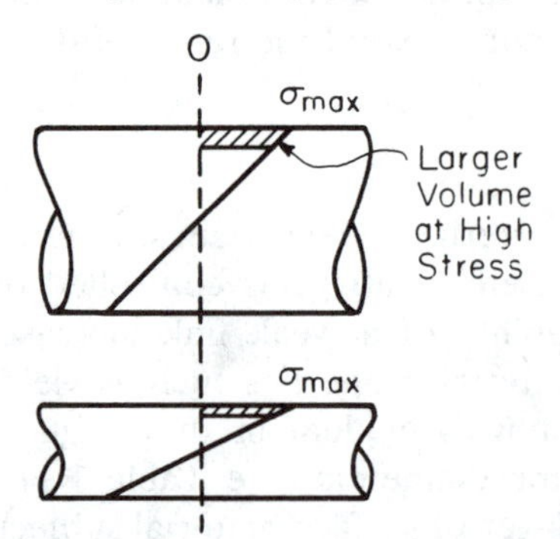

Figure 1.12 Stress gradient in large and small specimens.

The idea of critical volume can also be used to find a size modification factor for noncircular sections (see Ref. 5, p. 294).

1.4.2 Loading Effects

When relating the fatigue data from rotating bending and axial loading for a similar specimen, the volume idea discussed in the preceding section can be used. Since the axial specimen has no gradient, it has a greater volume of material subjected to the high stress. The ratio of endurance limits for a material found using axial and rotating bending tests ranges from 0.6 to 0.9. These test data may include some error due to eccentricity in axial loading. A conservative estimate is

$$S_e(\text{axial}) \approx 0.70 S_e(\text{bending}) \tag{1.15}$$

The ratio of endurance limits found using torsion and rotating bending tests ranges from 0.5 to 0.6. A theoretical value of 0.577 has been explained using the von Mises failure criterion. This relationship is discussed more thoroughly in Chapter 7. A reasonable estimate is

$$\tau_e(\text{torsion}) \approx 0.577 S_e(\text{bending}) \tag{1.16}$$

1.4.3 Surface Finish

The scratches, pits, and machining marks on the surface of a material add stress concentrations to the ones already present due to component geometry. Uniform fine-grained materials, such as high strength steel, are more adversely affected by a rough surface finish than a coarse-grained material such as cast iron.

The correction factor for surface finish is sometimes presented on graphs that use a qualitative description of surface finish such as "polished" or "machined" (Fig. 1.13). Some of the curves on this plot include effects other than

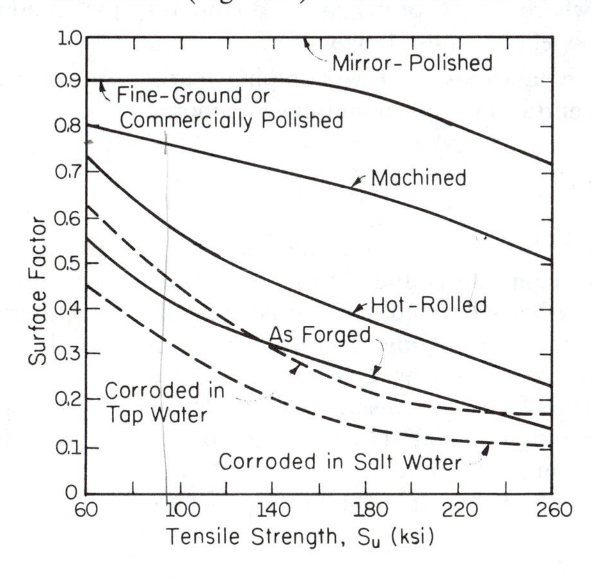

Figure 1.13 Surface finish factor: steel parts. (From Ref. 6.)

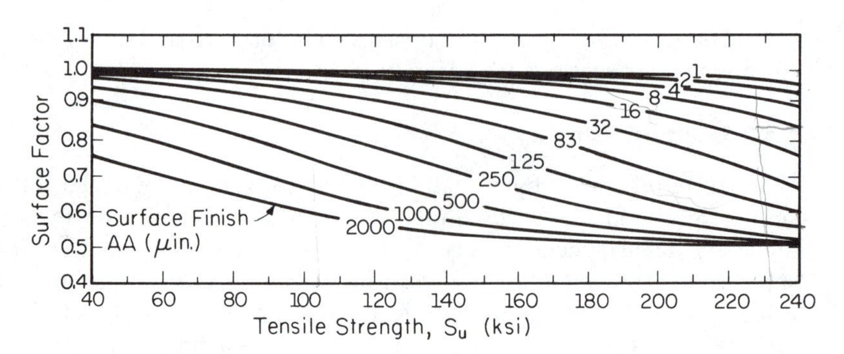

Figure 1.14 Surface finish factor versus surface roughness and strength: steel parts. (From Ref. 7.)

surface roughness. For example, the forged and hot-rolled curves include the effects of decarburization.

Other graphs, such as Fig. 1.14, use a quantitative measurement of surface roughness such as R_A, the root mean square or, *AA*, arithmetic average. The surface roughness resulting from various machining operations can be found in machining and manufacturing handbooks.

Some important points about the surface finish effect are:

1. The condition of the surface is more important for higher strength steels.
2. The residual surface stress caused by a machining operation can be important. An example is the residual tensile stress sometimes caused by some grinding operations.
3. At shorter lives, where crack propagation dominates, the condition of surface finish has less effect on the fatigue life.
4. Localized surface irregularities such as stamping marks can serve as very effective stress concentrations and should not be ignored.

1.4.4 Surface Treatment

Since fatigue cracks almost always initiate at a free surface, any surface treatment can have a significant effect on fatigue life. The effect of surface roughness from various forming operations was discussed in the preceding section. Other surface treatments can be categorized as plating, thermal, or mechanical. In all three cases the effect on fatigue life is due primarily to residual stresses.

As a review of residual stress, consider the unnotched beam (Fig. 1.15) subjected to a varying bending moment. The bending moment history is shown in Fig. 1.15d. If the simplifying assumption is made that the material is elastic–perfectly plastic, the history of the stress at the top surface of the beam is as shown in Fig. 1.15e.

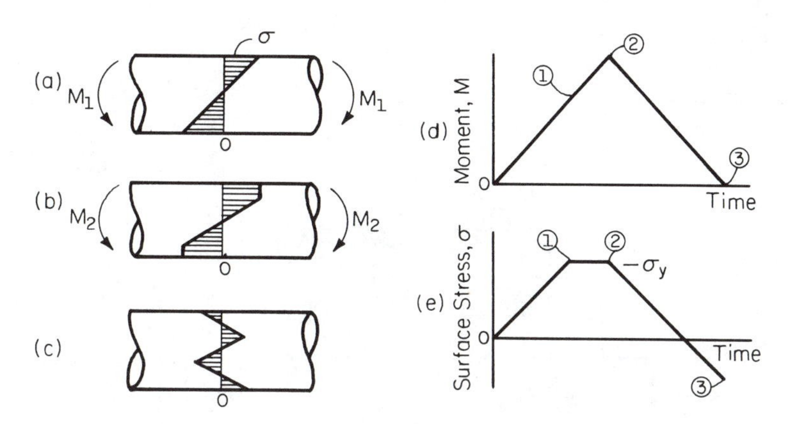

Figure 1.15 Residual stress in unnotched beam in bending.

1. At point 1 the surface of the beam is just at the point of yielding and the stress distribution is linear (Fig. 1.15a).
2. If the moment is increased to point 2, the outer layer of the beam begins to yield (Fig. 1.15b).
3. If the moment is reduced to point 3, the beam will have a residual stress distribution (Fig. 1.15c). When the outer layer of material yielded, it elongated and upon unloading the stresses and strains in the beam must meet compatibility and equilibrium requirements. Although the exact residual stress distribution is difficult to define, the important point is that the outer surface of the beam, which had yielded in tension, is now in residual compression.

Another example of residual stress is the notched member under axial loading, shown in Fig. 1.16. The loading history involves an initial tensile overload followed by fully reversed cyclic loads (Fig. 1.16d).

1. The initial overload (point 1) causes the material at the root of the notch to yield in tension (Fig. 1.16b) and when the load is released (point 2) this material will be in residual compression (Fig. 1.16c).
2. When the cyclic load is applied (points 3 and 4), the stress at the root of the notch will cycle between the limits shown on Fig. 1.16e.

Note that while the load is fully reversed, the stress at the root of the notch (where the fatigue crack will initiate) cycles about a compressive value. The residual stress in the material at the notch root has the same effect as an externally applied compressive mean stress of equal magnitude, and as pointed out in Section 1.3, this will increase life at a given alternating stress level. Remember that this discussion involves the residual mean stress in a notched member, not a mean stress due to an applied load.

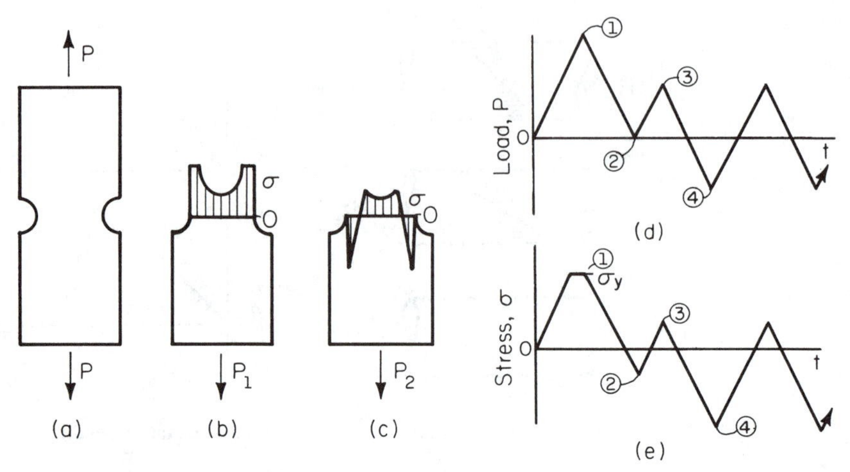

Figure 1.16 Residual stress in notched member under axial loading.

The method just described for producing a residual stress with an initial overload is called *prestressing* or *presetting*. An example of this is given in Table 1.2, which compares the endurance limit values of notched and unnotched plates of 4340 steel ($S_u = 130$ ksi). A comparison is also made of plates with and without an initial tensile overload. As can be seen, the preload sets up a residual stress which almost negates the effect of the notch.

Presetting is used on such components as coil and leaf springs. It should be noted that the initial overload on a component is favorable for future loading in the same direction as the overload, but unfavorable for loads in the opposite direction. For example, if a coil spring is preloaded in compression, it will have a beneficial effect only for future cyclic loading which is primarily in compression.

Presetting should not be used in cases of fully reversed loading. For example, the cold straightening of an axle shaft can reduce the endurance limit 20 to 50%.

In the following discussion on surface treatments it is important to keep these points in mind:

1. Since fatigue is a surface phenomenon, the residual stress at the surface of the material is critical.

TABLE 1.2 Endurance Limit of Plate with Hole under Axial Loading

	Endurance Limit (ksi)	
	Unnotched	Notched
No preload	58.0	23.0
With preload	56.6	53.7

Source: H. O. Fuchs and R. I. Stephens, *Metal Fatigue in Engineering,* John Wiley and Sons, New York, 1980. Reprinted with permission.

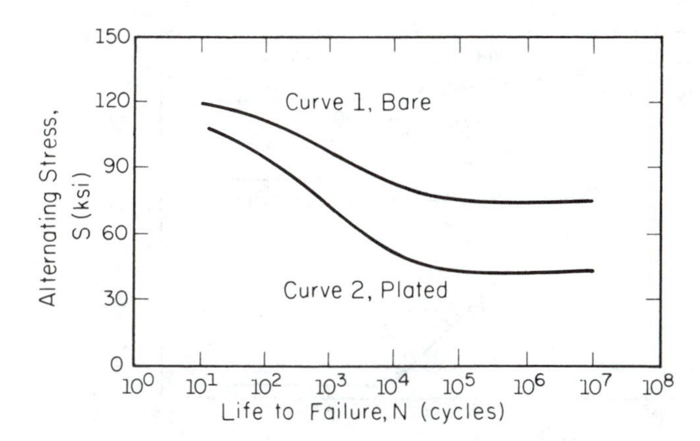

Figure 1.17 Effect of chrome plating on S–N curve of 4140 steel. (From Ref. 1.)

2. Compressive residual stresses are beneficial, and tensile stresses are detrimental to fatigue life.
3. Residual stresses are not always permanent, and various factors, such as high temperatures and overloads, may cause stress relaxation.

Plating. Chrome and nickel plating of steels can cause up to a 60% reduction in endurance limits (Figs. 1.17 and 1.18). This is due primarily to the high residual tensile stresses generated by the plating process. The following operations can help alleviate the residual tensile stress problem:

1. Nitride the part before plating.
2. Shot peen the part before or after plating.
3. Bake or anneal the part after plating.

Figure 1.19 shows the effect of shot peening a rotating beam specimen before and after a nickel-plating operation.

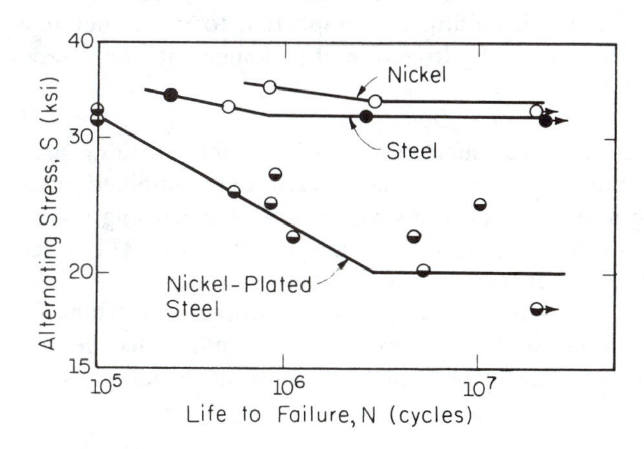

Figure 1.18 Effect of nickel plating on S–N curve of steel (S_u = 63 ksi). (From Ref. 9.)

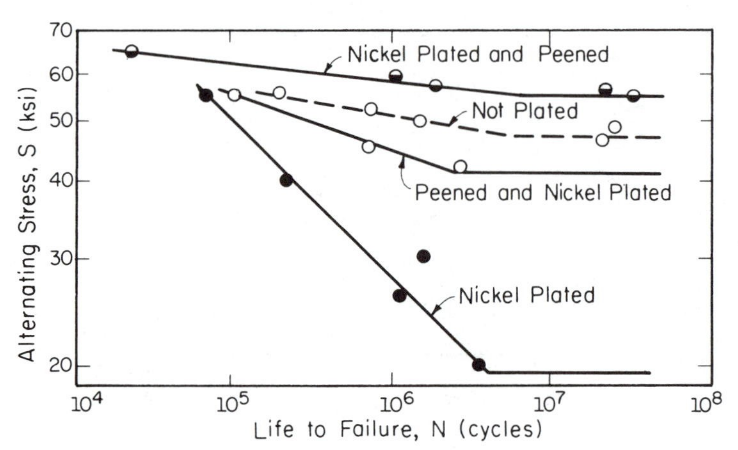

Figure 1.19 Effects of shot peening on S–N curve of nickel plated steel. (From Ref. 9.)

There are many factors involved with a plating operation which can affect fatigue life. The following are some general trends for chrome and nickel plating:

1. There is a greater reduction of fatigue strength as the yield strength of the material being plated increases.
2. The fatigue strength reduction due to plating is greater at longer lives.
3. The fatigue strength reduction is greater as the thickness of the plating increases.
4. It should also be noted that when fatigue occurs in a corrosive environment, the extra corrosion resistance offered by plating can more than offset the reduction in fatigue strength seen in a noncorrosive environment (see Table 1.7).

Plating with cadmium and zinc appear to have no effect on fatigue strength while still offering corrosion resistance. However, plating with these metals does not offer the wear resistance of chromium. It is important to remember that any electroplating operation can cause hydrogen embrittlement if the process is improperly controlled.

Thermal. Diffusion processes such as carburizing and nitriding are very beneficial for fatigue strength. These processes have the combined effect of producing a higher strength material on the surface as well as causing volumetric changes which produce residual compressive surface stresses. The effect of nitriding on notched steel members can be seen in Table 1.3.

Flame and induction hardening cause a phase transformation, which in turn causes a volumetric expansion. If these processes are localized to the surface, they produce a compressive residual stress which is beneficial for fatigue strength.

TABLE 1.3 Effect of Nitriding on Endurance Limit

	Endurance Limit (ksi)	
Geometry	Nitrided	Not Nitrided
Without notch	90	45
Half-circle notch	87	25
V notch	80	24

Source: Ref. 6.

Hot rolling and forging can cause surface decarburization. The loss of carbon atoms from the surface material causes it to have a lower strength and may also produce residual tensile stresses. Both of these factors are very detrimental to fatigue strength. The effect of decarburization on various high-strength steel alloys with notched and unnotched geometries can be seen in Table 1.4.

TABLE 1.4 Effect of Decarburization on Endurance Limit

		Endurance Limit (ksi)			
		Undecarburized		Decarburized	
Steel	S_u (ksi)	Smooth	Notched	Smooth	Notched
AISI 2340	250	122	69	35	25
AISI 2340	138	83	43	44	25
AISI 4140	237	104	66	31	22
AISI 4140	140	83	40	32	19

Source: Ref. 1.

Figure 1.20 shows the effect of forging on the endurance limit of steels with various tensile strengths. As can be seen, the endurance limit reduction for a low strength steel may only be a few percent, whereas there may be a five fold

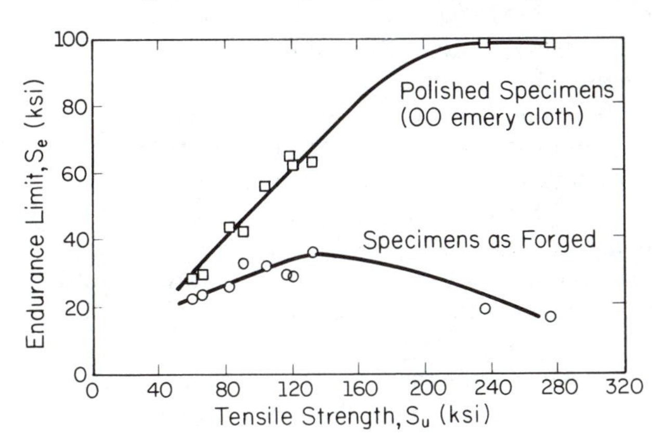

Figure 1.20 Effect of forging on the endurance limit of steels. (From Ref. 10.)

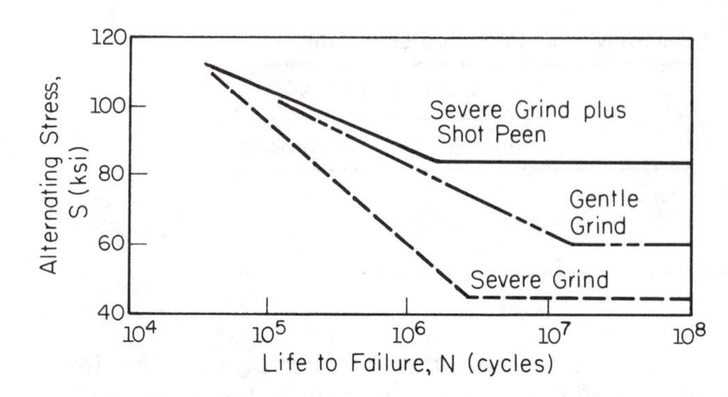

Figure 1.21 Effects of grinding on *S–N* curve of steel. (From Ref. 11.)

reduction for a high strength steel. This points out another general trend concerning the various surface factors that affect fatigue life. Almost all of these factors have a more pronounced influence as the strength of the base steel increases. In fact, for very low strength (i.e., lower carbon) steels, almost none of the factors will cause substantial increases or decreases in fatigue strength. In large part, this trend can be attributed to the ease with which residual stresses relax out of materials with low yield strengths.

It should also be noted that some manufacturing processes, such as welding, grinding, and flame cutting, can set up detrimental residual tensile stresses. Figure 1.21 shows the effect of gentle and severe grinding operations on the fatigue properties of a high strength steel (Rockwell C = 45). The figure also shows how shot peening can undo the damage caused by severe grinding.

Mechanical. There are several methods used to cold work the surface of a component to produce a residual compressive stress. The two most important are cold rolling and shot peening. Along with producing compressive residual stresses, these methods also work-harden the surface material. The great improvement in fatigue life is due primarily to the residual compressive stresses.

Cold rolling involves pressing steel rollers to the surface of a component which is usually rotated in a lathe. This method is used on large parts and can produce a deep residual stress layer. Figure 1.22 shows the effect of cold rolling. Another example of the benefits of cold rolling is the increased fatigue resistance of a bolt with rolled threads over one with cut threads (Table 1.5).

Shot peening is one of the most important methods of producing a residual compressive stress. This procedure involves blasting the surface of a component with high-velocity steel or glass beads. This puts the core of the material in residual tension and the skin in residual compression. The residual compressive stress layer is about 1 mm deep with a maximum value of about one-half the yield strength of the material. An example of the effect of shot peening is shown in Fig. 1.23.

One of the advantages of shot peening is the ease with which it can be used on oddly shaped parts such as coil springs. One disadvantage is that it leaves a

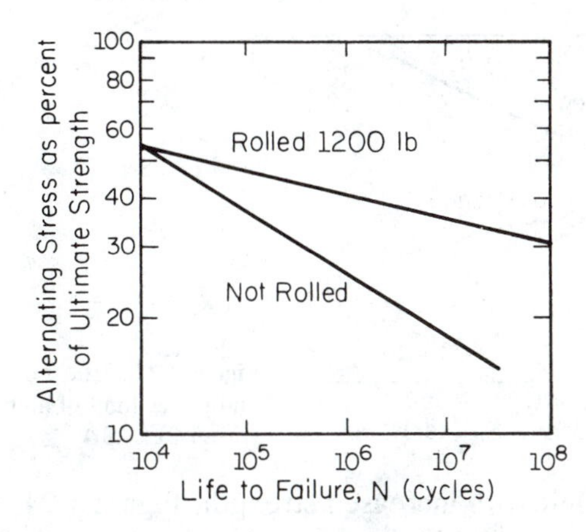

Figure 1.22 Effects of cold rolling on *S–N* curve of steel. (From Ref. 9.)

rough dimpled surface. If a smooth surface is required, a honing or polishing operation can be applied after the part is shot peened. This causes only a small reduction in fatigue strength.

Shot peening can be used to undo the deleterious effects caused by chrome

TABLE 1.5 Fatigue Strength at 10^5 cycles for Bolts (AISI 8635)[a]

	Fatigue Strength (ksi)
Rolled threads	74
Cut threads	44

[a] Bolts with the same thread design.

Source: H. O. Fuchs and R. I. Stephens, *Metal Fatigue in Engineering,* John Wiley and Sons, New York, 1980. Reprinted with permission.

and nickel plating (Fig. 1.19), decarburization, corrosion, and grinding (Fig. 1.21). It can also be used to great advantage on high-strength steels. As shown in Fig. 1.4, many steels with an ultimate strength above 200 ksi experience a

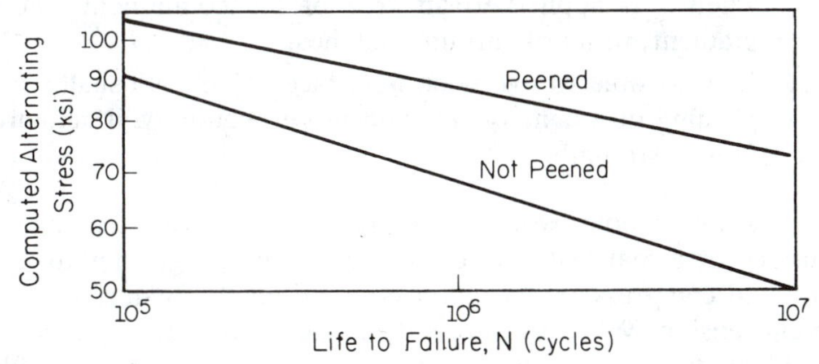

Figure 1.23 *S–N* curve of carburized gears in peened and unpeened conditions. (From Ref. 12.)

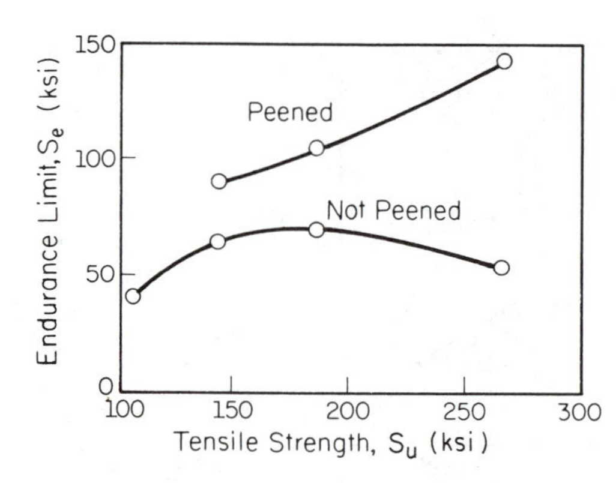

Figure 1.24 Effect of shot peening on endurance limit of high-strength steel. (From Ref. 13.)

reduction in endurance limit with increased strength. Figure 1.24 shows the effect of shot peening on the endurance limit of a high strength steel (0.45% C, 1.0% Mn, quenched and tempered). As can be seen, shot peening eliminates the roll-off in endurance limit, and a fatigue ratio of 0.5 is extended well beyond an ultimate strength of 200 ksi.

Some important points about cold working for residual compressive stresses are:

1. Cold rolling and shot peening have their greatest effect at long lives. At very short lives there is almost no improvement in the fatigue strength. At shorter lives the stress levels are high enough to cause yielding, which eliminates residual stresses.
2. Certain situations can cause the residual stresses to fade out or relax. These include high temperatures and overstressing. Approximate temperatures where this fading occurs are 500°F for steel and 250°F for aluminum.
3. Steels with yield strengths below 80 ksi are seldom cold rolled or shot peened. Due to their low yield points it is quite easy to introduce plastic strains that wipe out residual stresses.
4. A surface residual compressive stress has the greatest effect on fatigue life when it is applied to an area of the component where there is a stress gradient, primarily around notches.
5. It is possible to overpeen a surface. There is usually an optimum level for peening of a component, and more peening will actually begin to decrease fatigue strength.

Another point worth noting is that a high residual compressive stress at the surface of a material may cause subsurface fatigue failures. If the surface is in residual compression, to retain equilibrium the material below the surface must be in tension. When the stress distribution from the applied load is added to the residual stress distribution, the maximum tensile stress usually occurs below the surface. The fatigue failure may then initiate at this point of maximum stress.

This trend is especially true in carburized and nitrided parts, where the effect of the stress distribution is complemented by the change in material properties at the interface between the hard surface and the soft core.

A related trend is that residual compressive surface stresses will not significantly affect the fatigue strength of smooth axial specimens. This is because a smooth axial specimen has no stress gradient from applied loads. All the methods discussed for producing residual surface stresses will have the greatest effect in cases where there is a stress gradient due to applied loads. Examples of this are stress gradients due to bending, torsion, and notches.

Example 1.2

Several bars of high strength steel are to be used as leaf springs. These springs will be subjected to a zero-to-maximum ($R = 0$) three-point flexural loading. The bars are 1.50 in. wide and 0.192 in. thick. Half of the bars are in the "as-heat-treated" condition, while the other half have been shot peened. Determine the zero-to-maximum surface stress that will allow the bars to have an infinite life. The necessary data for the two sets of bars are given below. Use the Goodman relationship for these calculations.

As Heat Treated

Hardness = 48 Rockwell C ($\approx$ 465 BHN)

Residual surface stress = 0 ksi

Surface roughness (AA) = 24 μin.

Shot Peened

Hardness = 49 Rockwell C ($\approx$ 475 BHN)

Residual surface stress = −80 ksi

Surface roughness (AA) = 125 μin.

Solution First the calculations will be made for the as-heat-treated spring. The uncorrected endurance limit, S'_e, is found using Eq. (1.1). Since BHN > 400,

$$S'_e \approx 100 \text{ ksi}$$

The modification for size effect must take into account the fact that the cross-section of the bar is not round. Reference 5 suggests the following equation to determine the equivalent diameter, d_{eq}, for a rectangular section undergoing bending.

$$0.0766 d_{eq}^2 = 0.05 bh$$

where b is the width and h thickness. Then

$$0.0766 d_{eq}^2 = 0.05(0.192 \text{ in.})(1.5 \text{ in.})$$

$$d_{eq} = 0.43 \text{ in.}$$

Using Eq. (1.13) (since $d_{eq} > 0.3$ in.) yields

$$C_{size} = 0.869(0.43)^{-0.097}$$

The modification factor for size is 0.94.

The modification factor for loading effect is 1.0 since the loading is reversed bending. The modification factor for surface finish requires that ultimate strength be known. The ultimate strength, S_u, of the spring can be estimated as

$$S_u \approx 0.5(\text{BHN})$$

$$\approx 0.5(465)$$

$$\approx 232 \text{ ksi}$$

Reading from Fig. 1.14 at $S_u = 232$ ksi and AA = 24 μin. the modification for surface roughness is 0.75.

Determine the modified endurance limit, S_e, using Eq. (1.12):

$$S_e = S'_e \times \text{modification factors}$$

$$= (100 \text{ ksi})(0.94)(1.0)(0.75)$$

$$= 70.5 \text{ ksi}$$

The next step is to determine the allowable stress for the spring by using the Goodman relationship [Eq. (1.9)]:

$$\frac{\sigma_a}{S_e} + \frac{\sigma_m}{S_u} = 1$$

For this case where loading is zero to maximum ($R = 0$), the mean stress, σ_m, and alternating stress, σ_a, are equal. Solving for the unknown gives us

$$\frac{\sigma}{70.5 \text{ ksi}} + \frac{\sigma}{232 \text{ ksi}} = 1$$

$$\sigma = \sigma_a = \sigma_m = 54 \text{ ksi}$$

$$\sigma_{max} = \sigma_a + \sigma_m = 108 \text{ ksi}$$

This would mean that for an infinite life the outer surface of the spring could cycle between 0 and 108 ksi. Test results indicate that the actual value is 0 to 100 ksi. The analysis provides an answer with an 8% nonconservative error.

Next are the calculations for the shot peened spring. The uncorrected endurance limit, S'_e, and the modifications for size and loading would be the same as for the as-heat-treated spring.

The modification factor for surface finish requires a value for ultimate strength.

$$S_u \approx 0.5(\text{BHN})$$

$$\approx 0.5(475)$$

$$\approx 238 \text{ ksi}$$

Reading from Fig. 1.14 at $S_u = 238$ ksi and AA $= 124\,\mu$in. the modification for surface roughness is 0.58.

Determine the modified endurance limit, S_e.

$$S_e = (100\text{ ksi})(0.94)(1.0)(0.58)$$

$$= 54.5\text{ ksi}$$

It is necessary to include a modification for the residual surface stress. The residual surface stress (-80 ksi) can be accounted for in the Goodman equation since the residual stress can be combined with the imposed mean stress. The allowable stress is determined by using the relationship [Eq. (1.9)]

$$\frac{\sigma_a}{S_e} + \frac{\sigma_m}{S_u} = 1$$

For this case, where loading is zero to maximum ($R = 0$), the mean stress, σ_m, and alternating stress, σ_a, are equal, but the equation must also take into account the residual surface stress. This value will be combined with the mean stress. Solving for the unknown, we obtain

$$\frac{\sigma}{54.5\text{ ksi}} + \frac{\sigma - 80}{238\text{ ksi}} = 1$$

$$\sigma = \sigma_a = \sigma_m = 59.3\text{ ksi}$$

$$\sigma_{\max} = \sigma_a + \sigma_m = 118.6\text{ ksi}$$

Therefore, for an infinite life, the outer surface of the shot peened spring could cycle between 0 and 118.6 ksi. Test results indicate that the actual value is 0 to 140 ksi. The analysis provides an answer with a 15% conservative error.

One effect that was not considered in this analysis was that shot peening increases the uncorrected endurance limit (see Fig. 1.24). It should be noted that the beneficial effect of the compressive residual stress caused by shot peening more than offsets the detrimental increase in surface roughness. Another point which should be considered is that since the maximum surface stress is well below the yield strength of the material, there should be no problem with relaxation of residual stresses. (Data for this problem were taken from Ref. 14.)

1.4.5 Temperature

There is a tendency for the endurance limits of steels to increase at low temperatures. A more important design consideration, however, is that many materials experience a significant reduction in fracture toughness at low temperatures.

At high temperatures the endurance limit for steels disappears due to the mobilizing of dislocations. At temperatures beyond approximately one-half of the melting point of the material, creep becomes important. In this range the stress–life approach is no longer applicable. It is also important to note that high temperatures can cause annealing, which may remove beneficial residual compressive stresses.

1.4.6 Environment

When fatigue loading takes place in a corrosive environment the resulting detrimental effects are more significant than would be predicted by considering fatigue and corrosion separately. The interaction of fatigue and corrosion, which is called corrosion-fatigue, involves unique failure mechanisms which are very complex. The work in this area is still very much at the research stage and very little is available in the way of quantified data or useful theories.

The basic mechanism of corrosion-fatigue during the initiation stage can be explained this way. A corrosive environment attacks the surface of a metal and produces an oxide film. Usually, this oxide film would serve as a protective layer and prevent further corrosion of the metal. However, cyclic loading causes localized cracking of this layer, which exposes fresh metal surfaces to the corrosive environment. At the same time corrosion causes localized pitting of the surface, and these pits serve as stress concentrations. The mechanism of corrosion-fatigue during the crack propagation stage is very complicated and not well understood.

One of the main difficulties in trying to quantify corrosion-fatigue is the large number of variables involved in testing. Consider the corrosion-fatigue of the important combination of steel in water. Some of the variables that must be considered are alloying elements in the steel, chemical makeup of the water, temperature, degree of aeration, flow velocity, and salt content. One of the many trends is that corrosion-fatigue is much worse in a spray than when the metal is fully immersed. Another variable that is very critical is loading frequency. Fatigue tests in noncorrosive environments can be run at almost any frequency and similar data will be obtained. On the other hand, corrosion-fatigue data are greatly influenced by loading frequency. Low frequency tests allow more time for corrosion to take place, and resulting fatigue lives are shorter.

There are some general trends observed in corrosion-fatigue. Figure 1.25 shows the generalized S–N curves for steel in four different environments. The curves generated in room air and a vacuum show that even the humidity and oxygen in room air can slightly reduce fatigue strength. The curve for presoak involves the case where the steel is soaked in a corrosive environment and then the fatigue test is run in room air. The reduction of fatigue properties for this curve is due to the rough surface caused by corrosion pitting. The curve for corrosion-fatigue is significantly below the one for room air. Another trend is that corrosion-fatigue eliminates the endurance limit behavior seen in many steels.

Another important trend is shown in Fig. 1.26, which shows the endurance limit for various steels in room air and freshwater environments. The data for plain carbon steels indicate that higher strength steels have no advantage in a corrosive environment. Note that steels with a high chromium content have

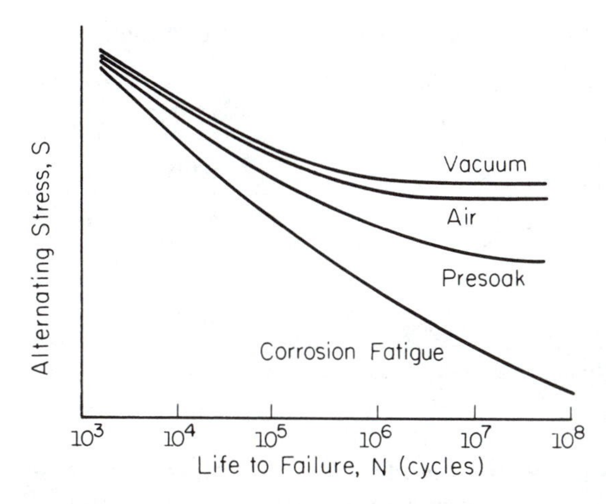

Figure 1.25 Effect of various environments on the *S–N* curve of steel. (H. O. Fuchs and R. I. Stephens, *Metal Fatigue in Engineering*, John Wiley and Sons, New York, 1980. Reprinted with permission.)

significantly better corrosion-fatigue resistance than that of plain carbon steels. A general trend is that materials which are resistant to corrosion alone will also have good corrosion-fatigue properties.

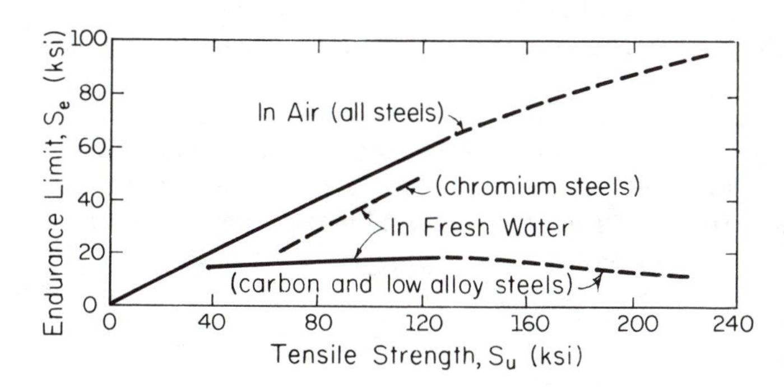

Figure 1.26 Influence of tensile strength and chemical composition on corrosion-fatigue strength of steels. (From Ref. 10.)

There are several methods that can be used to reduce the problems caused by corrosion-fatigue. Perhaps the most effective is to use a steel with a high chrome content. Table 1.6 compares the fatigue properties of a plain carbon and chromium steel in salt water.

TABLE 1.6 Fatigue Strength of Steels in Corrosive Environment[a]

Material	S_u (ksi)	Endurance Limit[b] (ksi) In Air	Endurance Limit[b] (ksi) In Salt Water	Percent Reduction
SAE 1050	116	53.8	22.6	58
5% Cr steel	116	66	47.2	28

[a] 6.8% Salt water, complete immersion.
[b] Basis for endurance limit in corrosive environment is 10^7 cycles.
Source: Ref. 10.

TABLE 1.7 Effect of Various Surface Treatments on the Corrosion Fatigue of Mild Steel

		Endurance Limit[a] (ksi)			
		In Air		In Fresh Water	
S_u (ksi)	Surface Treatment	Untreated	Treated	Untreated	Treated
53	Cold rolling	33	37	13	19
50	Nickel plate	28	20	14	20
50	Cadmium coat	28	29	14	22

[a] Basis for endurance limit in corrosive environment is 10^8 cycles.

Source: Ref. 10.

There are several surface treatments that will improve corrosion-fatigue resistance. Examples of these are shown in Table 1.7. Surface coatings such as paint, and platings using chrome, nickel, cadmium, or zinc, are useful. Note that nickel plating causes a reduction of fatigue strength in air but gives improvement in a corrosive environment. An advantage in using a softer metal for a coating is that it is more likely to remain intact when a crack forms in the base metal. One problem with surface coatings is that fatigue cracks can start at even the smallest break in a coating.

Surface treatments that produce compressive residual surface stresses (nitriding, shot peening, cold rolling, etc.) are also useful. These treatments cause the maximum tensile stress to occur below the surface. The reverse is also true and tensile residual surface stresses are very detrimental and promotes corrosion-fatigue.

1.5 IMPORTANT CONCEPTS

- Care should be taken when using the idea of an endurance limit, a "safe stress" below which fatigue will not occur. Only plain carbon and low-alloy steels exhibit this property, and it may disappear due to high temperatures, corrosive environments, and periodic overloads.
- As a general trend the following factors will reduce the value of endurance limit:
  Tensile mean stress
  Large section size
  Rough surface finish
  Chrome and nickel plating
  Decarburization (due to forging and hot rolling)
  Severe grinding
- The following factors tend to increase the endurance limit:
  Nitriding

Flame and induction hardening
Carburization
Shot peening
Cold rolling

1.6 IMPORTANT EQUATIONS

Endurance Limit Related to Hardness

$$S_e(\text{ksi}) \approx 0.25 \times \text{BHN} \quad \text{for BHN} \leq 400$$
$$\text{if} \quad \text{BHN} > 400,\ S_e \approx 100\ \text{ksi} \tag{1.1}$$

where BHN is the Brinell hardness number

Endurance Limit Related to Ultimate Strength

$$S_e \approx 0.5 \times S_u \quad \text{for } S_u \leq 200\ \text{ksi}$$
$$\text{if} \quad S_u > 200\ \text{ksi},\ Se \approx 100\ \text{ksi} \tag{1.2}$$

Alternating Stress Relationships

$$\Delta\sigma = \sigma_{\text{max}} - \sigma_{\text{min}} = \text{stress range}$$

$$\sigma_a = \frac{\sigma_{\text{max}} - \sigma_{\text{min}}}{2} = \text{stress amplitude}$$

$$\sigma_m = \frac{\sigma_{\text{max}} + \sigma_{\text{min}}}{2} = \text{mean stress}$$

$$R = \frac{\sigma_{\text{min}}}{\sigma_{\text{max}}} = \text{stress ratio}$$

$$A = \frac{\sigma_a}{\sigma_m} = \text{amplitude ratio}$$

R and A Values Corresponding to Common Loading Situations

Fully reversed:	$R = -1$	$A = \infty$
Zero to maximum:	$R = 0$	$A = 1$
Zero to minimum:	$R = \infty$	$A = -1$

Mean Stress Correction Relationships

Soderberg (USA, 1930): $$\frac{\sigma_a}{S_e} + \frac{\sigma_m}{S_y} = 1 \tag{1.8}$$

$$\text{Goodman (England, 1899):} \quad \frac{\sigma_a}{S_e} + \frac{\sigma_m}{S_u} = 1 \tag{1.9}$$

$$\text{Gerber (Germany, 1874):} \quad \frac{\sigma_a}{S_e} + \left(\frac{\sigma_m}{S_u}\right)^2 = 1 \tag{1.10}$$

$$\text{Morrow (USA, 1960s):} \quad \frac{\sigma_a}{S_e} + \frac{\sigma_m}{\sigma_f} = 1 \tag{1.11}$$

Relationship between S_e under Various Loading

$$S_e(\text{axial}) \approx 0.70 S_e(\text{bending}) \tag{1.15}$$

$$\tau_e(\text{torsion}) \approx 0.577 S_e(\text{bending}) \tag{1.16}$$

REFERENCES

1. C. C. Osgood, *Fatigue Design,* 2nd ed., Pergamon Press, Oxford, 1982.
2. U.S. Department of Defense, MIL-HBDK-5.
3. G. Sines and J. L. Waisman, (eds.), *Metal Fatigue,* McGraw-Hill, New York, 1959.
4. J. H. Faupel and F. E. Fisher, *Engineering Design,* Wiley-Interscience, New York, 1981.
5. J. E. Shigley and L. D. Mitchell, *Mechanical Engineering Design,* 4th ed., McGraw-Hill, New York, 1983.
6. R. C. Juvinall, *Engineering Considerations of Stress, Strain, and Strength,* McGraw-Hill, New York, 1967.
7. R. C. Johnson, *Mach. Des.,* Vol. 45, No. 11, 108, 1973.
8. H. O. Fuchs and R. I. Stephens, *Metal Fatigue in Engineering,* Wiley-Interscience, New York, 1980.
9. J. O. Almen and P. H. Black, *Residual Stresses and Fatigue in Metals,* McGraw-Hill, New York, 1963.
10. P. G. Forrest, *Fatigue of Metals,* Pergamon Press, Oxford, 1962.
11. L. P. Tarasov and H. J. Grover, "Effect of Grinding and Other Finishing Processes on the Fatigue Strength of Hardened Steel," *Am. Soc. Test. Mater. Proc.,* Vol. 50, 1950, p. 668.
12. J. C. Straub, "Shot-Peening," in *Metals, Engineering, Design,* 2nd ed., O. J. Horger (ed.), McGraw-Hill, New York, 1965, p. 258.
13. O. J. Horger, "Mechanical and Metallurgical Advantages of Shot-Peening," *Iron Age,* March 29 and April 5, 1945.
14. R. L. Mattson and J. G. Roberts, "The Effect of Residual Stresses Induced by Strain Peening upon Fatigue Strength," in *Internal Stresses and Fatigue in Metals,* G. M. Rossweiler and W. L. Grube, (eds.), Elsevier, New York, 1959, pp. 337–357.
15. H. F. Moore and J. B. Kommers, "An Investigation of the Fatigue of Metals," *Univ. Ill. Eng. Exp. Stn. Bull.,* 124, 1921.

16. H. F. Moore and T. M. Jasper, "An Investigation of the Fatigue of Metals," *Univ. Ill. Eng. Exp. Stn. Bull.*, 136, 1923.
17. J. A. Graham, (ed.), *Fatigue Design Handbook*, Society of Automotive Engineers, Warrendale, Pa., 1968.
18. J. O. Smith, "The Effect of Range of Stress on the Torsional Fatigue Strength of Steel," *Univ. Ill. Eng. Exp. Stn. Bull.*, 316, 1939.
19. R. Kuguel, "A Relation between Theoretical Stress Concentration Factor and Fatigue Notch Factor Deduced from the Concept of Highly Stressed Volume," *Am. Soc. Test. Mater. Proc.*, Vol. 61, 1961, pp. 732–748.
20. H. F. Moore and T. M. Jasper, "An Investigation of the Fatigue of Metals," *Univ. Ill. Eng. Exp. Stn. Bull.*, 142, 1924.
21. R. Cazaud, *Fatigue of Metals*, Philosophical Library, New York, 1953.
22. G. E. Dieter, *Mechanical Metallurgy*, 3rd ed., McGraw-Hill, New York, 1986.
23. A. F. Madayag, (ed.), *Metal Fatigue: Theory and Design*, Wiley, New York, 1969.
24. J. A. Collins, *Failure of Materials in Mechanical Design*, Wiley-Interscience, New York, 1981.
25. C. Lipson and N. J. Sheth, *Statistical Design and Analysis of Engineering Experiments*, McGraw-Hill, New York, 1973.
26. E. B. Haugen, *Probabilistic Mechanical Design*, Wiley-Interscience, New York, 1980.
27. L. Sors, *Fatigue Design of Machine Components*, Pergamon Press, Oxford, 1971.

References 1, 6, 8, 17, and 23 are general fatigue design texts or handbooks with several chapters devoted to the stress-life approach. Reference 6 is especially recommended.

References 4, 5, 22, and 24 are general design and materials texts with chapters devoted to fatigue.

References 3, 10, 21, and 27 give surveys of fatigue technology current to 1950, 1960, and 1970. Each has extensive lists of references. Reference 10 is a British perspective. References 21 and 27 were written by European authors.

Reference 9 deals with residual stresses, shot peening, and cold rolling.

References 25 and 26 deal with the statistical and probabilistic design aspects of fatigue.

PROBLEMS

SECTION 1.2

1.1. Given a steel with an ultimate strength of 100 ksi, estimate the allowable alternating stress for lives of 10^3, 10^4, 10^5, and 10^6 cycles. Solve this problem using the graphical method shown in Fig. 1.5 and Eq. (1.7). Repeat this procedure for a steel with an ultimate strength of 220 ksi.

1.2. Given below are the monotonic and rotating bending fatigue test data for three steels. Plot the fatigue data on log-log coordinates. Compare the test data to the estimate of the S–N curve made using the method shown in Fig. 1.5. (Data taken from Ref. 15.)

Material A		Material B		Material C	
S_u = 42 ksi BHN = 69		S_u = 102.6 ksi BHN = 209		S_u = 180 ksi BHN = 370	
Alternating Stress (ksi)	N_f (cycles)	Alternating Stress (ksi)	N_f (cycles)	Alternating Stress (ksi)	N_f (cycles)
32.3	4.5×10^4	81.4	4.4×10^4	110	2.4×10^4
30.3	2.4×10^5	74.7	8.5×10^4	105	3.1×10^4
27.9	8.0×10^5	71.6	1.4×10^5	100	4.5×10^4
25.9	1.5×10^6	64.7	6.3×10^5	98	8.7×10^4
25.4	2.7×10^6	62.1	1.9×10^6	95	1.5×10^5
24.4	7.8×10^6	59.6	2.9×10^6	92	1.0×10^{8}[a]
24.4	1.0×10^{7}[a]	58.8	6.4×10^5	90	1.0×10^{8}[a]
24.0	2.6×10^{7}[a]	58.7	1.4×10^6		
23.6	1.2×10^{7}[a]	57.2	1.0×10^{8}[a]		
23.5	2.2×10^{7}[a]	56.2	9.0×10^{7}[a]		

[a] Specimen did not fail.

1.3. Another method that can be used to estimate the S–N curve for steels is to construct on log–log coordinates a straight line between the true fracture stress, σ_f, at 1 cycle and the endurance limit, S_e, at 10^6 cycles. Compare this estimate to the fatigue data given in Problem 1.2. For steels the true fracture stress can be estimated as the ultimate strength, S_u, plus 50 ksi.

1.4 If a power law equation of the form $\sigma_a = AN_f^b$ is fit to each of the data sets contained in Problem 1.2, how do the constants A and b compare to the estimates shown in Eq. (1.7)?

1.5. Quite often the most difficult part of a fatigue design is to determine the required life of a component. As an exercise in estimating required life, determine the number of cycles for which the following components would need to be designed. Discuss the effect of estimate error on the life values. (*Hint*: Plot a range of values for each component on log coordinates.)
(a) Connecting rod in an automobile engine
(b) Motor shaft in an air conditioner
(c) Brake handle on a bicycle
(d) Latch assembly for a screen door

1.6. Estimate the minimum hardness (BHN) necessary for a steel to resist an alternating stress of ±100 ksi for 500,000 cycles. What would be the allowable alternating stress for this steel at 50,000 cycles?

1.7. A gearbox (containing steel gears) is currently being used in an application where it is driven by a 5-hp motor at 900 rpm. Tests indicate that the gearbox will last 500 hours under these conditions, with failure due to fatigue of the gear teeth. The capacity of the system is to be increased; a 6-hp motor will now be required and the rpm value will be decreased 10%. What effect will this have on the life (in hours) of the gearboxes?

SECTION 1.3

1.8. The values shown below correspond to long life tests ($>10^7$ cycles) for a nickel steel with an ultimate strength of 123 ksi. Determine the alternating stress, σ_a, mean stress, σ_m, stress ratio, R, and amplitude ratio, A, for each test. When these data are plotted on a Haigh diagram, how well does the diagram correlate with the Goodman, Gerber, and Morrow predictions? Estimate true fracture stress, σ_f as ultimate strength, S_u, plus 50 ksi. (Data taken from Ref. 16.)

Maximum Stress (ksi)	60	68	66	83	96	100	101
Minimum Stress (ksi)	−60	−54	−44	−33	0	29	51

1.9. Another method used to present mean stress fatigue data is to generate a family of curves on an S–N plot, with each curve representing a different stress ratio, R. Generate the curves for R values of −1, 0, and 0.5 for a steel with an ultimate strength of 100 ksi. For this example, use the Gerber relationship to generate these curves. Use the method shown in Fig. 1.5 to estimate the fully reversed ($R = -1$) fatigue behavior.

1.10. Another method used to present mean stress fatigue data is to generate a family of curves on an S–N plot, with each curve representing a different mean stress value, σ_m. Generate the curves for mean stress values of 0, 20, and 40 ksi for a steel with an ultimate strength of 100 ksi. For this example, use the Goodman relationship to generate these curves. Use the method shown in Fig. 1.5 to estimate the fully reversed ($\sigma_{\text{mean}} = 0$) fatigue behavior.

1.11. Draw on a Haigh diagram the estimated constant life lines at 10^3, 10^4, 10^5, and 10^6 cycles for a steel with an ultimate strength of 158 ksi. Use the Goodman relationship for these estimates. To estimate the fully reversed fatigue behavior, use the method shown in Fig. 1.5. How does the resulting Haigh diagram compare to the master diagram shown in Fig. 1.8? For the following questions, compare the results found using the estimated Haigh diagram and the master diagram (Fig. 1.8).

(a) At a life of 10^5 cycles, what is the allowable alternating stress for a mean stress of 50 ksi?

(b) If the maximum value for an alternating stress is 100 ksi, what is the allowable minimum stress for an infinite-life design?

(c) At a life of 10^4 cycles, what is the allowable alternating and mean stress for a stress ratio, R, of 0.2?

1.12. Given below are the monotonic and fully reversed fatigue properties of four steels. Also given are the results of mean stress fatigue tests. Compare the actual test results to life estimates made using the Goodman, Morrow, and Gerber relationships. (Data taken from Ref. 16.)

Monotonic and Fully Reversed Parameters

Material	S_u (ksi)	σ_f (ksi) (estimated)	S_{1000} (ksi) (estimated)	S_e (ksi)
A	91.5	140	68	33
B	123.3	170	110	64
C	117.5	165	105	64
D	101.6	150	85	48

Mean Stress Tests Results

Material	Mean Stress, σ_m (ksi)	Alternating Stress, σ_a (ksi)	Life to Failure, N_f (cycles)
A	10	40	1.0×10^5
A	9	35	7.0×10^5
A	36	24	3.2×10^5
A	52	17	4.4×10^4
B	15	75	1.2×10^5
B	13	66	2.2×10^5
B	26.5	62	5.9×10^5
B	52.5	52.5	4.8×10^5
C	7	63	2.2×10^5
C	16	63	6.5×10^5
C	30	69	2.1×10^5
C	40	60	1.2×10^5
D	5.2	47.3	1.0×10^6
D	5.2	46.8	8.0×10^{7}[a]
D	10.5	42	1.0×10^6
D	9.9	39.7	1.2×10^{9}[a]

[a] Specimen did not fail.

1.13. Given a material with an ultimate strength of 70 ksi, an endurance limit of 33 ksi, and a true fracture strength of 115 ksi, determine the allowable zero to maximum ($R = 0$) stress which can be applied for 10^3, 10^4, 10^5, and 10^6 cycles. Make predictions using the Goodman, Gerber, and Morrow relationships.

1.14. A component undergoes a cyclic stress with a maximum value of 75 ksi and a minimum value of −5 ksi. Determine the mean stress, stress range, stress amplitude, stress ratio, and amplitude ratio. If the component is made from a steel with an ultimate strength of 100 ksi, estimate its life using the Goodman relationship.

1.15. Another approach used to account for the mean stress effect is to consider the mean stress as an effective reduction of the monotonic true fracture strength in the Basquin equation (see Section 2.4). Stated in equation form, this is [17]

$$\sigma_a = (\sigma_f - \sigma_m)(2N_f)^b$$

where σ_a = stress amplitude
σ_m = mean stress
σ_f = true fracture strength
b = fatigue strength exponent
N_f = cycles to failure

For a material with the properties $S_u = 75$ ksi, $\sigma_f = 120$ ksi, and $b = -0.085$, determine the allowable alternating stress that could be superimposed over a tensile mean stress of 40 ksi to give a fatigue life of 5×10^5 cycles. Compare this value to predictions made using the Goodman and Gerber relationships.

1.16. Another relationship that is used to account for the effect of mean stress on infinite-life fatigue calculations is shown below. This method is developed in Ref. 8

and is a modification of the Smith–Watson–Topper parameter [Eq. (2.51)].

$$(\sigma_a + \sigma_m)\sigma_a = (S_e)^2$$

where σ_a = stress amplitude
σ_m = mean stress
S_e = endurance limit

Using the estimate that the endurance limit, S_e, is one-half of the ultimate strength, S_u, plot the relationship above on a Haigh diagram. Compare this method to the Goodman and Gerber relationships.

1.17. As discussed in Section 1.3, torsional fatigue tests on unnotched specimens often indicate that a mean shear stress, τ_m, has no effect on fatigue life when added to an alternating shear stress, τ_a. This relationship remains apparent until large maximum shear stress values, τ_{max} (where $\tau_{max} = \tau_m + \tau_a$) are reached.

The values shown below correspond to long life ($>10^7$ cycles) torsional tests on notch-free steel specimens. Determine the alternating shear stress, τ_a, mean shear stress, τ_m, stress ratio, R, and amplitude ratio, A, for each test. When these data are plotted on a Haigh diagram (τ_a versus τ_m), does it follow the relationship discussed earlier? (Data taken from Refs. 18 and 20.)

Monotonic properties

τ_y (yield point in shear) = 110.5 ksi

τ_u (modulus of rupture) = 137 ksi

S_y (in tension) = 153 ksi

S_u (in tension) = 162 ksi

Long-life fatigue test results

Maximum Shear Stress (ksi)	56	80	116	122
Minimum Shear Stress (ksi)	−56	−35	0	40

SECTION 1.4

1.18. The effect of size on fatigue behavior has been correlated with the volume of the thin layer of surface material subjected to high stresses. One such relationship was proposed by Kuguel [19]. He related the volume of material subjected to at least 95% of the maximum stress to the endurance limit using the relationship

$$S_e = S_{e0}\left(\frac{V}{V_0}\right)^{-0.034}$$

where S_{e0} is the endurance limit for a specimen of volume or size V_0, and S_e is the endurance limit at some other size V.

Using the relationship above, develop a modification factor for size, C_{size}, of the form shown in Eqs. (1.13) and (1.14). Use as the reference specimen an unnotched 0.3 in. diameter round bar undergoing rotating bending. (*Hint*: Assume that volume is proportional to the diameter cubed, d^3.)

1.19. Determine the diameter of a round bar undergoing rotating bending that will have the same highly stressed volume (subjected to at least 95% of the maximum stress) as a one inch by one inch square bar undergoing simple reversed bending. Assume that the two bars have the same maximum surface stress and the same effective length.

1.20. Given below are the material properties and fatigue test data for three steels. The fatigue data were generated using fully reversed axial loading on notch-free polished specimens. How do these data compare to the estimated S–N curve for rotating bending shown in Fig. 1.5? Determine a possible method or relationship to modify the estimated rotating bending S–N curve to account for axial loading. How well does the modification factor for endurance limit given in Eq. (1.15) work for these data? (Data taken from Ref. 20.)

Material A		Material B		Material C	
S_u = 42.4 ksi BHN = 69 S_e (bending) = 26 ksi		S_u = 117 ksi BHN = 224 S_e (bending) = 50 ksi		S_u = 133 ksi BHN = 262 S_e (bending) = 60 ksi	
Alternating Stress (ksi)	N_f (cycles)	Alternating Stress (ksi)	N_f (cycles)	Alternating Stress (ksi)	N_f (cycles)
27	2.0×10^4	33	4.0×10^4	62	6.0×10^3
26.5	2.4×10^4	31	7.5×10^4	50	1.5×10^4
24	3.8×10^4	29	3.7×10^5	46	1.7×10^4
22.5	7.0×10^4	28	4.0×10^5	40	1.8×10^5
20.5	1.6×10^5	27.5	3.8×10^{6}[a]	38	3.5×10^5
20	2.8×10^5	27.4	9.0×10^{6}[a]	37	2.5×10^5
19	5.0×10^5			36	2.6×10^5
17.5	1.5×10^6			34.5	3.4×10^{6}[a]
17	1.0×10^{7}[a]			34	7.6×10^{6}[a]

[a] Specimen did not fail.

1.21. Given below are the material properties and fatigue test data for three steels. The fatigue test data were generated using fully reversed torsional loading on notch-free polished specimens. Determine a possible method or relationship to approximate the alternating shear stress, τ_a, versus life, N, curve for steel using monotonic and/or rotating bending fatigue data. How well does the modification factor for endurance limit given in Eq. (1.16) work for these data? (Data taken from Ref. 15.)

Material A		Material B		Material C	
S_u = 42.4 ksi BHN = 69 S_e (bending) = 26 ksi		S_u = 113 ksi BHN = 247 S_e (bending) = 65 ksi		S_u = 188 ksi BHN = 380 S_e (bending) = 98 ksi	
Alternating Shear Stress (ksi)	N_f (cycles)	Alternating Shear Stress (ksi)	N_f (cycles)	Alternating Shear Stress (ksi)	N_f (cycles)
20	1.3×10^4	49.7	2.0×10^3	68.6	4×10^3
18	3.5×10^4	46.0	7.0×10^3	65.5	3.7×10^4
16	8.2×10^4	41.9	2.7×10^4	59.5	7.4×10^4
14.3	1.8×10^5	37.9	8.1×10^4	57.2	1.7×10^5
13.5	4.2×10^5	36.5	2.1×10^5	54.8	3.9×10^5
13.2	7.8×10^5	34.2	5.1×10^5	53.5	6.9×10^5
13.0	1.5×10^6	32.2	2.9×10^6 [a]	52.5	2.0×10^6 [a]
12.7	2.4×10^6 [a]	32.0	1.7×10^6	52.4	2.0×10^6 [a]

[a] Specimen did not fail.

1.22. Another method used to present the effect of surface roughness on the endurance limit of steel is to draw a family of curves corresponding to different surface finishes on a plot of endurance limit versus ultimate strength. Using the information contained on Fig. 1.13, draw on an S_e versus S_u plot the curves for mirror-polished, fine-ground, machined, hot-rolled, and forged surfaces. How do the forged and polished curves compare to Fig. 1.20? Use Eq. (1.2) to estimate the behavior of the steel with a mirror-polished surface.

1.23. Steel strips are to be used in service as one-directional bending members. Presetting has been suggested as a way to improve the fatigue resistance of these strips. Indicate with a sketch the direction that the strips should be flexed to give the proper preset. How would the zero-to-maximum ($R = 0$) fatigue strength of the preset and nonpreset strips compare at a life of 10^6 cycles? How would they compare at a life of 10^3 cycles?

1.24. It has been proposed to surface harden a steel leaf spring with a thickness, t, to improve its bending fatigue resistance. The steel is initially at 200 BHN and can be hardened to 350 BHN to whatever depth, d, is desired. However, the time required for deep hardening is excessive. Ignoring residual stresses, estimate the minimum d/t ratio to achieve the maximum bending fatigue strength at long lives ($N_f > 10^6$ cycles). What would the optimum value of d/t be if this member were to be subjected to cyclic axial loading?

1.25. Given below are the endurance limit values for a steel (S_u = 167 ksi) and the same steel after being nitrided. The endurance limit values correspond to axial, rotating bending and torsional loading. What conclusions can be drawn from these data regarding the effect of nitriding and various types of loading on fatigue behavior? (Data taken from Ref. 21.)

Condition of Steel	S_e(bending)	S_e(axial)	τ_e(torsion)
Nonnitrided	83 ksi	79 ksi	45.7 ksi
Nitrided	110 ksi	79 ksi	59.4 ksi

1.26. A steel alloy has an ultimate strength of 100 ksi, a true fracture strength of 150 ksi, and a completely reversed endurance limit of 50 ksi. Shot peening the surface of a specimen made from this steel induces a compressive residual stress of −50 ksi, increases the hardness from 200 to 250 BHN, and increases the surface roughness (AA) from 5 to 50 μin. Estimate the endurance limit for the specimen in the peened condition.

1.27. It is recommended in Ref. 6 that in the absence of specific data, a conservative method that can be used to account for the overall effect of shot peening is to use a surface finish factor of 1.0, regardless of the surface finish prior to peening. Referring to Fig. 1.13, discuss how this method, although conservative, will account for the trends outlined in Section 1.4.4.

1.28. High strength steel bars as described in Example 1.2 are to be used as leaf springs. The springs will be subjected to zero-to-maximum ($R = 0$) three-point flexural loading. The bars are 1.50 in. wide and 0.192 in. thick. The bars are in three different conditions:

1. *Preset.* The bars are initially overloaded in the same direction as the cyclic loading. This sets up a beneficial compressive residual stress.
2. *Shot peened and fully stress relieved.* This causes a rough surface finish and complete removal of any residual stresses.
3. *Strain peened.* In this process the spring is loaded in the same direction as the cyclic loading and then shot peened. This produces a much larger residual stress than would be caused by conventional shot peening.

Determine, for the three conditions, the zero-to-maximum surface stress that will allow the bars to have an infinite life. The necessary data for the three conditions are given below. Use the Goodman relationship for these calculations. (Data taken from Ref. 14.)

Preset

Hardness = 48 Rockwell C (BHN ≈ 465)
Residual surface stress = −35 ksi
Surface roughness (AA) = 12 μin.

Shot peened and stress relieved

Hardness = 50 Rockwell C (BHN ≈ 495)
Residual surface stress = 0 ksi
Surface roughness (AA) = 125 μin.

Strain peened

Hardness = 50 Rockwell C (BHN ≈ 495)
Residual surface stress = −140 ksi
Surface roughness (AA) = 125 μin.

Compare the predictions to the actual test results:

Preset:	0 to 128 ksi
Peened and relieved:	0 to 78 ksi
Strain peened:	0 to 176 ksi

1.29. You are handed a bar of AISI 4140 steel (S_u = 240 ksi) and asked to prepare specimens that will give the maximum endurance limit. What would you do and why? Calculate the endurance limit that you can expect to achieve with your samples.

1.30. Sketch the S–N curve for a steel specimen with an ultimate strength of 100 ksi, a 0.3 in. diameter, and a notch-free mirror-polished finish which is subjected to rotating bending. Sketch on the same plot the S–N curves resulting from each of the changes given below.

(a) 6-in.-diameter specimen
(b) Rough lathe finish (AA = 500 μin.)
(c) Initial axial overload of 80 ksi
(d) Chrome plated
(e) Decarburization of surface
(f) Shot peening
(g) Periodic peak 70-ksi overloads

Repeat this exercise for steels with ultimate strengths of 50 and 200 ksi.

1.31. Given below are the parameters for a notch-free round steel bar.

Diameter: 2.5 in.
Surface roughness: lathe finished (AA = 125 μin.)
Material: SAE 1035 CD, S_u = 92 ksi

Determine for an infinite fatigue life the allowable:

(a) Fully reversed axial load
(b) Fully reversed torsional load

1.32. Given below are the parameters describing two steel shafts. Shaft A, which is machined from cold-drawn stock, is to be replaced by shaft B, which is forged. Determine the diameter of shaft B that will allow the same fully reversed torsional load for 10^6 cycles as shaft A.

Shaft A

S_u = 80 ksi
Surface finish = machined (AA = 125 μin.)
Diameter = 1.5 in.

Shaft B

S_u = 90 ksi
Surface finish = as forged
Diameter to be determined

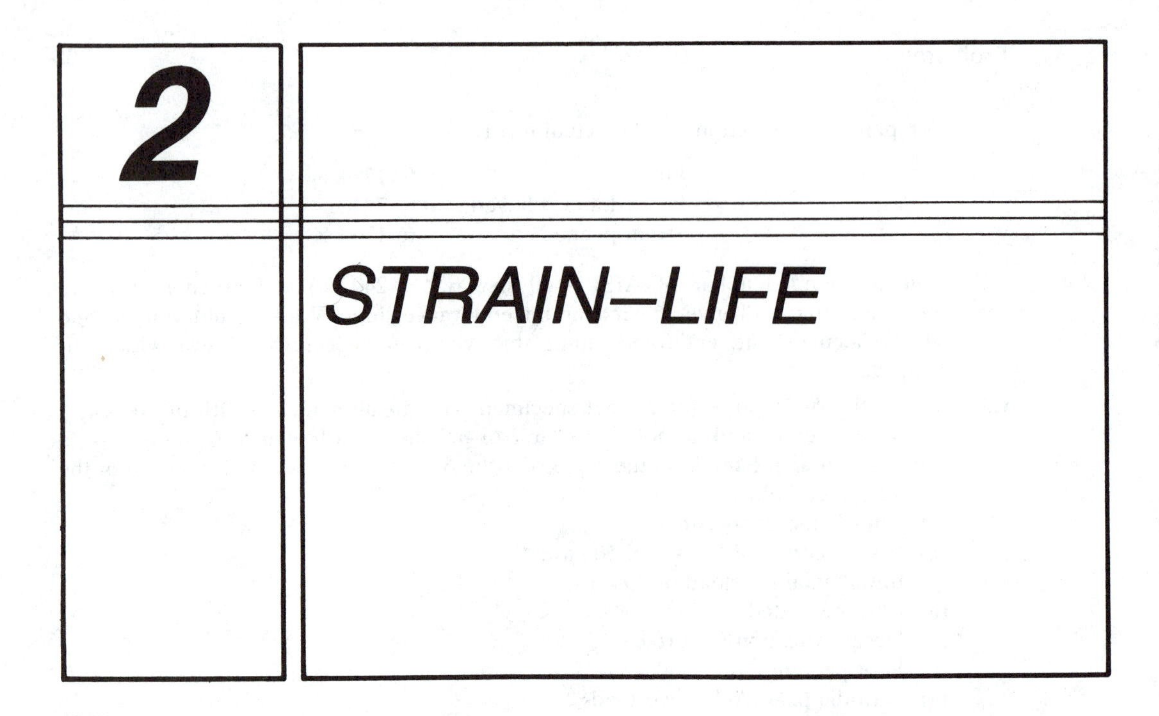

2.1 INTRODUCTION

The strain–life method is based on the observation that in many components the response of the material in critical locations (notches) is strain or deformation dependent. When load levels are low, stresses and strains are linearly related. Consequently, in this range, load-controlled and strain-controlled test results are equivalent. (Recall from Chapter 1 that stress–life data are generated from load-controlled tests.) At high load levels, in the low cycle fatigue (LCF) regime, the cyclic stress–strain response and the material behavior are best modeled under strain-controlled conditions.

Early fatigue research showed that damage is dependent on plastic deformation or strain. In the strain–life approach the plastic strain or deformation is directly measured and quantified. As discussed in Chapter 1, the stress–life approach does not account for plastic strain. At long lives, where plastic strain is negligible and stress and strain are easily related, the strain–life and stress–life approaches are essentially the same.

Although most engineering structures and components are designed such that the nominal loads remain elastic, stress concentrations often cause plastic strains to develop in the vicinity of notches. Due to the constraint imposed by the elastically stressed material surrounding the plastic zone, deformation at the notch root is considered strain-controlled. The strain–life method assumes that smooth specimens tested under strain-control can simulate fatigue damage at the notch root of an engineering component. Equivalent fatigue damage (and fatigue life) is assumed to occur in the material at the notch root and in the smooth

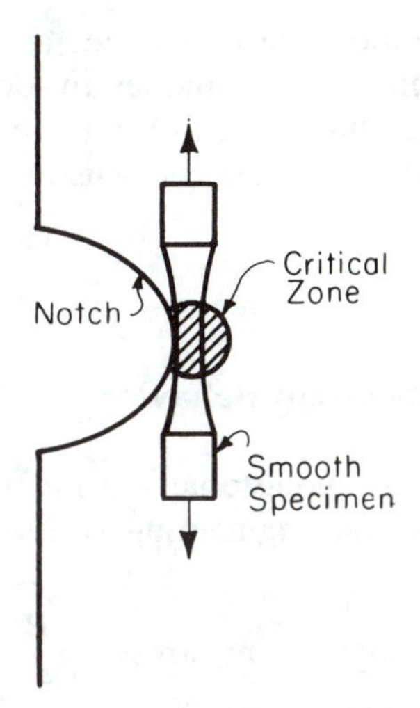

Figure 2.1 Equally stressed volume of material.

specimen when both are subjected to identical stress–strain histories. As seen in Fig. 2.1, the laboratory specimen models an equally stressed volume of material at the notch root.

Crack growth is not explicitly accounted for in the strain–life method. Rather, failure of the component is assumed to occur when the "equally stressed volume of material" fails. Because of this, strain–life methods are often considered "initiation" life estimates. For some applications the existence of a crack is an overly conservative criterion for component failure. In these situations, fracture mechanics methods may be employed to determine crack propagation life from an assumed initial crack size to a final crack length. Total lives are then reported as the sum of the initiation and propagation segments. The fracture mechanics approach is presented in Chapter 3.

The local strain–life approach has gained acceptance as a useful method of evaluating the fatigue life of a notched component. Both the American Society for Testing and Materials (ASTM) and the Society of Automotive Engineers (SAE) have recommended procedures and practices for conducting strain-controlled tests and using these data to predict fatigue lives [1–5].

Fatigue life predictions may be made using the strain–life approach with the following information:

1. Material properties obtained from smooth specimen strain-controlled laboratory fatigue data (cyclic stress–strain response and strain–life data)
2. Stress–strain history at the critical location (e.g., at a notch)
3. Techniques for identifying damaging events (cycle counting)
4. Methods to incorporate mean stress effects
5. Damage summation technique (e.g., Miner's rule)

In this chapter we review the necessary material behavior background needed for an understanding of the strain–life approach. Fundamental equations used in the strain–life method are presented, as well as modified equations to account for mean stress effects. Techniques to identify damaging events and to sum damage (items 3 and 5 above) are presented in Chapter 5. Fatigue analysis of notches using the strain–life approach is discussed in Chapter 4.

2.2 MATERIAL BEHAVIOR

2.2.1 Monotonic Stress–Strain Behavior

Basic definitions. A monotonic tension test of a smooth specimen is usually used to determine the engineering stress–strain behavior of a material where

$$S = \text{engineering stress} = \frac{P}{A_0} \tag{2.1}$$

$$e = \text{engineering strain} = \frac{l - l_0}{l_0} = \frac{\Delta l}{l_0} \tag{2.2}$$

The following terms are shown in Fig. 2.2:

P = applied load

l_0 = original length

d_0 = original diameter

A_0 = original area

l = instantaneous length

d = instantaneous diameter

A = instantaneous area

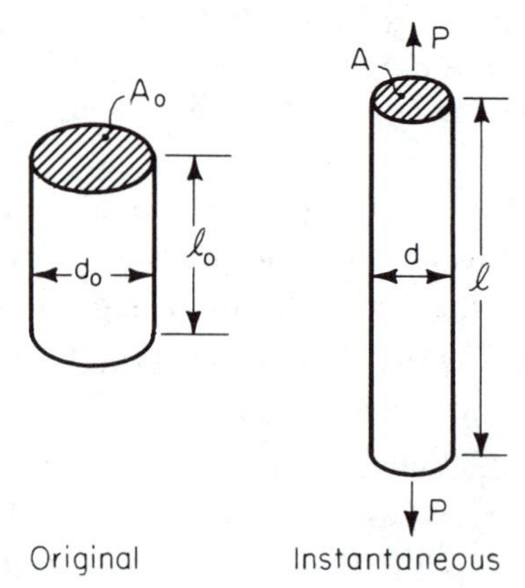

Figure 2.2 Original and deformed (instantaneous) configuration of test specimen.

In tension the true stress is larger than the engineering stress, due to changes in cross-sectional area during deformation.

$$\sigma = \text{true stress} = \frac{P}{A} \tag{2.3}$$

Similarly, until necking occurs in the specimen, true strain is smaller than engineering strain. *True* or *natural strain,* based on an instantaneous gage length l, is defined as

$$\epsilon = \text{true strain} = \int_{l_0}^{l} \frac{dl}{l} = \ln \frac{l}{l_0} \tag{2.4}$$

Figure 2.3 compares the monotonic tension stress–strain curve using true stress and strain and the engineering values.

True and engineering stress–strain. True stress and strain can be related to engineering stress and strain. The instantaneous length is

$$l = l_0 + \Delta l \tag{2.5}$$

Combining Eqs. (2.4) and (2.5), the true strain is

$$\epsilon = \ln \frac{l_0 + \Delta l}{l_0} \tag{2.6}$$

$$= \ln\left(1 + \frac{\Delta l}{l_0}\right) \tag{2.7}$$

From Eq. (2.2) the true strain in terms of engineering strain is

$$\epsilon = \ln(1 + e) \tag{2.8}$$

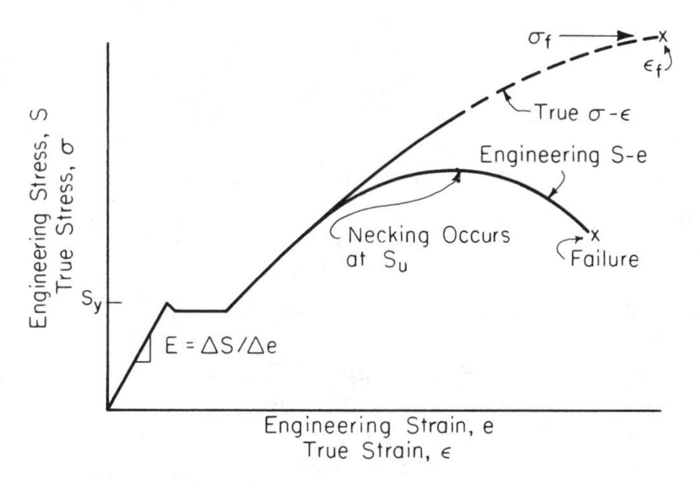

Figure 2.3 Comparison of engineering and true stress–strain.

Equation (2.8) is only valid up to necking. At necking the strain is no longer uniform throughout the gage length.

Assuming that the volume of the material remains constant during straining, we have

$$A_0 l_0 = Al = \text{constant} \tag{2.9}$$

Restating gives

$$\frac{A_0}{A} = \frac{l}{l_0} \tag{2.10}$$

True strain can be stated in terms of cross-sectional area

$$\epsilon = \ln\frac{l}{l_0} = \ln\frac{A_0}{A} \tag{2.11}$$

From Eq. (2.1),

$$P = SA_0 \tag{2.12}$$

and since

$$\sigma = \frac{P}{A} \tag{2.3}$$

true stress can be stated in terms of engineering stress:

$$\sigma = S\frac{A_0}{A} \tag{2.13}$$

Combining Eqs. (2.8) and (2.11) (valid only up to necking) gives us

$$\epsilon = \ln(1 + e) = \ln\frac{A_0}{A} \tag{2.14}$$

or

$$\frac{A_0}{A} = 1 + e \tag{2.15}$$

Therefore, true stress can be stated as a function of engineering stress and strain using Eqs. (2.13) and (2.15).

$$\sigma = S(1 + e) \tag{2.16}$$

This relation is valid only up to necking.

Stress–strain relationships. The total true strain ϵ_t in a tension test can be separated into elastic and plastic components:

1. *Linear elastic strain:* that portion of the strain which is recovered upon unloading, ϵ_e

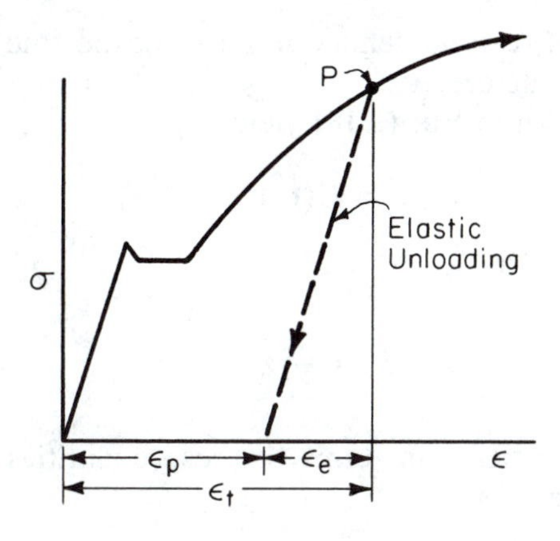

Figure 2.4 Elastic and plastic strain.

2. *Plastic strain (nonlinear):* that portion which cannot be recovered on unloading, ϵ_p (see Fig. 2.4)

Stated in equation form,

$$\epsilon_t = \epsilon_e + \epsilon_p \tag{2.17}$$

For most metals a log–log plot of true stress versus true plastic strain is modeled as a straight line. Consequently, this curve can be expressed using a power function

$$\sigma = K(\epsilon_p)^n \tag{2.18}$$

or

$$\epsilon_p = \left(\frac{\sigma}{K}\right)^{1/n} \tag{2.19}$$

where K is the *strength coefficient* and n is the *strain hardening exponent.*

At fracture two important quantities can be defined (see Fig. 2.3). These are true fracture strength and true fracture ductility. *True fracture strength,* σ_f, is the true stress at final fracture.

$$\sigma_f = \frac{P_f}{A_f} \tag{2.20}$$

where A_f is the area at fracture and P_f is the load at fracture.

True fracture ductility, ϵ_f, is the true strain at final fracture. This value can be defined in terms of the initial cross-sectional area and the area at fracture.

$$\epsilon_f = \ln\frac{A_0}{A_f} = \ln\frac{1}{1 - \text{RA}} \tag{2.21}$$

$$\text{RA} = \frac{A_0 - A_f}{A_0} = \text{reduction in area}$$

The strength coefficient, K, can be defined in terms of the true stress at fracture, σ_f, and the true strain at fracture, ϵ_f.

Substituting σ_f and ϵ_f into Eq. (2.18) yields

$$\sigma_f = K(\epsilon_f)^n \tag{2.22}$$

Rearranging gives

$$K = \frac{\sigma_f}{\epsilon_f^n} \tag{2.23}$$

We can also define plastic strain in terms of these quantities. Combining Eqs. (2.23) and (2.19), we have

$$\epsilon_p = \left(\frac{\sigma}{\sigma_f/\epsilon_f^n}\right)^{1/n} \tag{2.24}$$

$$= \left(\frac{\sigma\epsilon_f^n}{\sigma_f}\right)^{1/n} \tag{2.25}$$

$$= \epsilon_f\left(\frac{\sigma}{\sigma_f}\right)^{1/n} \tag{2.26}$$

The total strain can be expressed as

$$\epsilon_t = \epsilon_e + \epsilon_p \tag{2.17}$$

The elastic strain is defined as

$$\epsilon_e = \frac{\sigma}{E} \tag{2.27}$$

The expression for plastic strain is given in Eq. (2.19). Equation (2.17) may then be rewritten as

$$\epsilon_t = \frac{\sigma}{E} + \left(\frac{\sigma}{K}\right)^{1/n} \tag{2.28}$$

2.2.2 Cyclic Stress–Strain Behavior

Monotonic stress–strain curves have long been used to obtain design parameters for limiting stresses on engineering structures and components subjected to static loading. Similarly, cyclic stress–strain curves are useful for assessing the durability of structures and components subjected to repeated loading.

The response of a material subjected to cyclic inelastic loading is in the form of a hysteresis loop, as shown in Fig. 2.5. The total width of the loop is $\Delta\epsilon$ or the total strain range. The total height of the loop is $\Delta\sigma$ or the total stress range.

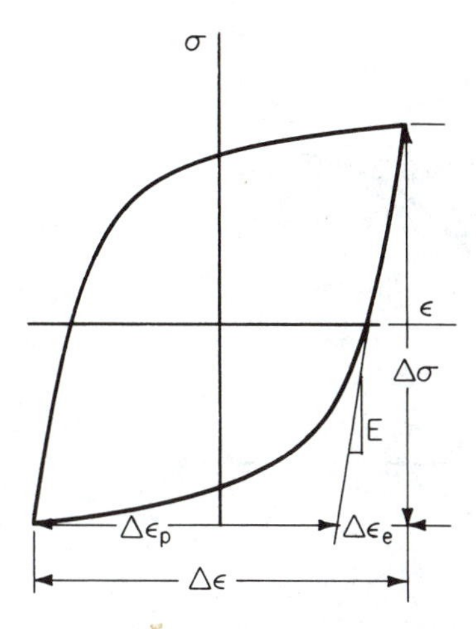

Figure 2.5 Hysteresis loop.

These can be stated in terms of amplitudes:

$$\epsilon_a = \frac{\Delta\epsilon}{2}$$

where ϵ_a is the strain amplitude and

$$\sigma_a = \frac{\Delta\sigma}{2}$$

where σ_a is the stress amplitude. The total strain is the sum of the elastic and plastic strain ranges,

$$\Delta\epsilon = \Delta\epsilon_e + \Delta\epsilon_p \tag{2.29}$$

or in terms of amplitudes,

$$\frac{\Delta\epsilon}{2} = \frac{\Delta\epsilon_e}{2} + \frac{\Delta\epsilon_p}{2} \tag{2.30}$$

Using Hooke's law, the elastic term may be replaced by $\Delta\sigma/E$.

$$\frac{\Delta\epsilon}{2} = \frac{\Delta\sigma}{2E} + \frac{\Delta\epsilon_p}{2} \tag{2.31}$$

The area within the loop is the energy per unit volume dissipated during a cycle. It represents a measure of the plastic deformation work done on the material.

The *Bauschinger effect* [6] is usually observed in most metals. This effect is described graphically in Fig. 2.6. Shown in Fig. 2.6a is the material response of a bar loaded past the yield strength, σ_y, to some value, $\sigma_{\max}$. In Fig. 2.6b, the material is unloaded and then loaded in compression to $-\sigma_{\max}$. Notice that under compressive loading, inelastic (plastic) strains develop before $-\sigma_y$ is reached. This behavior is known as the Bauschinger effect.

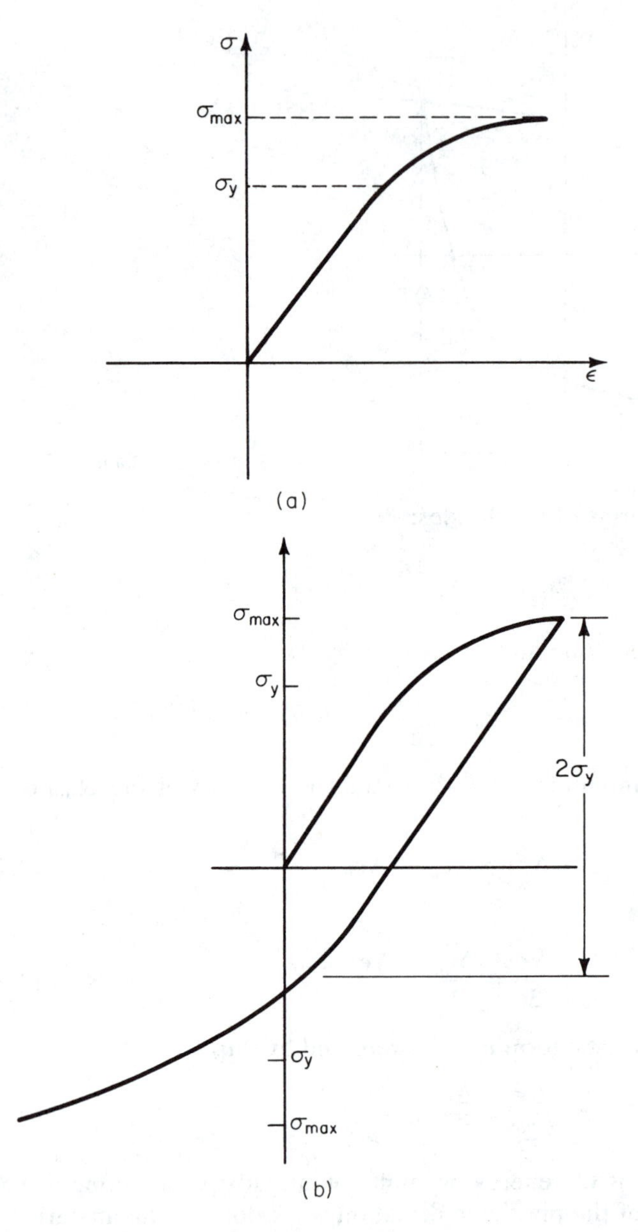

Figure 2.6 Bauschinger effect.

2.2.3 Transient Behavior: Cyclic Strain Hardening and Softening

The stress–strain response of metals is often drastically altered due to repeated loading. Depending on the initial conditions of a metal (i.e., quenched and

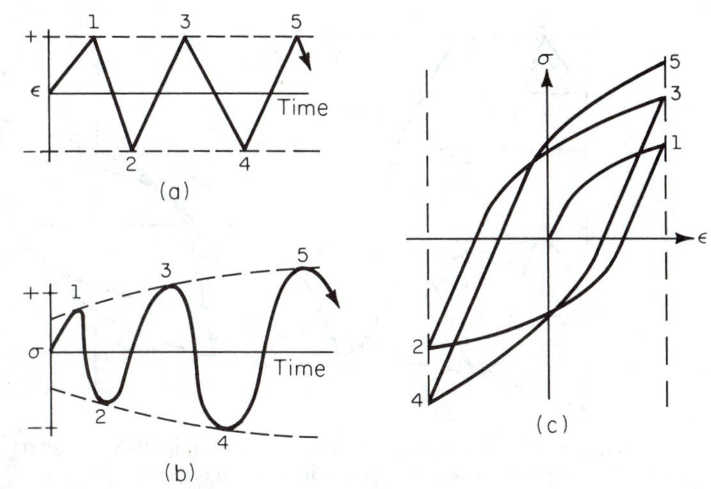

Figure 2.7 Cyclic hardening: (a) constant strain amplitude; (b) stress response (increasing stress level); (c) cyclic stress–strain response.

tempered, or annealed) and the test conditions, a metal may:

1. Cyclically harden
2. Cyclically soften
3. Be cyclically stable
4. Have mixed behavior (soften or harden depending on the strain range)

Figure 2.7b shows the stress response of a material loaded in strain-control. Figure 2.7c shows the hysteresis loops for the first two cycles. As seen, the maximum stress obtained increases with each cycle of strain. This is known as *strain hardening*. Conversely, if the maximum stress *decreases* with repeated straining, *strain softening* occurs as shown in Fig. 2.8.

The reason materials soften or harden appears to be related to the nature and stability of the dislocation substructure of the material [7]. Generally:

1. For a *soft material,* initially the dislocation density is low. The density rapidly increases due to cyclic plastic straining contributing to significant cyclic *strain hardening*.
2. For a *hard material* subsequent strain cycling causes a rearrangement of dislocations which offers less resistance to deformation and the material cyclically *softens*.

Manson [8] observed that the ratio of monotonic ultimate strength, σ_{ult}, to the 0.2% offset yield strength, σ_y, can be used to predict whether the material

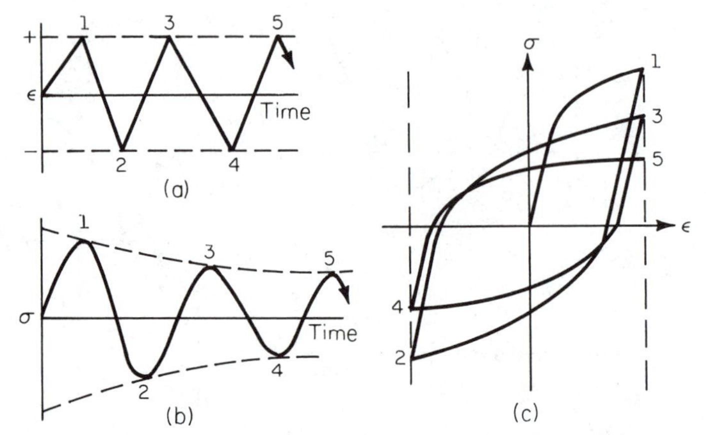

Figure 2.8 Cyclic softening: (a) constant strain amplitude; (b) stress response (decreasing stress level); (c) cyclic stress–strain response.

will soften or harden. If

$$\frac{\sigma_{ult}}{\sigma_y} > 1.4 \qquad \text{the material will cyclically harden}$$

$$\frac{\sigma_{ult}}{\sigma_y} < 1.2 \qquad \text{the material will cyclically soften}$$

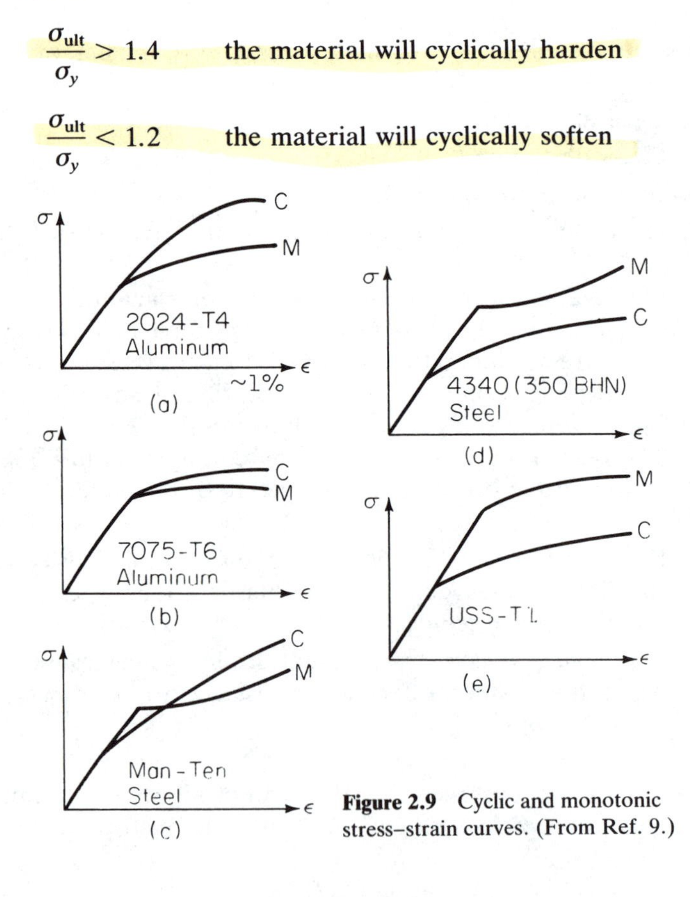

Figure 2.9 Cyclic and monotonic stress–strain curves. (From Ref. 9.)

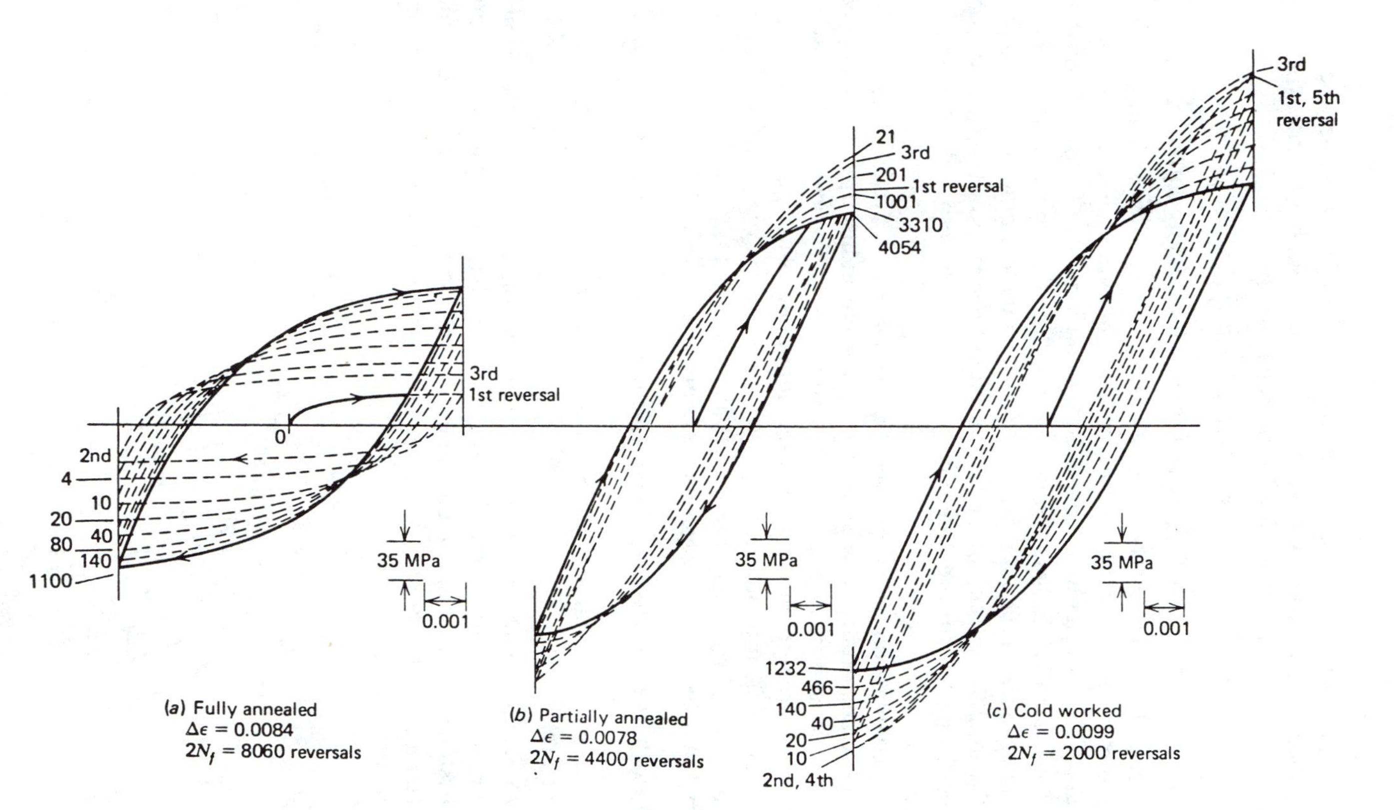

Figure 2.10 Hysteresis response of OFHC copper. (From Ref. 10.)

A large change in cyclic response is not expected for ratios between 1.2 and 1.4 and prediction is difficult. Also, the *monotonic* strain hardening exponent, n, can be used to predict the material's cyclic behavior. In general, if

$n > 0.20$ the material will cyclically harden

$n < 0.10$ the material will cyclically soften

Generally, transient behavior (strain hardening or softening) occurs only during the early fatigue life. After this, the material achieves a *cyclically stable condition.* This is usually achieved after approximately 20 to 40% of the fatigue life. Consequently, fatigue properties are usually specified at "half-life" (approximately 50% of the total fatigue life) when the material response is stabilized.

Figure 2.9 presents the cyclic and monotonic stress–strain curves for several materials. Figure 2.10 presents the hysteresis response of OFHC copper in three conditions.

A comparison between the monotonic and cyclic stress–strain curve provides a quantitative assessment of cyclically induced changes in mechanical behavior. As shown in Fig. 2.9e, a material that cyclically softens will have a cyclic yield strength lower than the monotonic. This points out the potential danger of using monotonic properties to predict cyclic strains. For example, monotonic properties may predict strains that are fully elastic, when in fact the material may experience large amounts of cyclic plastic strain.

2.2.4 Cyclic Stress–Strain Curve Determination

Cyclic stress–strain curves may be obtained from tests by several methods. Two of these are:

1. *Companion samples.* A series of companion samples are tested at various strain levels until the hysteresis loops become stabilized. The stable hysteresis loops are then superimposed and the tips of the loops are connected as shown in Fig. 2.11. This method is time consuming and requires many specimens.

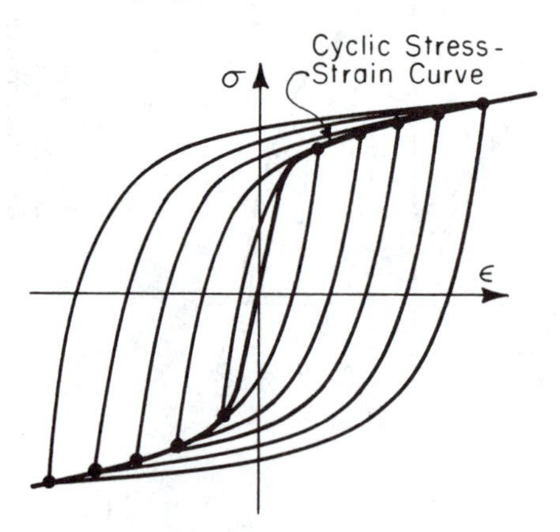

Figure 2.11 Cyclic stress–strain curve obtained by connecting tips of stabilized hysteresis loops.

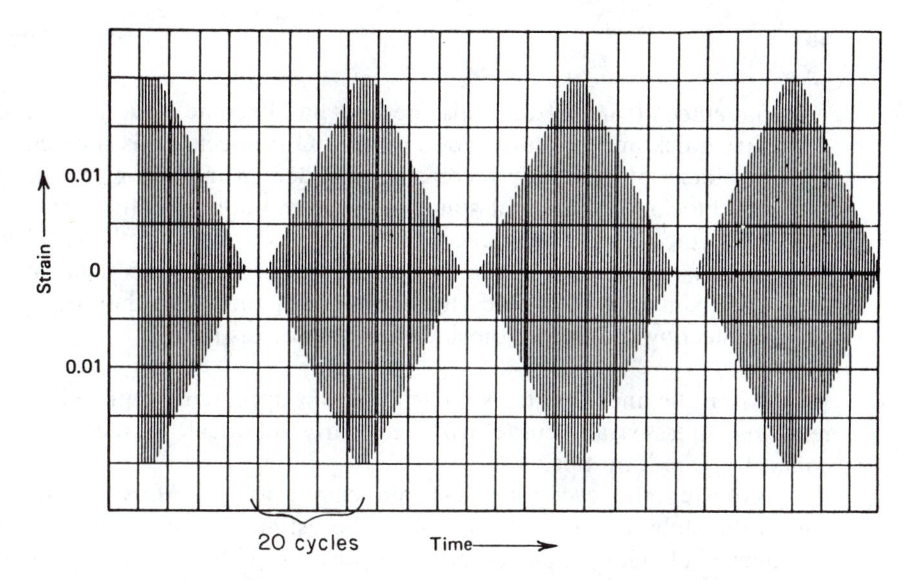

Figure 2.12 Incremental step test. (Data from Ref. 11.)

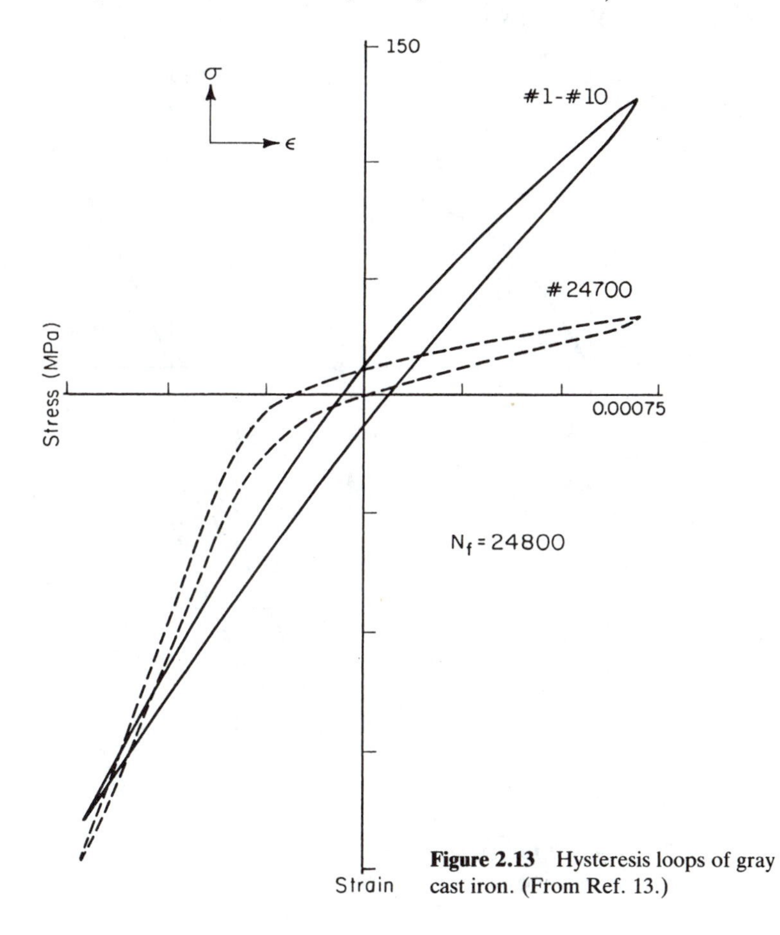

Figure 2.13 Hysteresis loops of gray cast iron. (From Ref. 13.)

2. *Incremental Step Test.* This method has become widely accepted, as it is very quick and produces good results. One specimen is subjected to a series of blocks of gradually increasing and decreasing strain amplitude. After a few blocks the material stabilizes. For example, for the test shown in Fig. 2.12, the loading block contains 20 cycles per half-block. The material response generally stabilizes after about three to four blocks and fails after approximately 20. The cyclic stress–strain curve can then be determined by connecting the tips of the stabilized hysteresis loops.

After the incremental step test, if the specimen is pulled to failure, the resulting stress–strain curve will be nearly identical to the one obtained by connecting the loop tips.

Knowing the cyclic stress–strain curve, use of *Massing's hypothesis* [12] allows the stabilized hysteresis loop to be estimated for a material that exhibits symmetric behavior in tension and compression. (The hysteresis loop of gray cast iron, for example, exhibits a different response in tension and compression, as shown in Fig. 2.13.)

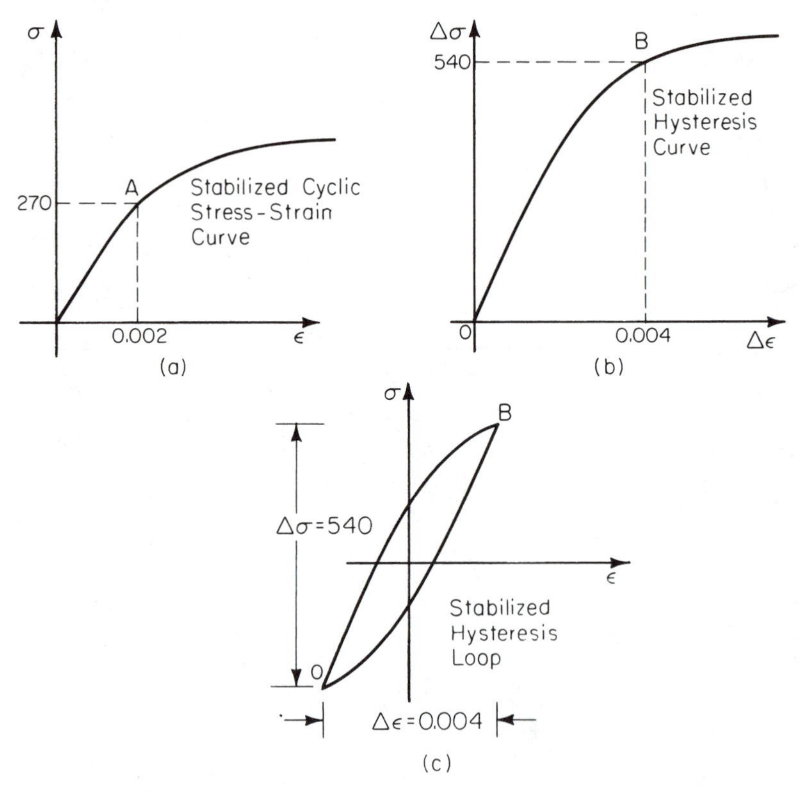

Figure 2.14 Development of stabilized hysteresis curve from cyclic stress–strain curve using Massing's hypothesis.

Massing's hypothesis states that the stabilized hysteresis loop may be obtained by doubling the cyclic stress–strain curve. By doubling the stress and strain value from the stabilized cyclic stress–strain curve, a corresponding point on the hysteresis loop is obtained as shown in Fig. 2.14. For example, by doubling the values corresponding to point A on the cyclic stress–strain curve in Fig. 2.14a, point *B* on the hysteresis loop (Fig. 2.14b) is obtained. Figure 2.14c shows the hysteresis loop for a fully reversed test. Note the location of point 0 on the hysteresis curve in Fig. 2.14b and c.

2.3 STRESS–PLASTIC STRAIN POWER LAW RELATION

Analogous to the monotonic stress–strain curve, a log–log plot of the completely reversed stabilized cyclic true stress versus true plastic strain can be approximated by a straight line as shown in Fig. 2.15.

Similar to the monotonic relationship, we can develop a power law function

$$\sigma = K'(\epsilon_p)^{n'} \tag{2.32}$$

where σ = cyclically stable stress amplitude
ϵ_p = cyclically stable plastic strain amplitude
K' = cyclic strength coefficient
n' = cyclic strain hardening exponent

For most metals the value of n' usually varies between 0.10 and 0.25, with an average value close to 0.15.

Rearranging Eq. (2.32) gives us

$$\epsilon_p = \left(\frac{\sigma}{K'}\right)^{1/n'} \tag{2.33}$$

The total strain is the sum of the elastic and plastic components. Using Eq. (2.33) and Hooke's law, the total strain can be written

$$\epsilon = \frac{\sigma}{E} + \left(\frac{\sigma}{K'}\right)^{1/n'} \tag{2.34}$$

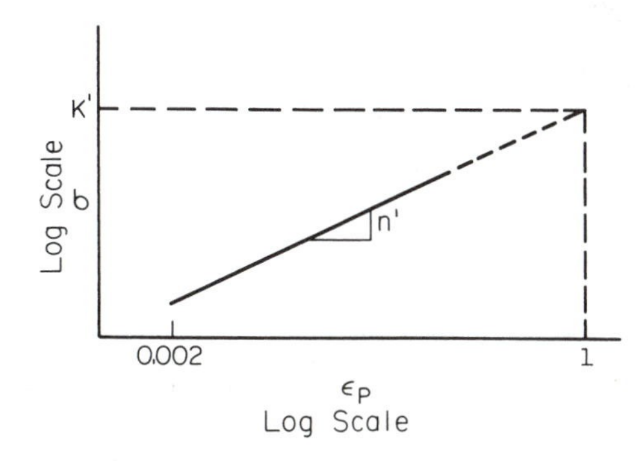

Figure 2.15 Log–log plot of true cyclic stress versus true cyclic plastic strain.

The equation of the hysteresis loop can be derived from the equation of the cyclic σ–ϵ curve [Eq. (2.34)] using Massing's hypothesis. Recall that Massing's hypothesis allows us to obtain the hysteresis loop by doubling the cyclic stress–strain curve as shown in Fig. 2.14. Given an arbitrary point, P_1, on the cyclic stress–strain curve, as shown in Fig. 2.16a, the corresponding values of stress and strain are σ_1 and ϵ_1, respectively. These values, σ_1 and ϵ_1, are related by the equation of the cyclic σ–ϵ curve. Equation (2.34) may be written

$$\epsilon_1 = \frac{\sigma_1}{E} + \left(\frac{\sigma_1}{K'}\right)^{1/n'} \tag{2.35}$$

From Massing's hypothesis a point corresponding to P_1 may be located on the hysteresis curve as shown in Fig. 2.16b. The coordinates of this point are $\Delta\sigma_1$ and $\Delta\epsilon_1$, where

$$\Delta\sigma_1 = 2\sigma_1$$
$$\Delta\epsilon_1 = 2\epsilon_1$$

Rearranging these equations, we obtain

$$\frac{\Delta\sigma_1}{2} = \sigma_1 \qquad \frac{\Delta\epsilon_1}{2} = \epsilon_1$$

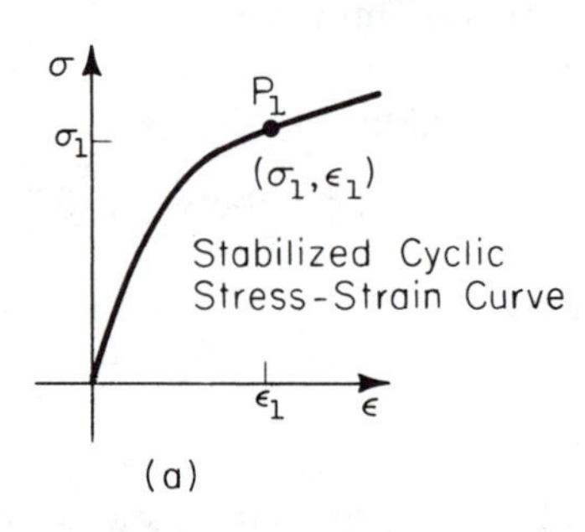

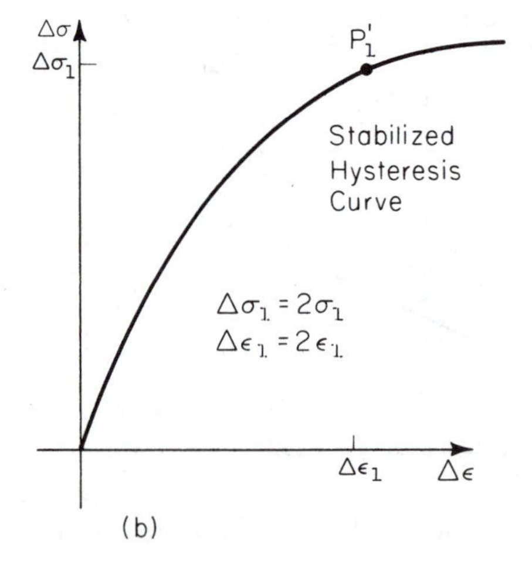

Figure 2.16 Stabilized cyclic stress–strain and hysteresis curves.

These can then be substituted into Eq. (2.35) to obtain the equation of the hysteresis loop:

$$\frac{\Delta\epsilon_1}{2} = \frac{\Delta\sigma_1}{2E} + \left(\frac{\Delta\sigma_1}{2K'}\right)^{1/n'}$$

Multiplying both sides by 2, the general hysteresis curve equation is

$$\Delta\epsilon = \frac{\Delta\sigma}{E} + 2\left(\frac{\Delta\sigma}{2K'}\right)^{1/n'} \tag{2.36}$$

(Since this equation was derived for an arbitrary point, P_1, the subscripts may be omitted.)

The relationship between the hysteresis stress–strain curve and the cyclic stress–strain curve can be made clearer with the following example.

Example 2.1

Consider a test specimen with the following material properties:

E = modulus of elasticity = 30×10^3 ksi
n' = cyclic strain hardening exponent = 0.202
K' = cyclic strength coefficient = 174.6 ksi

The specimen is subjected to a fully reversed cyclic strain with a strain range, $\Delta\epsilon$, of 0.04. Determine the stress–strain response of the material.

Solution Figure E2.1a shows the strain history. On the initial application of strain (point 1) the material response follows the cyclic stress–strain curve [Eq. (2.34)]:

$$\epsilon_1 = \frac{\sigma_1}{E} + \left(\frac{\sigma_1}{K'}\right)^{1/n'}$$

Substituting in the material properties and a strain value of 0.02 gives

$$0.02 = \frac{\sigma_1}{30 \times 10^3 \text{ ksi}} + \left(\frac{\sigma_1}{174.6 \text{ ksi}}\right)^{1/0.202}$$

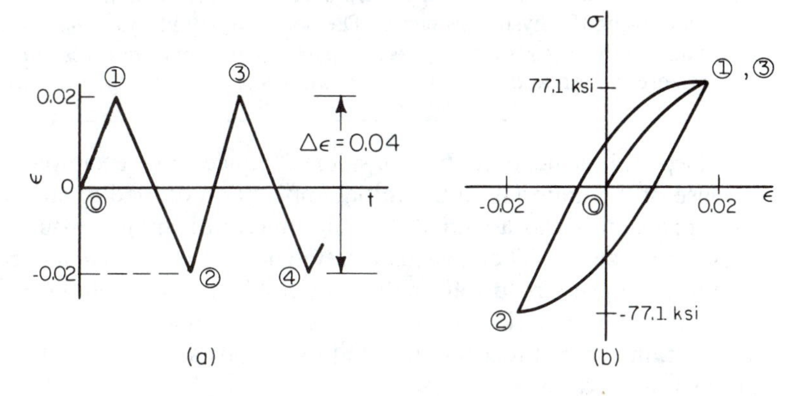

Figure E2.1 (a) Strain history; (b) stress–strain response.

The stress value at point 1 may be determined by solving this equation by iteration. As shown in Fig. E2.1b, this yields

$$\sigma_1 = 77.1 \text{ ksi}$$

The cyclic stress–strain curve is used only for the initial application of strain. On all successive strain reversals, the material response is modeled using the hysteresis curve [Eq. (2.36)]:

$$\Delta\epsilon = \frac{\Delta\sigma}{E} + 2\left(\frac{\Delta\sigma}{2K'}\right)^{1/n'}$$

Substituting in the material properties and a change in strain, $\Delta\epsilon$, of 0.04 gives

$$0.04 = \frac{\Delta\sigma}{30 \times 10^3} + 2\left(\frac{\Delta\sigma}{2 \times 174.6}\right)^{1/0.202}$$

The change in stress can be obtained by using an iterative method:

$$\Delta\sigma = 154.2 \text{ ksi}$$

The stress and strain values corresponding to point 2 can now be determined by subtracting the changes in stress and strain $(\Delta\sigma, \Delta\epsilon)$ from the values at point 1 (σ_1, ϵ_1).

$$\begin{aligned} \epsilon_2 &= \epsilon_1 - \Delta\epsilon \\ &= 0.02 - 0.04 \\ &= -0.02 \\ \sigma_2 &= \sigma_1 - \Delta\sigma \\ &= (77.1 - 154.2) \text{ ksi} \\ &= -77.1 \text{ ksi} \end{aligned}$$

(*Note:* The actual algebraic sign of the changes in stress and strain must be accounted for by considering the sign of the change in applied strain.)

The stress and strain values corresponding to point 3 can be determined by again using the hysteresis curve. The solution would show that the material response would return to point 1. As expected, the material response forms a closed hysteresis loop and all successive strain cycles would follow this loop.

Two points need to be considered concerning Example 2.1. First is the response of the material on the initial application of load or strain. In the solution of the problem it was assumed that the material response would follow the cyclic stress–strain curve. There is justification to an alternative approach which says that the response would follow the monotonic stress–strain curve. This argument states that the material response of a virgin material follows the monotonic stress–strain curve. Most fatigue life predictions are not greatly affected by which approach is used.

The second point that needs to be considered is that this analysis assumes that the material exhibits cyclically stable response from the initial loading. An exact analysis would account for the cyclically hardening or softening characteristics of the material. This type of analysis is very difficult and time consuming. It also requires that material properties be available to model this transient behavior. In general, there appears to be no significant effects on life predictions when this transient behavior is ignored.

The recommended procedure is to use the cyclic stress–strain curve to model material behavior on the initial load cycle and cyclically stable material properties during the entire analysis.

2.4 STRAIN–LIFE CURVE

In 1910, Basquin [14] observed that stress–life (S–N) data could be plotted linearly on a log–log scale. Using the true stress amplitude, the plot may be linearized by

$$\frac{\Delta\sigma}{2} = \sigma_f'(2N_f)^b \tag{2.37}$$

where $\frac{\Delta\sigma}{2}$ = true stress amplitude
$2N_f$ = reversals to failure (1 rev = $\frac{1}{2}$ cycle)
σ_f' = fatigue strength coefficient
b = fatigue strength exponent (Basquin's exponent)

σ_f' and b are fatigue properties of the material. The fatigue strength coefficient, σ_f', is approximately equal to the true fracture strength, σ_f. The fatigue strength exponent, b, will usually vary between -0.05 and -0.12.

Coffin [15] and Manson [16], working independently in the 1950s, found that plastic strain–life ($\epsilon_p - N$) data could also be linearized on log–log coordinates. Again, plastic strain can be related by a power law function

$$\frac{\Delta\epsilon_p}{2} = \epsilon_f'(2N_f)^c \tag{2.38}$$

where $\frac{\Delta\epsilon_p}{2}$ = plastic strain amplitude
$2N_f$ = reversals to failure
ϵ_f' = fatigue ductility coefficient
c = fatigue ductility exponent

ϵ_f' and c are also fatigue properties of the material. The fatigue ductility coefficient, ϵ_f', is approximately equal to true fracture ductility, ϵ_f. The fatigue ductility exponent, c, varies between -0.5 and -0.7.

An expression may now be developed that relates total strain range to life to failure. As discussed with reference to Eq. (2.30), the total strain is the sum of the elastic and plastic strains. In terms of strain amplitude [repeating Eq. (2.30)],

$$\frac{\Delta\epsilon}{2} = \frac{\Delta\epsilon_e}{2} + \frac{\Delta\epsilon_p}{2} \tag{2.30}$$

The elastic term can be written as

$$\frac{\Delta\epsilon_e}{2} = \frac{\Delta\sigma}{2E} \tag{2.39}$$

Using Eq. (2.37) we can now state this in terms of life to failure:

$$\frac{\Delta\epsilon_e}{2} = \frac{\sigma_f'}{E}(2N_f)^b \tag{2.40}$$

From Eq. (2.38) the plastic term is

$$\frac{\Delta\epsilon_p}{2} = \epsilon_f'(2N_f)^c \tag{2.38}$$

Using Eq. (2.30), the total strain can now be rewritten using Eqs. (2.38) and (2.40):

$$\frac{\Delta\epsilon}{2} = \underbrace{\frac{\sigma_f'}{E}(2N_f)^b}_{\text{elastic}} + \underbrace{\epsilon_f'(2N_f)^c}_{\text{plastic}} \tag{2.41}$$

Equation (2.41) is the basis of the strain–life method and is termed the *strain–life relation.*

Equation (2.41) can be explained graphically. Recalling that the elastic and plastic relations are both straight lines on a log–log plot, the total strain amplitude, $\Delta\epsilon/2$, can be plotted simply by summing the elastic and plastic values as shown in Fig. 2.17. At large strain amplitudes the strain–life curve approaches the plastic line, and at low amplitudes, the curve approaches the elastic line.

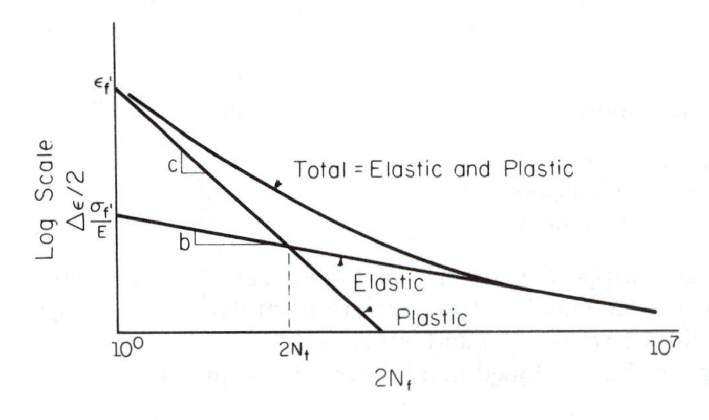

Figure 2.17 Strain–life curve.

In Fig. 2.17, the transition fatigue life, $2N_t$, represents the life at which the elastic and plastic curves intersect. Note that this is the life at which the stabilized hysteresis loop has equal elastic and plastic strain components. By equating elastic and plastic terms the following expression is derived for the transition life:

$$\frac{\Delta\epsilon_e}{2} = \frac{\Delta\epsilon_p}{2}$$

$$\frac{\sigma_f'}{E}(2N_f)^b = \epsilon_f'(2N_f)^c \qquad \text{at } N_f = N_t$$

$$2N_t = \left(\frac{\epsilon_f' E}{\sigma_f'}\right)^{1/(b-c)} \tag{2.42}$$

A schematic representation of the shape of the hysteresis loop at different lives is shown in Fig. 2.18 in relationship to the transition life. As seen, at shorter lives more plastic strain is present and the loop is wider. At long lives the loop is narrower, representing less plastic strain.

As shown in Fig. 2.19, the transition life of steel decreases with increasing hardness.

As the ultimate strength of the material increases, the transition life decreases, and elastic strains dominate for a greater portion of the life range.

Figure 2.20 presents the strain–life curves for a medium carbon steel in two different heat treated conditions. The material in a normalized (soft) ductile condition has a transition life of 90,000 cycles, while the material in a quenched (high strength) condition has a transition life of 15 cycles. As shown, for a given strain the high strength material (quenched) provides longer fatigue lives in the high cycle regime. At short lives or high strains the ductile (normalized) material exhibits better fatigue resistance.

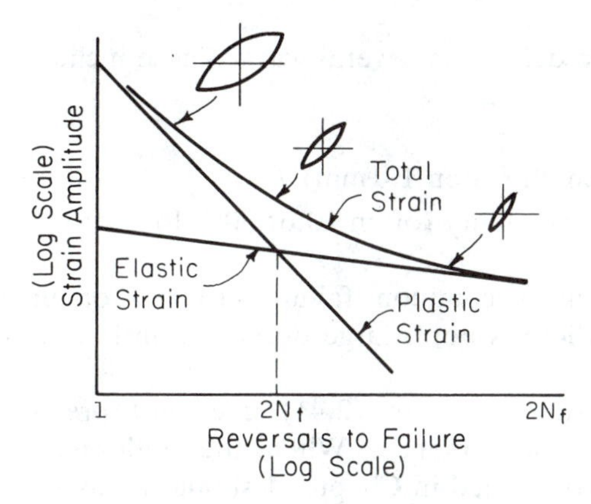

Figure 2.18 Shape of the hysteresis curve in relation to the strain–life curve. (From Ref. 17.)

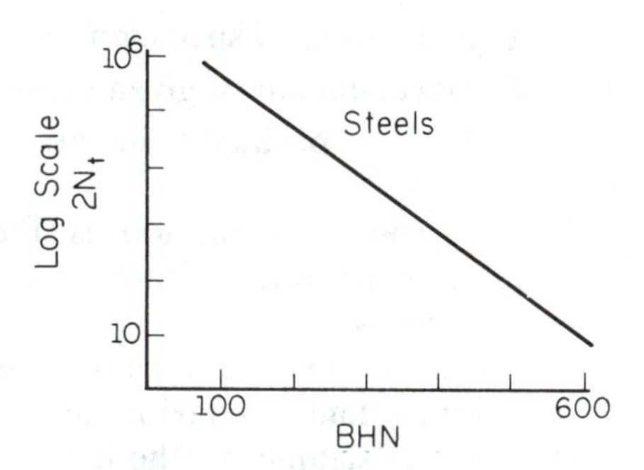

Figure 2.19 Relationship between transition life and hardness for steels. (From Ref. 17.)

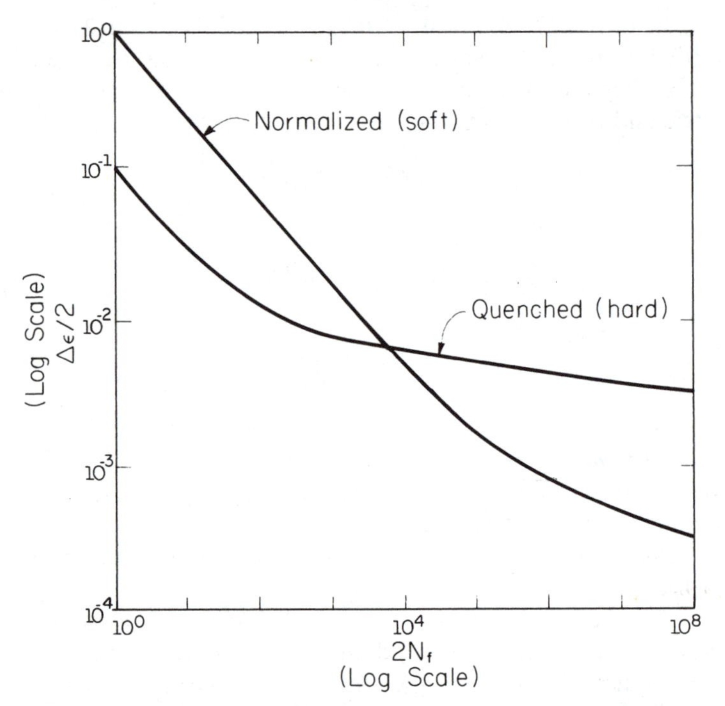

Figure 2.20 Strain–life curves for a medium carbon steel in a quenched and normalized condition.

The optimum material would be one that has *both* high ductility and high strength. Unfortunately, there is usually a trade-off between these two properties and a compromise must be made for the expected load or strain conditions being considered.

Note that life to failure may be defined in several ways. These include:

1. Separation of specimen
2. Development of given crack length (often 1.0 mm)
3. Loss of specified load carrying capability (often 10 or 50% load drop)

Specimen separation is the most common failure criteria for uniaxial loading. However, in many cases, there is not a large difference in life between these criteria.

The strain–life equation as stated in Eq. (2.41) does not predict the endurance limit behavior seen in some metals. When the endurance limit behavior is significant, the methods described in Chapter 1 should be used.

Before continuing it is worthwhile to consider the "factor of 2" problem found in the strain–life analysis. There are three cases where it is very easy to lose track of a factor of 2 and cause errors in a strain–life fatigue analysis. All of these cases have been discussed earlier, but to reemphasize, they are:

1. *Cycles versus reversal.* The strain–life approach measures life in terms of reversals ($2N$), whereas the stress–life method uses cycles (N). A reversal is one-half of a full cycle.
2. *Amplitude versus range.* The strain–life approach uses both strain range, $\Delta\epsilon$, and amplitude, ϵ_a, which differ by a factor of 2 (i.e., $\Delta\epsilon/2 = \epsilon_a$).
3. *Cyclic σ–ϵ curve versus hysteresis curve.* Massing's hypothesis states that the hysteresis curve can be modeled as twice the cyclic stress–strain curve.

Although the preceding points may seem trivial to some, it is the experience of the authors that the "factor of 2" phantom can cause extreme hardship for the unwary.

2.5 DETERMINATION OF FATIGUE PROPERTIES

The strain–life equation [Eq. (2.41)] requires four empirical constants (b, c, σ'_f, ϵ'_f). Several points must be considered in attempting to obtain these constants from fatigue data.

1. Not all materials may be represented by the four-parameter strain–life equation. (Examples of these are some high strength aluminum alloys and titanium alloys.)
2. The four fatigue constants may represent a curve fit to a limited number of data points. The values of these constants may be changed if more data points are included in the curve fit.
3. The fatigue constants are determined from a set of data points over a given range. Gross errors may occur when extrapolating fatigue life estimates outside this range.
4. The use of power law relationships in Eqs. (2.32), (2.37), and (2.38) is strictly a matter of mathematical convenience and is not based on a physical phenomenon.

From Eqs. (2.34) and (2.41) the following properties may be related:

$$K' = \frac{\sigma'_f}{(\epsilon'_f)^{n'}} \tag{2.43}$$

$$n' = \frac{b}{c} \tag{2.44}$$

Although these relationships may be useful, K' and n' are usually obtained from a curve fit of the cyclic stress–strain data using Eq. (2.32). Due to the approximate nature of the curve fits, values obtained from Eqs. (2.32), (2.43), and (2.44) may not be equal.

Fatigue properties may be approximated from monotonic properties. Currently, due to the available data, these techniques are no longer used extensively. Nevertheless, the following approximate methods may be useful.

Fatigue strength coefficient, σ'_f. A fairly good approximation is

$$\sigma'_f \approx \sigma_f \qquad \text{(corrected for necking)} \tag{2.45}$$

For steels with hardnesses below 500 BHN:

$$\sigma_f \approx S_u + 50\,\text{ksi} \tag{2.46}$$

Fatigue strength exponent, b. b varies from -0.05 to -0.12 for most metals with an average of -0.085. (Note that this corresponds to the approximate slope of the S–N curve discussed in Section 1.2.)

Fatigue ductility coefficient, ϵ'_f. A fairly good approximation is

$$\epsilon'_f \approx \epsilon_f \tag{2.47}$$

$$\text{where } \epsilon_f = \ln \frac{1}{1 - \text{RA}}$$

where RA is the reduction in area.

Fatigue ductility exponent, c. c is not as well defined as the other parameters. A rule-of-thumb approach must be followed rather than an empirical equation.

Coffin found c to be about -0.5.
Manson found c to be about -0.6.
Morrow found that c varied between -0.5 and -0.7.

Fairly ductile metals (where: $\epsilon_f \approx 1$) have average values of $c = -0.6$. For strong metals (where: $\epsilon_f \approx 0.5$) a value of $c = -0.5$ is probably more reasonable.

Example 2.2

Given below are the monotonic and cyclic strain–life data for smooth steel specimens. Determine the cyclic stress–strain and strain–life constants (K', n', σ'_f, b, ϵ'_f, c) for this material.

Monotonic data

$S_y = 158$ ksi $\qquad E = 28.4 \times 10^3$ ksi

$S_u = 168$ ksi $\qquad \sigma_f = 228$ ksi

$\%\,\text{RA} = 52 \qquad \epsilon_f = 0.734$

Smooth Specimen-Cyclic Data

Total Strain Amplitude, $\Delta\epsilon/2$	Stress Amplitude, $\Delta\sigma/2$ (ksi)	Plastic Strain Amplitude, $\Delta\epsilon_p/2$[a]	Reversals to Failure, $2N_f$
0.0393	162.5	0.0336	50
0.0393	162	0.0336	68
0.02925	155	0.0238	122
0.01975	143.5	0.0147	256
0.0196	143.5	0.0145	350
0.01375	136.5	0.00894	488
0.00980	130.5	0.00521	1,364
0.00980	126.5	0.00534	1,386
0.00655	121	0.00229	3,540
0.00630	119	0.00211	3,590
0.00460	114	0.00059	9,100
0.00360	106	0.00000	35,200
0.00295	84.5	0.00000	140,000

[a] $\dfrac{\Delta\epsilon_p}{2} = \dfrac{\Delta\epsilon}{2} - \dfrac{\Delta\epsilon_e}{2} = \dfrac{\Delta\epsilon}{2} - \dfrac{\Delta\sigma}{2E}.$

Solution Determine the fatigue strength coefficient, σ_f', and the fatigue strength exponent, b, by fitting a power law relationship to the stress amplitude, $\Delta\sigma/2$, versus reversals to failure, $2N_f$, data.

$$\frac{\Delta\sigma}{2} = \sigma_f'(2N_f)^b$$

Determine the fatigue ductility coefficient, ϵ_f', and the fatigue ductility exponent, c, by fitting a power law relationship to the plastic strain amplitude, $\Delta\epsilon_p/2$, versus reversals to failure, $2N_f$, data.

$$\frac{\Delta\epsilon_p}{2} = \epsilon_f'(2N_f)^c$$

The curve fits to the strain–life data are shown in Fig. E2.2. The resulting cyclic

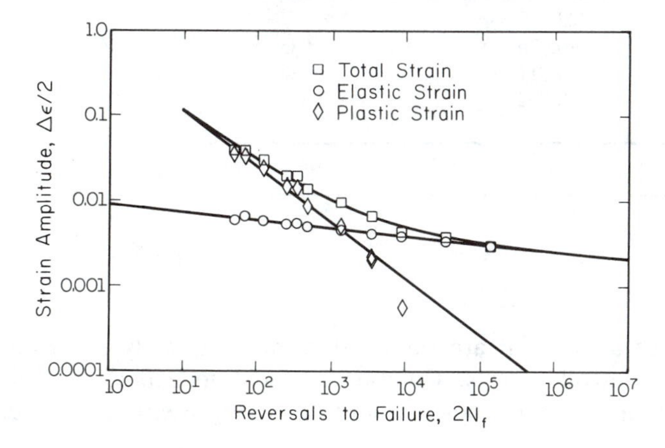

Figure E2.2 Strain–life curve.

properties are

$$\sigma_f' = 222 \text{ ksi} \qquad b = -0.076$$

$$\epsilon_f' = 0.811 \qquad c = -0.732$$

Determine the cyclic strength coefficient, K', and the cyclic strain hardening exponent, n'. These can be found using two different procedures. First, these terms can be found by fitting a power law relationship to the stress amplitude, $\Delta\sigma/2$, versus plastic strain amplitude, $\Delta\epsilon_p/2$, data.

$$\sigma = K'(\epsilon_p)^{n'}$$

If this is done, the resulting values are

$$K' = 216 \text{ ksi} \qquad n' = 0.094$$

These terms may also be determined using the relationships stated in Eqs. (2.43) and (2.44).

$$K' = \frac{\sigma_f'}{(\epsilon_f')^{n'}}$$

$$n' = \frac{b}{c}$$

Using these relationships, the resulting values are

$$K' = 227 \text{ ksi} \qquad n' = 0.104$$

Note the difference between the predictions for K' and n' found using the two different methods. In general, the first procedure (curve fit) is the preferred method.

For comparison purposes the following table lists the strain–life constants determined from strain–life data and the values found using the approximations discussed in Section 2.5 [Eqs. (2.45) and (2.47)]. Note the difference between these values.

Value	Determined from Strain–Life Data	Determined Using Approximations
σ_f'	222	228
b	−0.076	−0.085
ϵ_f'	0.811	0.734
c	−0.732	−0.6

2.6 MEAN STRESS EFFECTS

Cyclic fatigue properties of a material are obtained from completely reversed, constant amplitude strain-controlled tests. Components seldom experience this type of loading, as some mean stress or mean strain is usually present. The effect

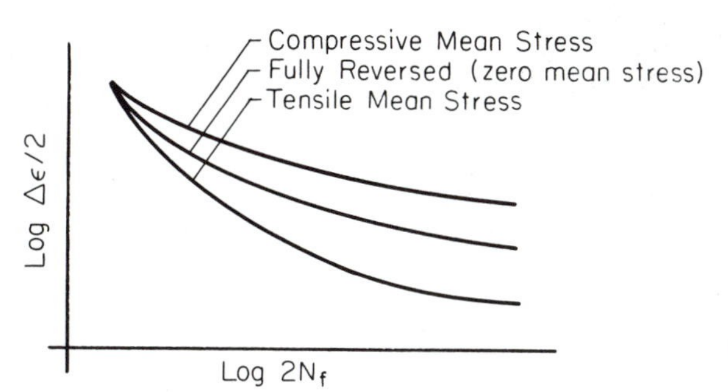

Figure 2.21 Effect of mean stress on strain–life curve.

of mean strain is, for the most part, negligible on the fatigue life of a component. Mean stresses, on the other hand, may have a significant effect on the fatigue life.

Mean stress effects are seen predominantly at longer lives. They can either increase the fatigue life with a nominally compressive load or decrease it with a nominally tensile value, as shown schematically in Fig. 2.21.

At high strain amplitudes (0.5% to 1% or above), where plastic strains are significant, mean stress relaxation occurs and the mean stress tends toward zero (see Fig. 2.22.) Note that this is not cyclic softening. Mean stress relaxation can occur in materials that are cyclically stable.

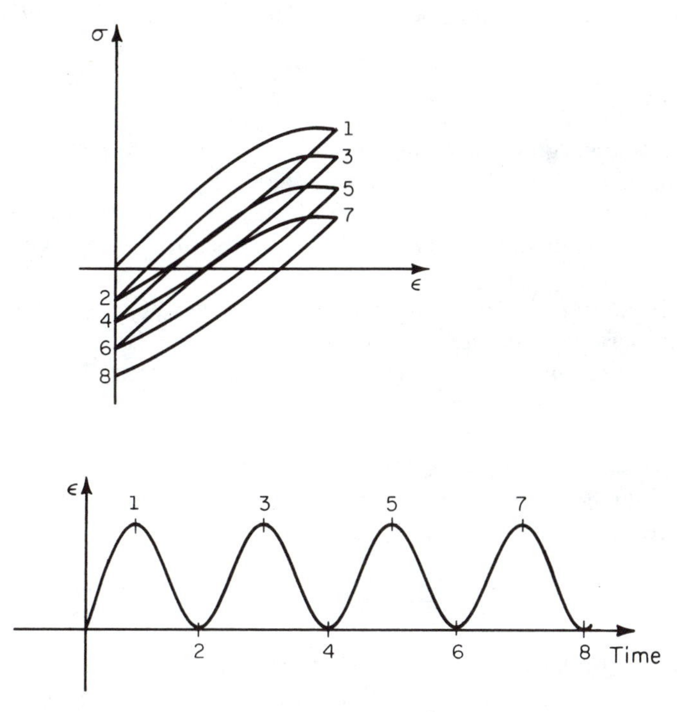

Figure 2.22 Mean stress relaxation.

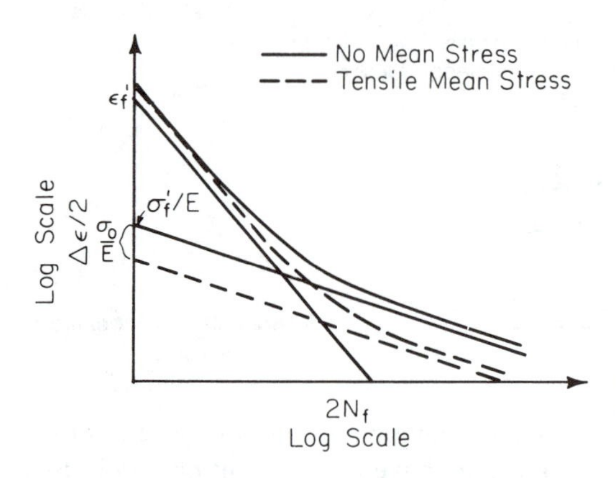

Figure 2.23 Morrow's mean stress correction to the strain–life curve for a tensile mean.

Modifications to the strain–life equation have been made to account for mean stress effects. Morrow [18] suggested that the mean stress effect could be taken into account by modifying the elastic term in the strain–life equation [Eq. (2.41)] by the mean stress, σ_0.

$$\frac{\Delta\epsilon_e}{2} = \frac{\Delta\sigma}{2E} = \frac{\sigma_f' - \sigma_0}{E}(2N_f)^b \tag{2.48}$$

The strain–life equation, accounting for mean stresses is, then,

$$\frac{\Delta\epsilon}{2} = \frac{\sigma_f' - \sigma_0}{E}(2N_f)^b + \epsilon_f'(2N_f)^c \tag{2.49}$$

This is shown graphically in Fig. 2.23. The predictions made with this equation are consistent with the observations that mean stress effects are significant at low values of plastic strain, where the elastic strain dominates. They also reflect the trend that mean stresses have little effect at shorter lives, where plastic strains are large.

Equation (2.49), though, incorrectly predicts that the ratio of elastic to plastic strain is dependent on mean stress. This is clearly not true, as demonstrated in Fig. 2.24. The two smaller hysteresis loops have the same strain range and the same ratio of elastic to plastic strain, while they have vastly different mean stresses.

Manson and Halford [19] modified both the elastic and plastic terms of the strain–life equation to maintain the independence of the elastic–plastic strain ratio from mean stress. This equation,

$$\frac{\Delta\epsilon}{2} = \frac{\sigma_f' - \sigma_0}{E}(2N_f)^b + \epsilon_f'\left(\frac{\sigma_f' - \sigma_0}{\sigma_f'}\right)^{c/b}(2N_f)^c \tag{2.50}$$

is shown graphically in Fig. 2.25. (Note that the transition life remains constant.) This equation tends to predict too much mean stress effect at short lives or where plastic strains dominate. At high plastic strains, mean stress relaxation occurs.

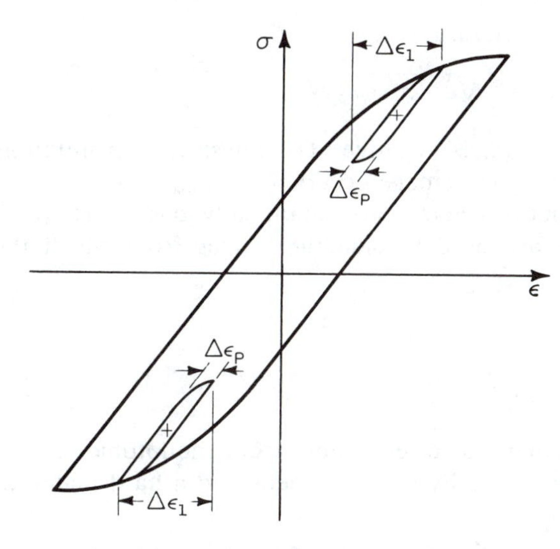

Figure 2.24 Independence of elastic/plastic strain ratio from mean stress. (Note that in this figure the plastic strain portion of the small hysteresis loops has been exaggerated for clarity.)

Although Eq. (2.49) violates the constitutive relationship, it generally does a better job predicting mean stress effects.

Smith, Watson, and Topper (SWT) [20] have proposed another equation to account for mean stress effects. Recalling Eq. (2.37), for completely reversed loading

$$\sigma_{\max} = \frac{\Delta\sigma}{2} = \sigma_f'(2N_f)^b \tag{2.51}$$

and multiplying the strain–life equation by this term, results in

$$\sigma_{\max}\frac{\Delta\epsilon}{2} = \frac{(\sigma_f')^2}{E}(2N_f)^{2b} + \sigma_f'\epsilon_f'(2N_f)^{b+c} \tag{2.52}$$

For application of this equation, the term $\sigma_{\max}$ is evaluated as

$$\sigma_{\max} = \frac{\Delta\sigma}{2} + \sigma_0 \tag{2.53}$$

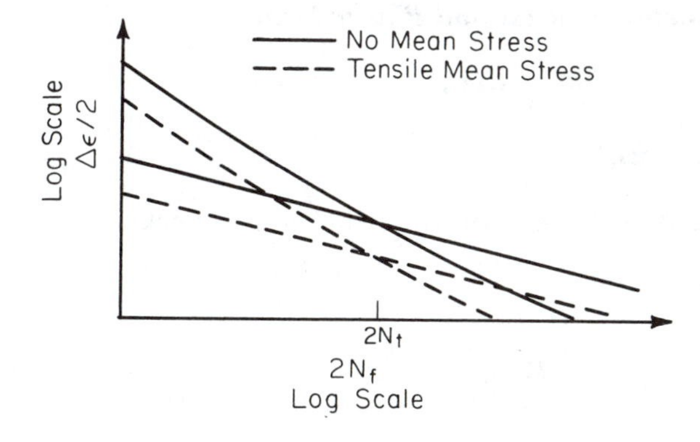

Figure 2.25 Mean stress correction for independence of elastic/plastic strain ratio from mean stress.

Since this equation is in the general form

$$\sqrt{\sigma_{max}\,\Delta\epsilon} \propto N_f \qquad (2.54)$$

it becomes undefined when σ_{max} is negative. The physical interpretation of this approach assumes that no fatigue damage occurs when $\sigma_{max} < 0$.

The mean stress equations above are empirically based. Therefore, care must be taken when they are used outside the ranges from which they were developed.

2.7 IMPORTANT CONCEPTS

- Cyclic response of a material often differs from monotonic response. For instance, a soft material usually strain hardens and a hard material usually strain softens.
- Fatigue properties, σ'_f, ϵ'_f, b, c, n', and K', are curve fit to the experimental data and must be used for the range of life from which they were determined.
- Tensile mean stresses are detrimental to fatigue life, while compressive mean stresses are beneficial.
- At high strain levels, cyclic plastic strains tend to cause the mean stresses to relax to zero.
- Three cases where it is easy to lose a factor of 2 are:
  a. Cycles versus reversals
  b. Amplitude versus range
  c. Cyclic stress–strain curve versus hysteresis curve

2.8 IMPORTANT EQUATIONS

Power Relationship between Stress and Plastic Strain

$$\sigma = K(\epsilon_p)^n \qquad (2.18)$$

Strain Hardening Exponent

$$n = \text{slope of } \log\sigma \text{ vs. } \log\epsilon_p \quad \text{or} \quad n = \ln(1 + e \text{ at necking})$$

Strength Coefficient

$$K = \frac{\sigma_f}{\epsilon_f^n} \qquad (2.23)$$

Total Strain = Elastic Strain + Plastic Strain

$$\epsilon_t = \epsilon_e + \epsilon_p = \frac{\sigma}{E} + \left(\frac{\sigma}{K}\right)^{1/n} \tag{2.17}$$

Strain and Stress Amplitude

$$\epsilon_a = \frac{\Delta\epsilon}{2}$$

$$\sigma_a = \frac{\Delta\sigma}{2}$$

$$\Delta\epsilon = \Delta\epsilon_e + \Delta\epsilon_p \tag{2.29}$$

$$\frac{\Delta\epsilon}{2} = \frac{\Delta\sigma}{2E} + \frac{\Delta\epsilon_p}{2} \tag{2.31}$$

Cyclic Stress–Plastic Strain Relationship

$$\sigma = K'(\epsilon_p)^{n'} \tag{2.32}$$

Cyclic Stress–Total Strain Relationship

$$\epsilon = \frac{\sigma}{E} + \left(\frac{\sigma}{K'}\right)^{1/n'} \tag{2.34}$$

Hysteresis Curve

$$\Delta\epsilon = \frac{\Delta\sigma}{E} + 2\left(\frac{\Delta\sigma}{2K'}\right)^{1/n'} \tag{2.36}$$

Strain–Life Relationship

$$\frac{\Delta\epsilon}{2} = \underbrace{\frac{\sigma_f'}{E}(2N_f)^b}_{\text{elastic}} + \underbrace{\epsilon_f'(2N_f)^c}_{\text{plastic}} \tag{2.41}$$

Transition Life

$$2N_t = \left(\frac{\epsilon_f' E}{\sigma_f'}\right)^{1/(b-c)} \tag{2.42}$$

Cyclic Strength Coefficient

$$K' = \frac{\sigma_f'}{(\epsilon_f')^{n'}} \tag{2.43}$$

Cyclic Strain Hardening Exponent

$$n' = \frac{b}{c} \tag{2.44}$$

REFERENCES

1. American Society for Testing and Materials, *Manual on Low Cycle Fatigue Testing,* ASTM STP 465, ASTM, Philadelphia, 1969.
2. American Society for Testing and Materials, ASTM Standard E606-80, *Annual Book of ASTM Standards,* ASTM, Philadelphia, 1980.
3. J. A. Graham (ed.), *SAE Fatigue Design Handbook,* Vol. 4, Society of Automotive Engineers, Warrendale, Pa., 1968.
4. R. M. Wetzel (ed.), *Fatigue under Complex Loading: Analysis and Experiments,* Advances in Engineering, Vol. 6, Society of Automotive Engineers, Warrendale, Pa., 1977.
5. J. Morrow and D. F. Socie, "The Evolution of Fatigue Crack Initiation Life Prediction Methods," in *Materials, Experimentation and Design in Fatigue,* F. Sherratt and J. B. Sturgeon (eds.), Westbury House, Warwick, England, 1981, p. 3.
6. J. Bauschinger, *Mitt. Mech.-Tech., Lab München,* Vol. 13, No. 1, 1886.
7. R. W. Hertzberg, *Deformation and Fracture Mechanics of Engineering Materials,* 2nd ed., Wiley, New York, 1983.
8. S. S. Manson and M. H. Hirschberg, *Fatigue: An Interdisciplinary Approach,* Syracuse University Press, Syracuse, N.Y., 1964, p. 133.
9. R. W. Landgraf, J. Morrow, and T. Endo, "Determination of the Cyclic Stress–Strain Curve," *J. Mater.,* Vol. 4, No. 1, 1969, p. 176.
10. J. Morrow, "Cyclic Plastic Strain Energy and Fatigue of Metals," in *Internal Friction, Damping, and Cyclic Plasticity,* ASTM STP 378, American Society for Testing and Materials, Philadelphia, 1965, p. 45.
11. J. F. Martin, "Cyclic Stress–Strain Behavior and Fatigue Resistance of Two Structural Steels," Fracture Control Program Report No. 9, University of Illinois at Urbana–Champaign, 1973.
12. G. Massing, *Proc. 2nd Int. Cong. Appl. Mech.,* Zurich, 1926.
13. D. Weinacht, "Fatigue Behavior of Gray Cast Iron under Torsional Loads," Report No. 126, College of Engineering, University of Illinois at Urbana–Champaign, May 1986.
14. O. H. Basquin, "The Exponential Law of Endurance Tests," *Am. Soc. Test. Mater. Proc.,* Vol. 10, 1910, pp. 625–630.
15. L. F. Coffin, Jr., "A Study of the Effects of Cyclic Thermal Stresses on a Ductile Metal," *Trans. ASME,* Vol. 76, 1954, pp. 931–950.
16. S. S. Manson, "Behavior of Materials under Conditions of Thermal Stress," *Heat Transfer Symposium,* University of Michigan Engineering Research Institute, 1953, pp. 9–75 (also published as NACA TN 2933, 1953).
17. R. W. Landgraf, "The Resistance of Metals to Cyclic Deformation," in *Achievement*

of *High Fatigue Resistance in Metals and Alloys,* ASTM STP 467, American Society for Testing and Materials, Philadelphia, 1970, pp. 3–36.

18. J. Morrow, *Fatigue Design Handbook,* Advances in Engineering, Vol. 4, Society of Automotive Engineers, Warrendale, Pa., 1968, Sec. 3.2, pp. 21–29.
19. S. S. Manson and G. R. Halford, "Practical Implementation of the Double Linear Damage Rule and Damage Curve Approach for Treating Cumulative Fatigue Damage," *Int. J. Fract.,* Vol. 17, No. 2, 1981, pp. 169–172, R35–R42.
20. K. N. Smith, P. Watson, and T. H. Topper, "A Stress–Strain Function for the Fatigue of Metals," *J. Mater.,* Vol. 5, No. 4, 1970, pp. 767–778.
21. T. Fugger, Jr., "Service Load Histories Analyzed by the Local Strain Approach," Report No. 120, College of Engineering, University of Illinois at Urbana–Champaign, May 1985.
22. D. F. Socie, N. E. Dowling, and P. Kurath, "Fatigue Life Estimation of Notched Members," in *Fracture Mechanics: Fifteenth Symposium,* ASTM STP 833, R. J. Sanford (ed.), American Society for Testing and Materials, Philadelphia, 1984, pp. 284–299.
23. R. W. Landgraf, "Cyclic Deformation and Fatigue Behavior of Hardened Steels," Report No. 320, Department of Theoretical and Applied Mechanics, University of Illinois at Urbana–Champaign, Nov. 1968.
24. T. Endo and J. Morrow, "Cyclic Stress–Strain and Fatigue Behavior of Representative Aircraft Metals," *J. Mater.,* Vol. 4, No. 1, 1969, pp. 159–175.
25. S. S. Manson, "Fatigue: A Complex Subject—Some Simple Approximations," *Exp. Mech.,* Vol. 5, No. 7, 1965, p. 193.
26. J. H. Crews, Jr., "Crack Initiation at Stress Concentrations as Influenced by Prior Local Plasticity," in *Achievement of High Fatigue Resistance in Metals and Alloys,* ASTM STP 467, American Society for Testing and Materials, Philadelphia, 1970, p. 37.
27. J. L. Koch, "Proportional and Non-Proportional Biaxial Fatigue of Inconel 718," Report No. 121, College of Engineering, University of Illinois at Urbana–Champaign, July 1985.

PROBLEMS

SECTION 2.2.1

2.1. During the initial stages of a tension test of an engineering material, the engineering stress and strain (S, e) were determined to be 62.2 ksi and 0.0098. Later, after some plastic deformation had occurred, the engineering stress and strain were determined to be 90.8 ksi and 0.0898. Calculate the true stress and strain (σ, ϵ) values at these two points. Discuss the relationship between the engineering and true values as plastic strain increases.

2.2. A cylindrical bar of structural steel with an initial diameter of 50 mm is loaded in tension. (*Note:* The general behavior of structural steel in a tension test is shown in Fig. 2.3.) The following deflection measurements are made over a 250-mm gage length:

Load (MN)	Deflection (mm)	Comment
0	0	Start of test
0.33	0.20	Behavior is elastic
0.37	0.25	Sudden drop in load
0.33	1.00	Load nearly constant
0.53	50.00	Maximum load (necking starts)
0.30	64.00	Fracture (final diameter = 30 mm)

Determine:
(a) Modulus of elasticity, E
(b) Upper yield point
(c) Lower yield point
(d) 0.2% offset yield strength, S_y
(e) Ultimate strength, S_u
(f) Engineering and true strain at start of necking
(g) Strain hardening exponent, n
(h) Percent reduction in area, % RA
(i) True fracture ductility, ϵ_f
(j) True fracture strength, σ_f
(k) Strength coefficient, K

2.3. Shown below is the load–deflection curve ($P - \delta$) for an engineering brass. The modulus of elasticity, E, of the material is 100 GPa. Given the following values for the test specimen:

Initial length, $l_0 = 167$ mm
Initial diameter, $d_0 = 3.17$ mm
Final diameter at necked section, $d_f = 2.55$ mm

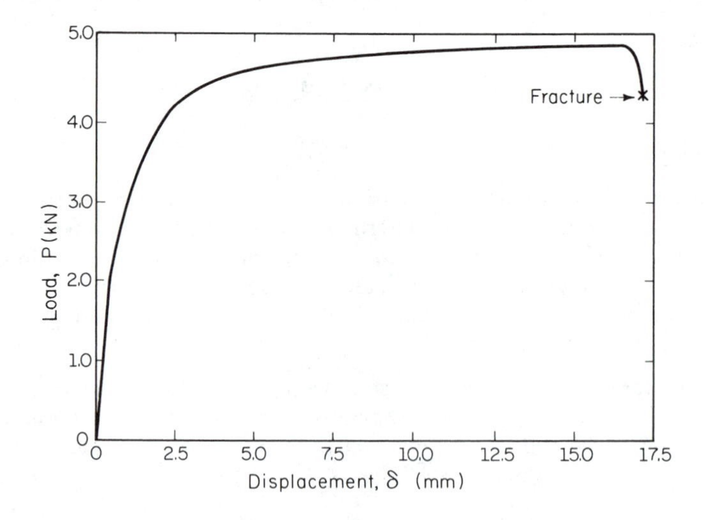

Determine:
(a) 0.2% offset yield strength, S_y
(b) Ultimate strength, S_u
(c) Percent reduction in area, % RA
(d) True fracture ductility, ϵ_f
(e) True fracture strength, σ_f
(f) Strength coefficient, K
(g) Strain hardening exponent, n
(h) True stress at ultimate load
(i) True strain at ultimate load
Also plot the engineering and true stress–strain curves.

2.4. The following true stress–strain properties of a metal were obtained in a monotonic tension test:

$$E = 30 \times 10^3 \text{ ksi} \qquad n = 0.2 \qquad \epsilon_f = 1.0 \qquad \sigma_f = 270 \text{ ksi}$$

Determine:
(a) Strength coefficient, K
(b) 0.2% offset yield strength, S_y
(c) Percent reduction in area, % RA
Plot the monotonic true stress–strain curve.

2.5. It has been determined that a certain steel ($E = 30 \times 10^3$ ksi) follows the following true stress, σ, true plastic strain, ϵ_p, relation:

$$\sigma = (360 \text{ ksi})(\epsilon_p)^{0.11}$$

The true plastic strain at fracture was found to be 0.48. Determine:
(a) True fracture strength, σ_f
(b) Total true strain at fracture
(c) Strength coefficient, K
(d) Strain hardening exponent, n
(e) Strength at 0.2% offset, S_y
(f) Percent reduction in area, % RA
(g) True fracture ductility, ϵ_f

2.6. A cylindrical metal specimen is tested in tension and found to have the following properties:

$$E = 10 \times 10^3 \text{ ksi} \qquad n = 0.5 \qquad K = 70 \text{ ksi} \qquad \% \text{ RA} = 86\%$$

A second specimen is prestrained in tension to an elongation of 20%. (*Note:* This does not cause necking.) It is then regarded as a new material and tested in tension. Determine the following properties for the original and for the new (20% prestrained) metal.
(a) True fracture ductility, ϵ_f
(b) True fracture strength, σ_f
(c) Ultimate tensile strength, S_u

2.7. A high strength steel is tested in tension and observed to fail without necking. The following engineering properties were determined:

$$S_y \text{ (0.2\% offset)} = 700 \text{ MPa}$$

$$S_u = 1000 \text{ MPa}$$

$$\% \text{ RA} = 10\%$$

Determine the following true stress–strain properties:

(a) True fracture ductility, ϵ_f

(b) True fracture strength, σ_f

(c) Strain hardening exponent, n

(d) Strength coefficient, K

(*Note:* To solve this problem, you may need the equations given in Problem 2.8.)

2.8. Values for yield strength, S_y, are usually determined using a small offset, typically 0.2% (or 0.002), on the stress–strain curve. Prove that for small offsets,

$$\frac{S_u}{S_y} = \left(\frac{n}{\text{offset}}\right)^n \exp(-n)$$

and

$$\frac{\sigma_f}{S_y} = \left(\frac{\epsilon_f}{\text{offset}}\right)^n$$

2.9. If the relationships given in Problem 2.8 are true, show that all the true stress–strain properties (σ_f, ϵ_f, n, K) can be determined from the engineering monotonic properties (S_y, S_u, % RA).

2.10. Prove that if a material follows the power relationship for true stress and true plastic strain,

$$\sigma = K(\epsilon_p)^n$$

The true plastic strain at the beginning of necking, ϵ_{pn}, is equal to the strain hardening exponent, n.

SECTIONS 2.2.2 through 2.3

2.11. For a material with the following properties:

$$E = 30 \times 10^3 \text{ ksi} \qquad n' = 0.202 \qquad K' = 174.6 \text{ ksi}$$

draw the cyclic stress–strain (σ–ϵ) curve and the hysteresis curve (see Fig. 2.16). Draw the cyclic curve up to $\epsilon = 0.02$ and the hysteresis curve up to $\Delta\epsilon = 0.04$. Use the same scales to draw both figures.

The cyclic stress–strain response of a material can be determined graphically by using templates made from the cyclic stress–strain curve and hysteresis curve. These templates can be used to track the material behavior on σ–ϵ coordinates. Use the curves developed in the first part of this problem to construct templates. Then use these templates to verify the calculations carried out in Example 2.1.

Compare the graphical and analytical solutions at strain amplitudes of 0.001, 0.002, 0.005, 0.01, and 0.015. Discuss the shape of the hysteresis loops at these different levels.

2.12. The table below lists the stress range ($\Delta\sigma$) and strain range ($\Delta\epsilon$) values for several stabilized hysteresis loops for a stainless steel. From these data determine the cyclic strength coefficient, K', and cyclic strain hardening exponent, n'. The modulus of elasticity for this material is 29.4×10^3 ksi.

Total Strain Range, $\Delta\epsilon$	Stabilized Stress Range, $\Delta\sigma$ (ksi)
0.00084	24.27
0.00200	50.55
0.00400	72.61
0.00610	80.89
0.00800	83.64
0.01010	88.24
0.01200	91.92
0.01390	96.51
0.01600	101.11
0.01780	106.62
0.01900	111.22

If the monotonic properties for this material were

$$S_y \text{ (0.2\% offset)} = 47 \text{ ksi} \qquad S_u = 94 \text{ ksi} \qquad n = 0.193$$

did the material harden or soften?

2.13. A metal has the following monotonic tension properties:

$$E = 193 \text{ GPA} \qquad S_y \text{ (0.2\% offset)} = 325 \text{ MPa}$$

$$S_u = 650 \text{ MPa} \qquad \sigma_f = 1400 \text{ MPa}$$

$$\epsilon_f = 1.731 \qquad \%\,\text{RA} = 80\%$$

$$n = 0.193$$

Under cyclic loading will the material harden or soften? Calculate the strain reached on the first half-cycle for a stress amplitude of 200 MPa.

Given that the material has the following cyclic properties:

$$K' = 1660 \text{ MPa} \qquad n' = 0.287$$

determine the stable total strain and plastic strain amplitudes for a stress amplitude of 200 MPa.

Repeat the calculations above, but now determine the stress response for a strain amplitude ($\Delta\epsilon/2$) of 0.01.

2.14. Given below are the monotonic properties for several engineering metals.

Material	E (ksi)	S_y (ksi)	S_u (ksi)	% RA	n	σ_f (ksi)
Pure aluminum	10×10^3	14	48	88	0.3	110
High strength aluminum	10×10^3	68	84	33	0.11	108
Low carbon steel	30×10^3	38	50	80	0.16	123
Medium carbon steel	30×10^3	92	105	65	0.13	178
High strength steel	30×10^3	235	355	6	0.06	375
Gray iron	15×10^3	25	28	Nil	—	28

Which of these materials will cyclically harden, soften, or be stable?

Indicate which of these materials would be the best choice to obtain:

(a) Maximum tensile load on smooth rods
(b) Maximum uniform elongation before necking in tension
(c) Maximum work required to load smooth rods to necking
(d) Minimum tensile load required on smooth rods to cause a strain of 0.001
(e) Most likely to cyclically harden
(f) Most likely to cyclically soften
(g) Maximum work required to cause fracture
(h) Minimum elastic strain at necking
(i) Maximum total strain at necking

2.15. In Section 2.2.3 two methods were given to determine if a material cyclically softens or hardens. The first uses the ratio of monotonic ultimate strength, S_u, and 0.2% offset yield, S_y. The second method uses the monotonic strain hardening exponent, n. The following relationship can be used to relate these values:

$$\frac{S_u}{S_y} = \left(\frac{n}{\text{offset}}\right)^n \exp(-n)$$

where the offset is 0.002. Plot this relationship for S_u/S_y versus n and mark the three regions corresponding to hardening, softening, and mixed behavior. Plot on this graph the values corresponding to the materials listed in Problems 2.13 and 2.14.

SECTION 2.4

2.16. The strain–life values for a hardened steel are

$$E = 30 \times 10^3 \text{ ksi}$$

$$\sigma_f' = 300 \text{ ksi} \qquad b = -0.1$$

$$\epsilon_f' = 0.1 \qquad c = -0.6$$

(a) Draw on log–log coordinates the elastic strain–life, plastic strain–life, and total strain–life curves.
(b) Determine the transition life ($2N_t$) both graphically and analytically.
(c) Determine the total strain amplitude ($\Delta\epsilon/2$) at the transition fatigue life ($2N_t$).
(d) Determine the cyclic strength coefficient (K') and cyclic strain hardening exponent (n') for this material.
(e) A highly strained component made from this steel is failing in service after about 100 reversals. Suggest what could be done to the fatigue properties to increase the life. What processes would alter the necessary properties?
(f) In another application failure occurs after about 10^6 reversals. Suggest methods to increase the fatigue life under these circumstances.

2.17. The following stress–strain and strain–life properties are given for a steel:

$$E = 30 \times 10^3 \text{ ksi} \qquad K' = 137 \text{ ksi} \qquad n' = 0.22$$

$$\sigma_f' = 120 \text{ ksi} \qquad b = -0.11$$

$$\epsilon_f' = 0.95 \qquad c = -0.64$$

(a) Draw on log–log coordinates the elastic strain–life, plastic strain–life, and total strain–life curves. Determine the transition life ($2N_t$).

(b) Draw the hysteresis loops corresponding to strain amplitude ($\Delta\epsilon/2$) values of 0.05, 0.00125, and 0.0007. Determine the fatigue life in reversals at these three strain levels.

(c) Determine the elastic, plastic, and total strain amplitude for a life ($2N_f$) of 2×10^6 reversals.

(d) Determine the elastic, plastic, and total strain amplitude for a life ($2N_f$) of 500 reversals.

(e) Determine the cyclic stress amplitude corresponding to fatigue lives of 500 and 2×10^6 reversals.

(f) A component made from this material is required to have a life of no less than 10^4 reversals. The loading on the component causes a total strain amplitude of 0.008. Determine if the component will meet the life requirements.

2.18. Listed below are the strain–life properties for a high and low strength steel.

	σ'_f (MPa)	ϵ'_f	b	c	E (GPa)
Low strength	800	1.0	−0.10	−0.50	200
High strength	2700	0.1	−0.08	−0.70	200

(a) Determine the transition life ($2N_t$) for the two steels.

Which steel would allow:

(b) The largest completely reversed strain for a life of 200 reversals?

(c) The largest completely reversed stress for a life of 200 reversals?

(d) The largest completely reversed strain for a life of 2×10^6 reversals?

(e) The largest completely reversed stress for a life of 2×10^6 reversals?

2.19. It is useful to understand how sensitive a calculated fatigue life value is to various material and loading parameters. Determine how sensitive the fatigue life, $2N_f$, is to the properties and strain levels shown below. Make reasonable assumptions on the variability of the parameters. (If unsure, use a variability of ±10%.)

$$E = 185 \text{ GPa}$$

$$\sigma'_f = 1000 \text{ MPa} \qquad b = -0.114$$

$$\epsilon'_f = 0.171 \qquad c = -0.402$$

$$\Delta\epsilon/2 = 0.001 \text{ and } 0.01$$

2.20. At the transition fatigue life, $2N_t$, determine the stress and strain amplitude ($\Delta\sigma/2$, $\Delta\epsilon/2$) in terms of the cyclic stress–strain properties of a metal (E, K', n').

2.21. A medium-carbon steel (E = 200 GPa) has the following fatigue properties in the normalized (soft) condition:

$$\sigma'_f = 600 \text{ MPa} \qquad \epsilon'_f = 1.0 \qquad b = -0.12 \qquad c = -0.6$$

When hardened to form martensite and then tempered at 200°C, this steel has the following fatigue properties:

$$\sigma'_f = 2400 \text{ MPa} \qquad \epsilon'_f = 0.1 \qquad b = -0.07 \qquad c = -0.7$$

(a) Draw on log–log coordinates the completely reversed cyclic strain amplitude, $\Delta\epsilon/2$, versus fatigue life, $2N_f$, curves for these two steels. Determine the strain amplitude where they have the same life.

(b) A composite member is fabricated by securely bonding a rod of the soft steel inside a tube of the hardened steel. This member is then subjected to completely reversed cyclic strain in the axial direction. On the same plot used in part (a), show the expected $\Delta\epsilon/2$ versus $2N_f$ plot for this composite member. Indicate on the plot whether fatigue cracks first initiate in the rod or tube. Does the fatigue response of this member depend on the relative areas of the rod and tube?

(c) Plot the cyclic σ–ϵ curves for the steels in the soft and hard conditions up to a strain of 0.02. On the same plot show the cyclic stress–strain curve for the composite member for the case where the rod and tube are of equal areas.

(d) On log–log coordinates, plot the completely reversed stress amplitude ($\Delta\sigma/2$) versus fatigue life ($2N_f$) line for the soft and hard steels. Also plot the stress amplitude versus fatigue life curve for the composite member from part (c).

2.22. Derive the following relationships:

$$K' = \frac{\sigma_f'}{(\epsilon_f')^{n'}} \qquad n' = \frac{b}{c}$$

2.23. Another parameter that can be related to fatigue resistance is the cyclic plastic strain energy (area inside the hysteresis loop). Derive an expression relating the plastic strain energy per cycle, ΔW_p, to fatigue life, $2N_f$. Assuming that a metal is cyclically stable during its entire fatigue life, derive an expression for the total plastic strain energy required for fatigue failure, W_f, as a function of fatigue life.

SECTION 2.5

2.24. Given below are the results of constant amplitude strain-controlled tests. The material has a modulus of elasticity, E, of 200 GPa.

Total Strain Amplitude, $\Delta\epsilon/2$	Stress Amplitude, $\Delta\sigma/2$ (MPa)	Reversals to Failure, $2N_f$
0.00202	261	416,714
0.00510	372	15,894
0.0102	428	2,671
0.0151	444	989

Determine:

(a) The cyclic stress–strain properties (K', n')

(b) The strain–life properties (ϵ_f', σ_f', b, c)

(c) The transition life ($2N_t$)

(d) The fatigue life at strain amplitude, $\Delta\epsilon/2$, of 0.0075

2.25. The following equations can be used to estimate the fatigue strength coefficient, σ_f', and fatigue strength exponent, b, for steels.

$$\sigma_f' \approx \sigma_f \approx S_u + 50\,\text{ksi}$$

$$b \approx -\frac{1}{6}\log\frac{2\sigma_f}{S_u}$$

where σ_f is the true fracture strength and S_u is the ultimate strength. How well do these approximations work for the following steels?

S_u (ksi)	σ_f (ksi)	σ_f' (ksi)	b
52	104	84	−0.09
77	117	117	−0.07
105	178	178	−0.095
136	155	180	−0.07
150	220	230	−0.09
180	240	240	−0.076
195	270	230	−0.074
238	258	258	−0.067
255	290	290	−0.08
270	310	310	−0.071
295	300	300	−0.082
325	385	385	−0.089
355	375	375	−0.075

2.26. The following relationships can be used to estimate the fatigue ductility coefficient, ϵ_f', and fatigue ductility exponent, c, for steels.

$$\epsilon_f' \approx \epsilon_f = \ln\frac{1}{1 - \text{RA}}$$

where RA is the reduction in area. For ductile steels (where $\epsilon_f \approx 1$), $c \approx -0.6$, and for strong steels (where $\epsilon_f \approx 0.5$), $c \approx -0.5$. How well do these approximations work for the following steels?

% RA	ϵ_f	ϵ_f'	c
73	1.30	0.15	−0.43
65	1.04	1.00	−0.66
25	0.29	0.27	−0.53
59	0.89	0.45	−0.68
42	0.54	0.40	−0.73
20	0.22	0.20	−0.77
27	0.31	0.07	−0.76
11	0.12	0.18	−0.56
49	0.68	0.68	−0.65
43	0.56	0.66	−0.69
72	1.27	0.85	−0.61
55	0.79	0.89	−0.69
35	0.43	0.09	−0.61

2.27. The following equations can be used to estimate the fatigue strength exponent, b, and fatigue ductility exponent, c, using the cyclic strain hardening exponent, n'.

$$b = \frac{-n'}{1 + 5n'} \qquad c = \frac{-1}{1 + 5n'}$$

How well do these approximations work for the following steels?

n'	b	c
0.07	−0.077	−0.740
0.08	−0.070	−0.750
0.11	−0.073	−0.410
0.12	−0.073	−0.600
0.13	−0.081	−0.610
0.14	−0.070	−0.690
0.15	−0.071	−0.470
0.16	−0.110	−0.670
0.17	−0.080	−0.750
0.18	−0.120	−0.510
0.21	−0.102	−0.420
0.22	−0.110	−0.480
0.26	−0.150	−0.570

2.28. Given below are the results of constant amplitude strain-controlled tests on four different steels ($E = 30 \times 10^3$ ksi).

Total Strain Amplitude, $\Delta\epsilon/2$	Stress Amplitude, $\Delta\sigma/2$ (ksi)	Reversals to Failure, $2N_f$	Total Strain Amplitude, $\Delta\epsilon/2$	Stress Amplitude, $\Delta\sigma/2$ (ksi)	Reversals to Failure, $2N_f$
	Steel A[a]			Steel B[b]	
0.0100	49.5	2,174	0.0264	103.5	250
0.0080	48.5	2,246	0.0153	91.5	824
0.0070	42.8	3,034	0.00885	77.5	4,128
0.0040	39.0	15,880	0.00540	69.5	11,900
0.0030	33.9	27,460	0.00274	61.0	72,000
0.0020	29.5	106,700	0.00253	63.5	972,000
0.0018	27.6	171,700	0.00204	62.5	456,000
0.0014	26.2	536,500	0.00187	56.5	636,000
0.0012	25.3	426,200	0.00187	56.5	818,000
			0.00197	59.5	1,914,000
	Steel C[c]			Steel D[c]	
0.0160	148	240	0.0135	190	178
0.0100	135	722	0.0125	184	258
0.0090	128	1,020	0.0110	185	266
0.0084	125	1,250	0.0100	185	488
0.0080	125	1,350	0.0083	184	584
0.0072	122	1,760	0.0075	166	956
0.0060	120	3,000	0.0058	152	2,350
0.0052	115	6,000	0.0050	146	6,880
0.0042	112	15,000	0.0040	120	63,400
0.0033	100	82,000	0.0033	100	785,000

[a] Data taken from Ref. 21.
[b] Data taken from Ref. 22.
[c] Data taken from Ref. 23.

Material	S_y (ksi)	S_u (ksi)	% RA	σ_f (ksi)	ϵ_f
Steel A	38	57	55	108	0.81
Steel B	94	114	68	225	1.14
Steel C	185	195	59	270	0.89
Steel D	230	255	42	320	0.54

Determine for each of these steels:

(a) The cyclic stress–strain properties (K', n') using a regression fit of stress amplitude, $\Delta\sigma/2$, versus plastic strain amplitude, $\Delta\epsilon_p/2$.

(b) The strain life properties $(\sigma_f', \epsilon_f', b, c)$. Compare these values to the approximations discussed in Section 2.5.

(c) The transition life, $2N_t$.

(d) The cyclic stress–strain properties using Eqs. (2.43) and (2.44). Compare these values to the ones found in part (a).

On the same set of log–log coordinates, draw the strain–life curves for the four steels. Compare the relative behavior of these four steels at various strain amplitudes.

2.29. Given below are the results of constant amplitude strain-controlled tests on a high strength aluminum ($E = 10.5 \times 10^3$ ksi).

Total Strain Amplitude $\Delta\epsilon/2$	Stress Amplitude, $\Delta\sigma/2$ (ksi)	Reversals to Failure $2N_f$
0.0725	88.5	10
0.0445	86.0	28
0.0290	81.0	90
0.0182	76.0	284
0.0123	68.0	620
0.0082	64.5	2,000
0.0056	57.0	8,400
0.0047	49.0	24,800

Determine the strain–life properties $(\sigma_f', \epsilon_f', b, c)$ for this material.

It has been found that some high strength aluminums cannot be represented well by the four-parameter strain–life equation. Plot the data above and discuss how well this material follows the four-parameter model. (Data taken from Ref. 24.)

2.30. An engineering rule of thumb is that a strain amplitude of 1% $(\Delta\epsilon/2 = 0.01)$ corresponds to a life of 1000 cycles (or 2000 reversals). Does this rule hold for the materials in Problems 2.24, 2.28, and 2.29?

2.31. A method of approximating the strain–life relationship was proposed by Manson [25]. This procedure is referred to as the method of universal slopes:

$$\Delta\epsilon = 3.5\frac{S_u}{E}(N)^{-0.12} + \epsilon_f^{0.6}(N)^{-0.6}$$

where S_u = ultimate strength
E = modulus of elasticity
ϵ_f = true fracture ductility

$$= \ln \frac{1}{1 - \text{RA}}$$

This approximation only requires monotonic tensile properties. Discuss how this method compares to the approximations discussed in Section 2.5. Note that the universal slopes equation is stated in terms of strain range, $\Delta\epsilon$, and cycles, N.

2.32. Restate the equation for transition life for steel:

$$2N_t = \left(\frac{\epsilon_f' E}{\sigma_f'}\right)^{1/(b-c)}$$

in terms of the monotonic properties, ultimate strength, S_u, and reduction in area, % RA. Use the approximations from Section 2.5 and note that

$$\frac{1}{b - c} \approx 2$$

SECTION 2.6

2.33. The stabilized hysteresis loop for a material under cyclic loading is determined to have the following values:

Strain amplitude ($\Delta\epsilon/2$) = 0.002
Mean stress (σ_0) = −9.5 ksi
Maximum stress (σ_{max}) = 30.5 ksi

Determine the life to failure, $2N_f$, using the Morrow [Eq. (49)], Manson–Halford [Eq. (50)], and Smith–Watson–Topper [Eq. (52)] relationships.

The strain–life properties for the material are

$$E = 30 \times 10^3 \text{ ksi}$$

$$\sigma_f' = 133 \text{ ksi} \qquad b = -0.095$$

$$\epsilon_f' = 0.26 \qquad c = -0.47$$

2.34. Smooth aluminum specimens are subjected to two series of cyclic load-controlled tests. The first test (level A) varies between a maximum stress value, σ_{max}, of 21.3 ksi and a minimum value, σ_{min}, of −30.1 ksi. The second test (level B) varies between 61.5 and 10.1 ksi. Predict the life to failure, in reversals, at the two levels. Use the Morrow [Eq. (2.49)], Manson–Halford [Eq. (2.50)], and Smith–Watson–Topper [Eq. (2.52)] relationships for the predictions. Assume that there is no mean stress relaxation. The material properties for the aluminum are

$$E = 10.6 \times 10^3 \text{ ksi} \qquad K' = 95 \text{ ksi} \qquad n' = 0.065$$

$$\sigma_f' = 160 \text{ ksi} \qquad b = -0.124$$

$$\epsilon_f' = 0.22 \qquad c = -0.59$$

Listed below are actual test results at the two levels. Three tests were run at each of the levels. Compare the predictions to these values. (Data taken from Ref. 26.)

Level	Test Results: Lives in Reversals, $2N_f$		
A	5.4×10^5	5.5×10^5	7.2×10^5
B	5.6×10^4	6.4×10^4	6.8×10^4

2.35. Determine the lives to failure for a nickel alloy under the following load histories:

History	Description	Strain Amplitude $\Delta\epsilon/2$	Mean Strain ϵ_0
A	Fully reversed ($R = -1$)	0.005	0
B	Fully reversed ($R = -1$)	0.010	0
C	Zero to maximum ($R = 0$)	0.005	0.005
D	Zero to maximum ($R = 0$)	0.010	0.010

Use the Morrow [Eq. (2.49)], Manson–Halford [Eq. (2.50)], and Smith–Watson–Topper [Eq. (2.52)] relationships for these predictions. Compare the predictions made using the three methods.

The stress–strain and strain–life properties for the alloy are

$$E = 208.5 \text{ GPA} \qquad K' = 1530 \text{ MPa} \qquad n' = 0.073$$

$$\sigma_f' = 1640 \text{ MPa} \qquad b = -0.06$$

$$\epsilon_f' = 2.67 \qquad c = -0.82$$

Listed below are actual test results for the four histories. Two tests were run for each history. Compare the predictions to these values. Discuss the effect of mean strain on fatigue life at high and low strain amplitudes. (*Hint:* Refer to Fig. 2.22.) (Data taken from Ref. 27.)

History	Test Results: Lives in Reversals, $2N_f$	
A	2.8×10^4	2.6×10^4
B	2.4×10^3	2.6×10^3
C	1.6×10^4	1.4×10^4
D	1.8×10^3	1.9×10^3

2.36. Plot on log–log coordinates the allowable alternating strain amplitude, $\Delta\epsilon/2$, versus fatigue life, $2N_f$, for a mean stress, σ_0, of 50 ksi. Draw curves using the Morrow [Eq. (2.48)], Manson–Halford [Eq. (2.50)], and Smith–Watson–Topper [Eq. (2.52)] relationships. Compare the three curves. The stress–strain and strain–life parameters are

$$E = 30 \times 10^3 \text{ ksi}$$

$$K' = 182 \text{ ksi} \qquad n' = 0.208$$

$$\sigma_f' = 137.5 \text{ ksi} \qquad b = -0.0924$$

$$\epsilon_f' = 0.26 \qquad c = -0.445$$

Repeat the foregoing procedure for a mean stress, σ_0, of -50 ksi.

2.37. A metal specimen is shot peened, which produces a residual surface stress of -40 ksi. Plot on log–log coordinates the strain–life ($\Delta\epsilon/2$ versus $2N_f$) curves for both the peened and unpeened specimens. The cyclic stress–strain and strain–life parameters for the material are

$$E = 30 \times 10^3 \text{ ksi}$$

$$K' = 154 \text{ ksi} \qquad n' = 0.123$$

$$\sigma_f' = 169 \text{ ksi} \qquad b = -0.081$$

$$\epsilon_f' = 1.142 \qquad c = -0.670$$

Using the strain–life relationship and cyclic stress–strain properties, generate the stress amplitude, $\Delta\sigma/2$, versus fatigue life, $2N_f$, curves for the material in the peened and unpeened conditions.

Use the Morrow relationship [Eq. (2.49)] for these calculations.

2.38. Given below are the cyclic stress–strain and strain–life parameters for a steel:

$$E = 30 \times 10^3 \text{ ksi}$$

$$K' = 174.6 \text{ ksi} \qquad n' = 0.202$$

$$\sigma_f' = 133 \text{ ksi} \qquad b = -0.095$$

$$\epsilon_f' = 0.26 \qquad c = -0.47$$

Determine the life of this material under the following strain histories. History A is constant amplitude, while histories B and C have initial overloads which set up

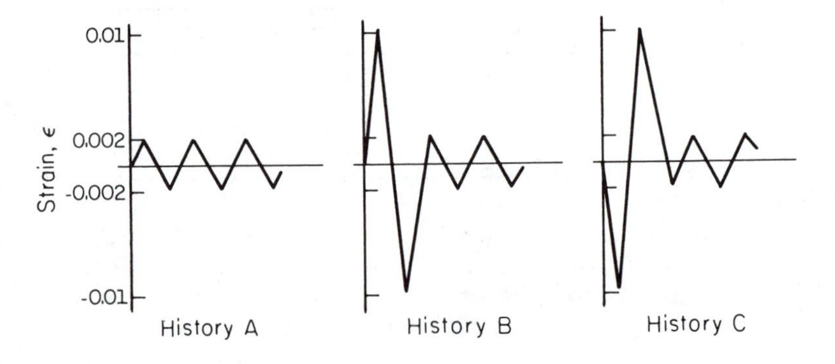

residual stresses. Use the Morrow [Eq. (2.49)], Manson–Halford [Eq. (2.50)], and Smith–Watson–Topper [Eq. (2.52)] relationships for these calculations. Compare the predictions made using the three methods.

2.39. Plot on log–log coordinates the strain amplitude, $\Delta\epsilon/2$, versus life to failure, $2N_f$, curve for fully reversed ($R = -1$) strain-controlled tests. Use the material properties given below. The plot should go from 1 to 10^6 reversals. On the same graph, plot the strain amplitude versus life to failure curve for zero-to-maximum ($R = 0$) strain-controlled tests. Assume that there will be no mean stress relaxation. Use the Morrow relationship [Eq. (2.49)] for these calculations.

$$E = 200 \text{ GPa} \qquad K' = 1484.2 \text{ MPa} \qquad n' = 0.150$$

$$\sigma'_f = 1600 \text{ MPa} \qquad b = -0.09$$

$$\epsilon'_f = 1.65 \qquad c = -0.60$$

Discuss possible errors in the fully reversed curve ($R = 0$) at very short lives.

2.40. As discussed in Section 2.3, there is justification for using monotonic, as opposed to cyclic, properties to model the initial loading when performing a strain–life analysis. To determine the difference between using monotonic or cyclic properties, compare the life predictions for a zero-to-maximum ($R = 0$) strain loading of 0.0 to 0.005. In the first case use monotonic properties (K, n) to model the initial loading and cyclic properties (K', n') on all successive cycles. For the second case use cyclic properties on all loadings. Use the Morrow relationship [Eq. (2.49)] for these calculations. The material properties are

$$E = 185 \text{ GPA}$$

$$K = 1210 \text{ MPa} \qquad n = 0.193$$

$$K' = 1660 \text{ MPa} \qquad n' = 0.287$$

$$\sigma'_f = 1000 \text{ MPa} \qquad b = -0.114$$

$$\epsilon'_f = 0.171 \qquad c = -0.402$$

3 FRACTURE MECHANICS

3.1 INTRODUCTION

The fatigue life of a component is made up of initiation and propagation stages. This is illustrated schematically in Fig. 3.1. The size of the crack at the transition from initiation to propagation is usually unknown and often depends on the point of view of the analyst and the size of the component being analyzed. For example, for a researcher equipped with microscopic equipment it may be on the order of a crystal imperfection, dislocation, or a 0.1 mm-crack, while to the inspector in the field it may be the smallest crack that is readily detectable with nondestructive inspection equipment. Nevertheless, the distinction between the initiation life and propagation life is important. At low strain amplitudes up to 90% of the life may be taken up with initiation, while at high amplitudes the majority of the fatigue life may be spent propagating a crack. Fracture mechanics approaches are used to estimate the propagation life.

Fracture mechanics approaches require that an initial crack size be known or assumed. For components with imperfections or defects (such as welding porosities, inclusions and casting defects, etc.) an initial crack size may be known. Alternatively, for an estimate of the total fatigue life of a defect-free material, fracture mechanics approaches can be used to determine propagation. Strain–life approaches may then be used to determine initiation life, with the total life being the sum of these two estimates.

In this chapter we briefly review the fundamentals of fracture mechanics and discuss the use of these concepts in applications to constant amplitude fatigue crack propagation analyses. In Chapter 4 we review fracture mechanics ap-

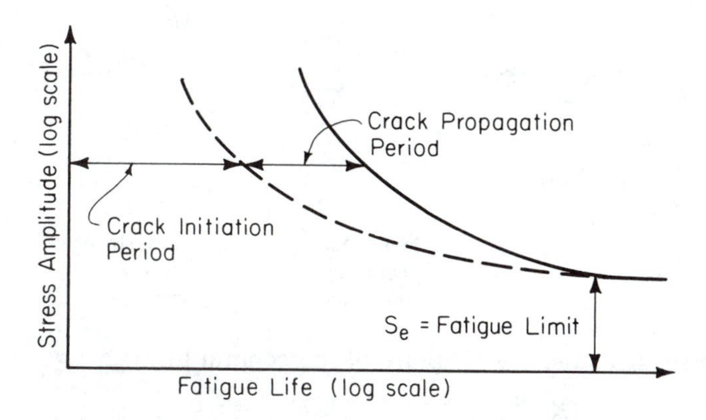

Figure 3.1 Initiation and propagation portions of fatigue life.

proaches for the analysis of notched components and in Chapter 5 discuss fracture mechanics approaches used to predict fatigue crack growth under variable amplitude loading.

3.2 LINEAR ELASTIC FRACTURE MECHANICS BACKGROUND

Linear elastic fracture mechanics (LEFM) principles are used to relate the stress magnitude and distribution near the crack tip to:

- Remote stresses applied to the cracked component
- The crack size and shape
- The material properties of the cracked component

3.2.1 Historical Overview

In the 1920s, Griffith [1] formulated the concept that a crack in a component will propagate if the total energy of the system is lowered with crack propagation. That is, if the change in elastic strain energy due to crack extension is larger than the energy required to create new crack surfaces, crack propagation will occur.

Griffith's theory was developed for brittle materials. In the 1940s, Irwin [2] extended the theory for ductile materials. He postulated that the energy due to plastic deformation must be added to the surface energy associated with the creation of new crack surfaces. He recognized that for ductile materials, the surface energy term is often negligible compared to the energy associated with plastic deformation. Further, he defined a quantity, G, the strain energy release rate or "crack driving force," which is the total energy absorbed during cracking per unit increase in crack length and per unit thickness.

In the mid-1950s, Irwin [3] made another significant contribution. He

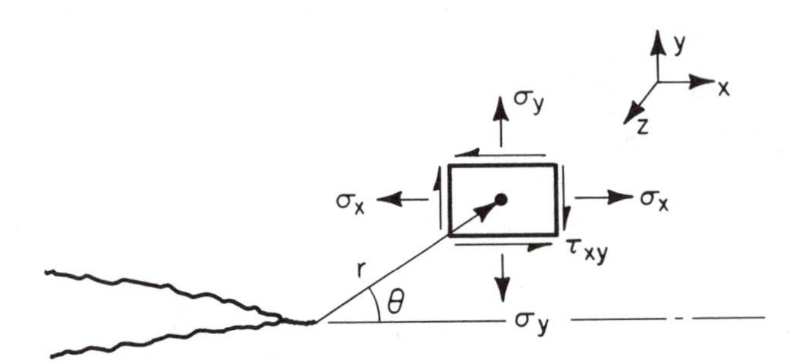

Figure 3.2 Location of local stresses near a crack tip in cylindrical coordinates.

showed that the local stresses near the crack tip are of the general form

$$\sigma_{ij} = \frac{K}{\sqrt{2\pi r}} f_{ij}(\theta) + \cdots \tag{3.1}$$

where r and θ are cylindrical coordinates of a point with respect to the crack tip (see Fig. 3.2) and K is the stress intensity factor. He further showed that the energy approach (the "G" approach above) is equivalent to the stress intensity approach (described in Section 3.2.4) and that crack propagation occurs when a critical strain energy release rate, G_c (or in terms of a critical stress intensity, K_c) is achieved.

3.2.2 LEFM Assumptions

Linear elastic fracture mechanics (LEFM) is based on the application of the theory of elasticity to bodies containing cracks or defects. The assumptions used in elasticity are also inherent in the theory of LEFM: namely, small displacements and general linearity between stresses and strains.

The general form of the LEFM equations is given in Eq. (3.1). As seen, a singularity exists such that as r, the distance from the crack tip, tends toward zero, the stresses go to infinity. Since materials plastically deform as the yield stress is exceeded, a plastic zone will form near the crack tip. The basis of LEFM remains valid, though, if this region of plasticity remains small in relation to the overall dimensions of the crack and cracked body.

3.2.3 Loading Modes

There are generally three modes of loading, which involve different crack surface displacements (see Fig. 3.3). The three modes are:

Mode I: opening or tensile mode (the crack faces are pulled apart)

Mode II: sliding or in-plane shear (the crack surfaces slide over each other)

Mode III: tearing or anti-plane shear (the crack surfaces move parallel to the leading edge of the crack and relative to each other)

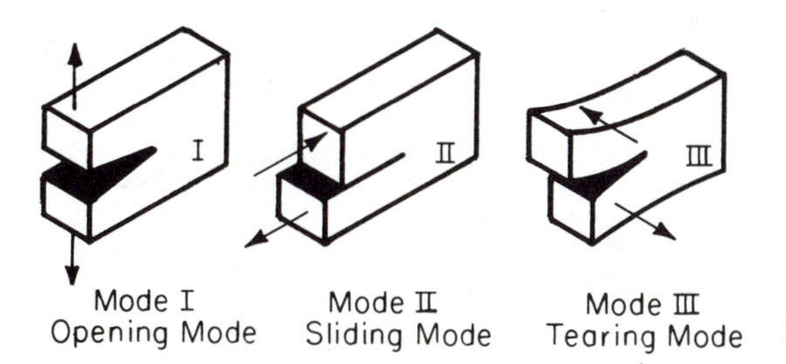

Figure 3.3 Three loading modes.

The following discussion deals with Mode I since this is the predominant loading mode in most engineering applications. Similar treatments can readily be extended to Modes II and III. Equations and additional details are found in Refs. 4 to 6.

3.2.4 Stress Intensity Factor

The stress intensity factor, K, which was introduced in Eq. (3.1), defines the magnitude of the local stresses around the crack tip. This factor depends on loading, crack size, crack shape, and geometric boundaries, with the general form given by

$$K = f(g)\sigma\sqrt{\pi a} \tag{3.2}$$

where σ = remote stress applied to component [not to be confused with the local stresses, σ_{ij}, in Eq. (3.1)]
a = crack length
$f(g)$ = correction factor that depends on specimen and crack geometry

Stress intensity factor solutions have been obtained for a wide variety of problems and published in handbook form [7–9]. Figure 3.4 gives the stress intensity relationships for a few of the more common loading conditions.

Stress intensity factors for a single loading mode can be added algebraically. Consequently, stress intensity factors for complex loading conditions of the same mode can be determined from the superposition of simpler results, such as those readily obtainable from handbooks.

One superposition method, the compounding technique, has been used to obtain relatively accurate approximations. The technique consists of reducing a complicated problem into a number of simpler configurations with known solutions. By superposition of these simpler K solutions, a stress intensity factor may be obtained for the complicated geometry. In equation form,

$$K_{\text{tot}} = K_0 + \left[\sum_{n=1}^{N} (K_n - K_0)\right] + K_e \tag{3.3}$$

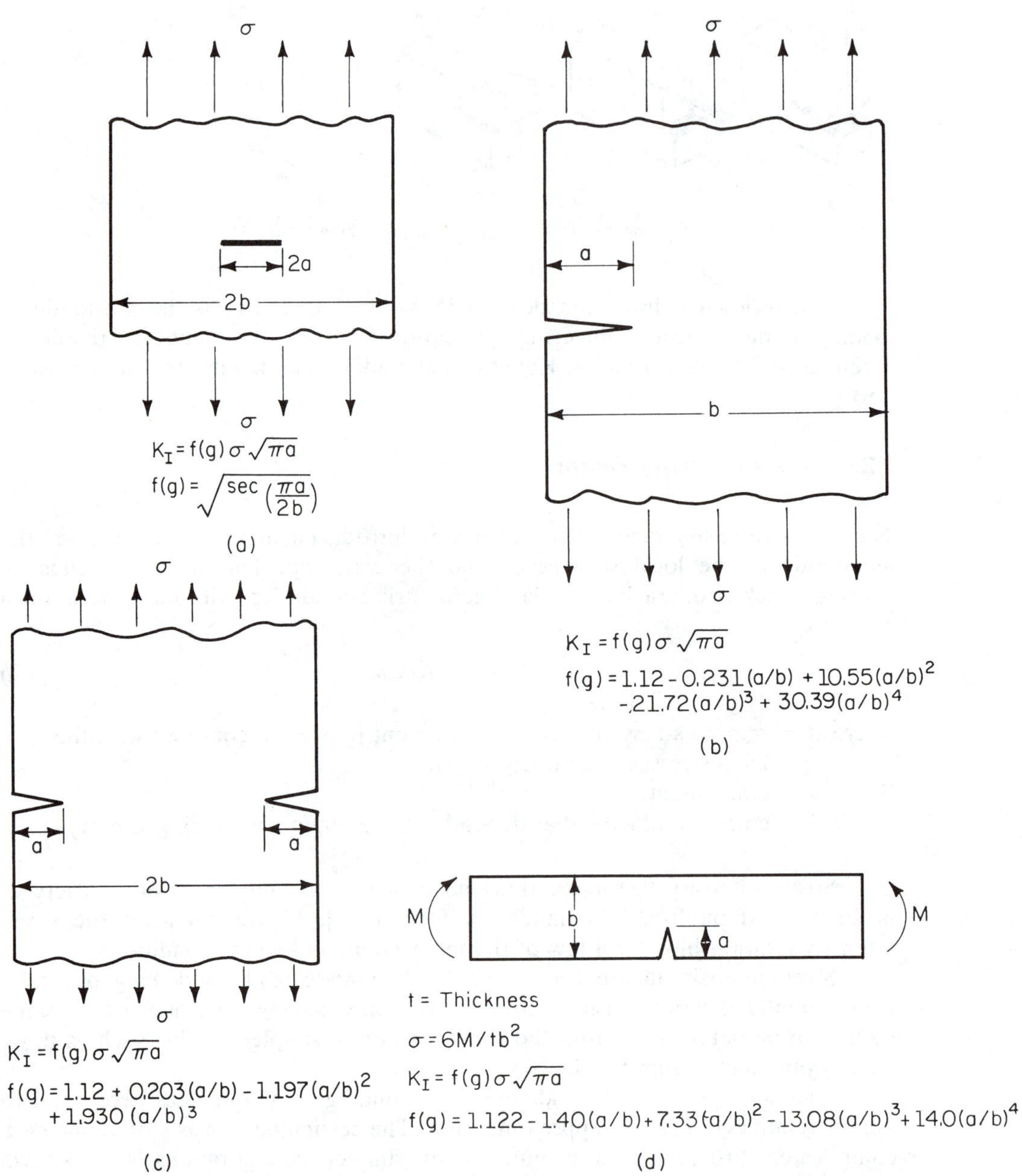

Figure 3.4 Stress intensity factor for (a) Center-cracked plate loaded in tension (from Ref. 10), (b) Edge-cracked plate loaded in tension (from Ref. 11), (c) Double-edge-cracked plate loaded in tension (from Ref. 12), (d) Cracked beam in pure bending (from Ref. 12), (e) Circular (penny shaped) crack embedded in infinite body subjected to tension (from Ref. 13), (f) Elliptical crack embedded in infinite body subjected to tension (from Ref. 14–15).

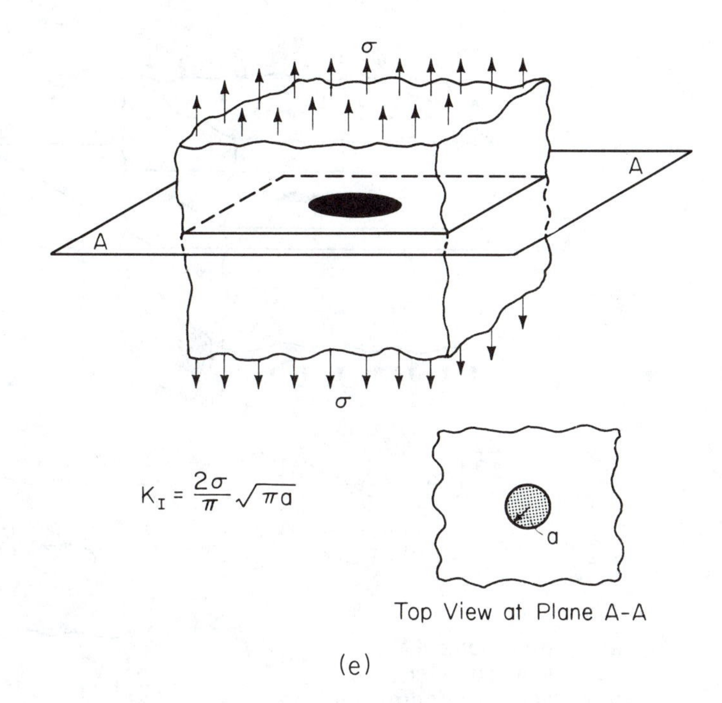

Figure 3.4 (*Continued*)

where K_{tot} = stress intensity factor for complicated geometry
K_0 = stress intensity factor in the absence of all boundaries of a form applicable to the loading (i.e., $K_I = \sigma\sqrt{\pi a}$)
K_n = stress intensity factor for the nth simpler configuration
K_e = factor that accounts for the effect of the interaction between boundaries

K_e is the only unknown. Neglecting this term will lead to underestimates of less than 10% [16]. References 17 to 21 give more details of this approach.

Another approximate method is simply to multiply the individual correction factors for the various geometric effects, such as

$$K = f_1 \cdot f_2 \cdot f_3 \cdots \sigma\sqrt{\pi a} \tag{3.4}$$

Correction factors, f_i, are used to account for

- Finite width (back wall) effect
- Front wall effect
- Crack shape (i.e., elliptical flaw)

Other superposition methods that are employed include the alternating method and the weight function method [22–26].

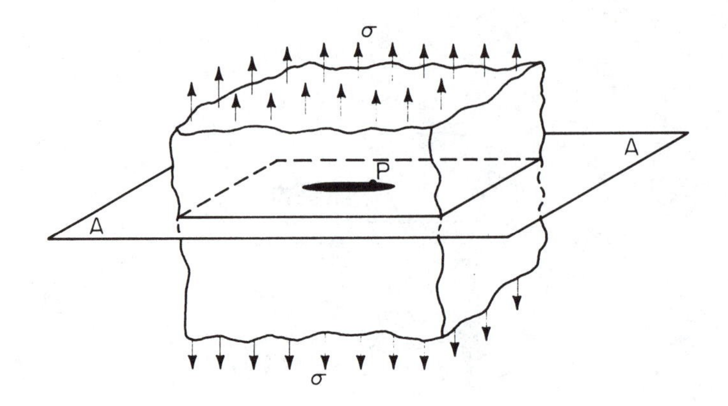

$$K_{I_{at\,P}} = \frac{\sigma\sqrt{\pi a}}{\Phi}\left\{\sin^2\theta + \frac{a^2}{c^2}\cos^2\theta\right\}^{1/4}$$

$$\Phi = \int_0^{\frac{\pi}{2}} \sqrt{1 - \left(1 - \frac{a^2}{c^2}\right)\sin^2\phi}\; d\phi$$

Top View at Plane A-A

P θ a c

$$\frac{x^2}{c^2} + \frac{y^2}{a^2} = 1 \quad (c > a)$$

2c = major diameter
2a = minor diameter
Φ = Elliptical Integral

When $\theta = 90°$, K_I is Maximum
When $\theta = 0°$, K_I is Minimum
(Note: when a=b, this becomes the solution for a circular (penny shaped) crack: $K_I = \frac{2\sigma\sqrt{\pi a}}{\pi}$)

The value for the elliptical integral, Φ, for a number of a/c values is:

a/c	0	0.1	0.2	0.3	0.4	0.5	0.6	0.7	0.8	0.9	1.0
Φ	1.000	1.016	1.051	1.097	1.151	1.211	1.277	1.345	1.418	1.493	1.571

(f)

Figure 3.4 (*Continued*)

In determining K, numerical methods (including finite element methods) have been widely used in recent years. In fact, many commercially available finite element computer programs include subroutines to calculate K. References 27 to 33 review numerical techniques used to determine K.

Determination methods for K tend to be approximate. In general, values for $f(g)$ in Eq. (3.2) tend to be between 1 and 1.4, with the value for many engineering situations being between 1 and 1.2. Errors in K may be small compared to uncertainties in a fatigue analysis, such as material properties, load levels, load history, and service environment.

Example 3.1

As discussed in this section, an approximate method to obtain the stress intensity factor for a complicated geometry is simply to approximate the geometric correction factor by the product of the individual correction factors for the various geometric effects.

For the semi-elliptical surface crack in a finite thickness plate subjected to Mode I loading:

$$K_I = f_1 f_2 \cdots f_n \frac{\sigma\sqrt{\pi a}}{\Phi}\left(\sin^2\theta + \frac{a^2}{c^2}\cos^2\theta\right)^{1/4}$$

where $f_1, f_2, \ldots, f_n$ are the individual correction factors for the various geometric effects.

Using this method, determine an estimate for the stress intensity factor for a semi-circular crack in a thick plate (see Fig. E3.1a). Also using this method, estimate the stress intensity factor, K_I, for the circular corner crack in the plate shown in Fig. E3.1b.

Solution For the stress intensity factor for a semi-circular crack in a thick plate,

$$K_I \approx 1.12\left(\frac{2\sigma\sqrt{\pi a}}{\pi}\right)$$

where $f_1 = 1.12$ is the free edge correction factor and $2\sigma\sqrt{\pi a}/\pi$ is the stress intensity for a circular crack embedded in an infinite body subjected to tension (see Fig.

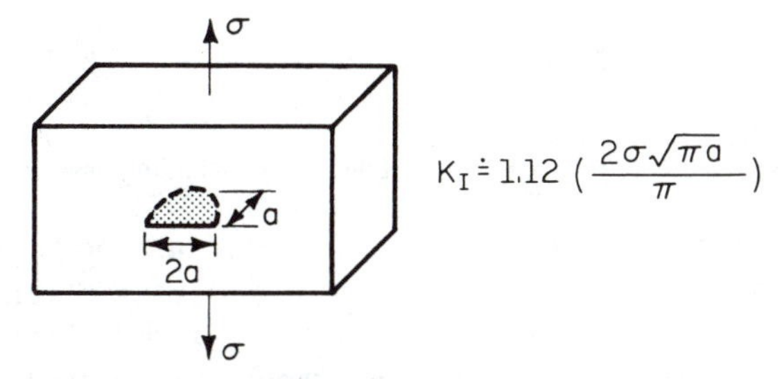

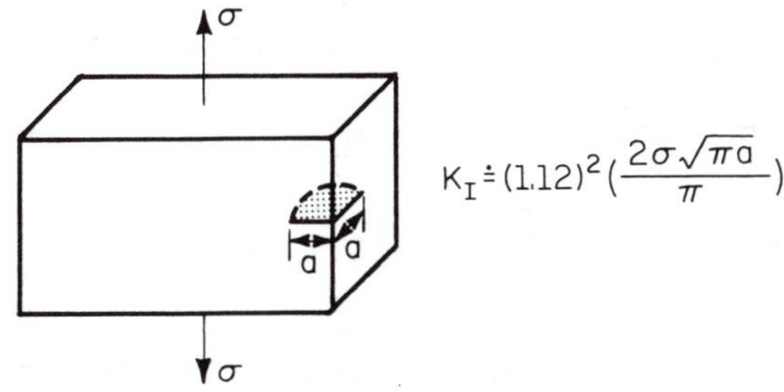

Figure E3.1 Stress intensity factors for: (a) a semi-circular crack in a thick plate; (b) a circular corner crack in a thick plate.

3.4e). The stress intensity factor for a circular corner crack in a thick plate

$$K_I \approx (1.12)^2(2\sigma)\frac{\sqrt{\pi a}}{\pi}$$

where f_1 = 1.12, the free edge correction factor for one face of the plate

f_2 = 1.12, the free edge correction factor the other face of the plate

$\frac{2\sigma\sqrt{\pi a}}{\pi}$ = Mode I stress intensity factor for a circular crack embedded in an infinite body subjected to tension

3.2.5 Plastic Zone Size

As mentioned previously, materials develop plastic strains as the yield stress is exceeded in the region near the crack tip (see Fig. 3.5). The amount of plastic deformation is restricted by the surrounding material, which remains elastic. The size of this plastic zone is dependent on the stress conditions of the body.

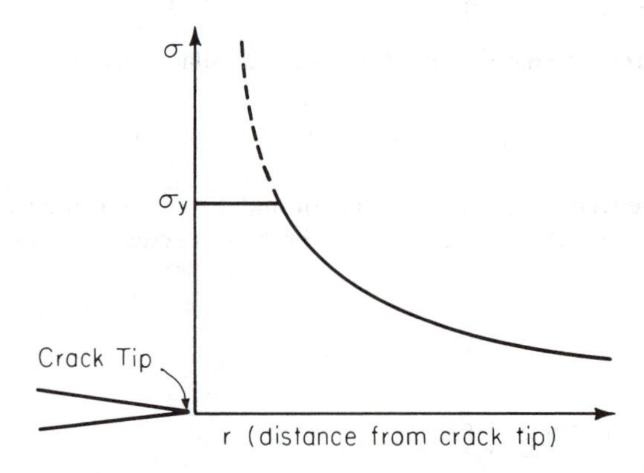

Figure 3.5 Yielding near crack tip.

Plane stress and plane strain conditions. In a thin body, the stress through the thickness (σ_z) cannot vary appreciably due to the thin section. Because there can be no stresses normal to a free surface, $\sigma_z = 0$ throughout the section and a biaxial state of stress results. This is termed a *plane stress condition* (see Fig. 3.6).

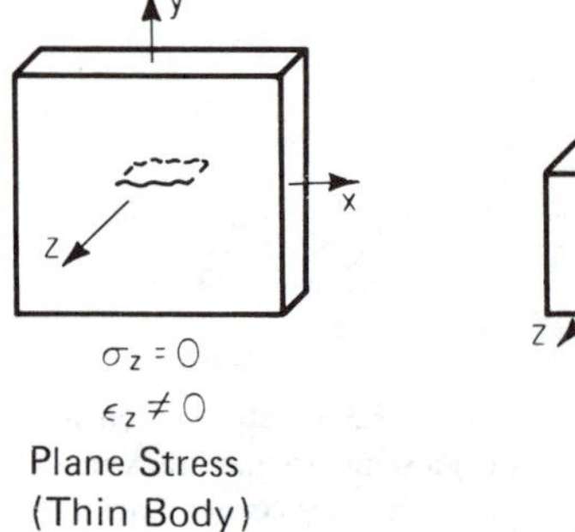

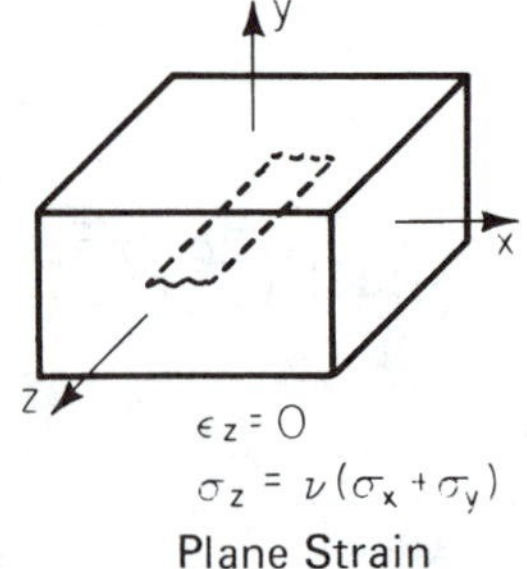

Figure 3.6 Plane stress and plane strain conditions.

In a thick body, the material is constrained in the z direction due to the thickness of the cross section and $\epsilon_z = 0$, resulting in a *plane strain condition.* Due to Poisson's effect, a stress, σ_z, is developed in the z direction. Maximum constraint conditions exist in the plane strain condition, and consequently the plastic zone size is smaller than that developed under plane stress conditions.

Monotonic plastic zone size. The plastic zone sizes under monotonic loading have been estimated to be

$$r_y = \begin{cases} \dfrac{1}{2\pi}\left(\dfrac{K}{\sigma_y}\right)^2 & \text{plane stress} \quad (3.5a) \\[2ex] \dfrac{1}{6\pi}\left(\dfrac{K}{\sigma_y}\right)^2 & \text{plane strain} \quad (3.5b) \end{cases}$$

where r is defined as shown in Fig. 3.7.

Cyclic plastic zone size. The reversed or cyclic plastic zone size is four times smaller than the comparable monotonic value. As the nominal tensile load is reduced, the plastic region near the crack tip is put into compression by the surrounding elastic body. As shown in Fig. 3.8, the change in stress at the crack tip due to the reversed loading is twice the value of the yield stress.

Equations (3.5a) and (3.5b) become

$$r_y = \begin{cases} \dfrac{1}{2\pi}\left(\dfrac{K}{2\sigma_y}\right)^2 = \dfrac{1}{8\pi}\left(\dfrac{K}{\sigma_y}\right)^2 & \text{plane stress} \quad (3.6a) \\[2ex] \dfrac{1}{6\pi}\left(\dfrac{K}{2\sigma_y}\right)^2 = \dfrac{1}{24\pi}\left(\dfrac{K}{\sigma_y}\right)^2 & \text{plane strain} \quad (3.6b) \end{cases}$$

The cyclic plastic zone size is smaller than the monotonic and more characteristic of a plane strain state even in thin plates. Thus LEFM concepts can

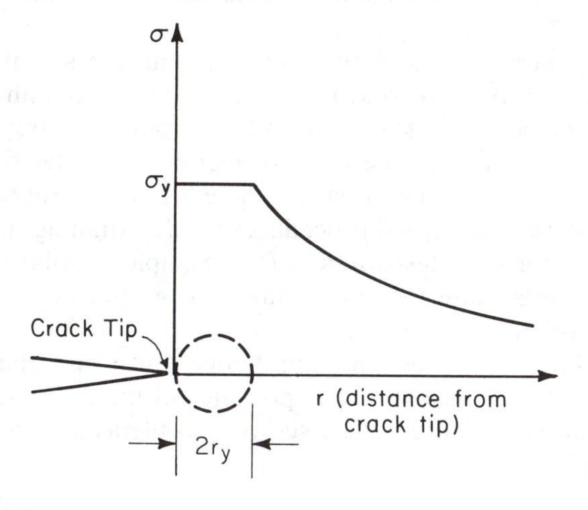

Figure 3.7 Monotonic plastic zone size.

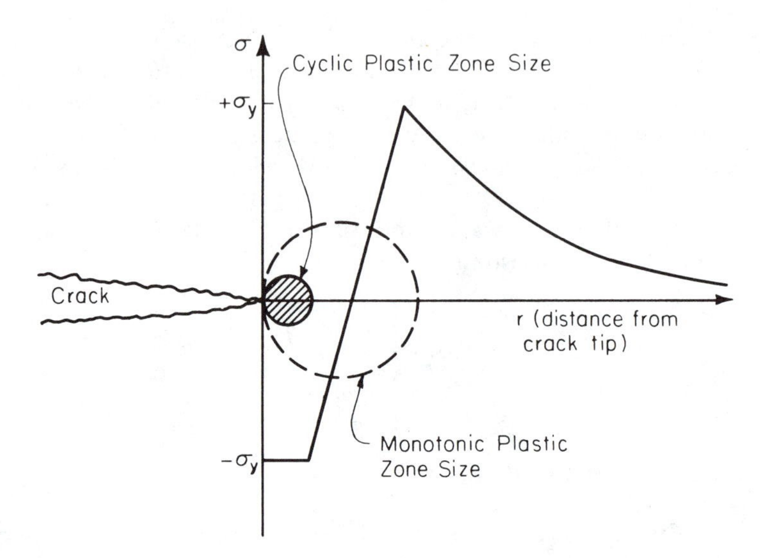

Figure 3.8 Reversed plastic zone size.

often be used in the analysis of fatigue crack growth problems even in materials that exhibit considerable amounts of ductility. The basic assumption that the plastic zone size is small in relationship to the crack and the cracked body usually remains valid.

3.2.6 Fracture Toughness

As the stress intensity factor reaches a critical value, K_c, unstable fracture occurs. This critical value of the stress intensity factor is known as the fracture toughness of the material. The fracture toughness can be considered the limiting value of stress intensity just as the yield stress might be considered the limiting value of applied stress.

The fracture toughness varies with specimen thickness until limiting conditions (maximum constraint) are reached. Recall that maximum constraint conditions occur in the plane strain state. The plane strain fracture toughness, K_{Ic}, is dependent on specimen geometry and metallurgical factors. ASTM Designation E-399, Standard Method of Test for Plane Strain Fracture Toughness of Metallic Materials, sets forth accepted procedures for determining this value. It is often difficult to perform a valid test for K_{Ic}. For example, a valid test using a thin plate of high toughness material often cannot be performed. Rather the value, K_c, at the given conditions is obtained.

The fracture toughness depends on both temperature and the specimen thickness. The following example shows the importance of the fracture toughness in designing against unstable fracture. (Also see the problems at the end of the chapter.)

Example 3.2

A company is building a 3 ft diameter pressure vessel from material that has a fracture toughness of 60 ksi $\sqrt{\text{in.}}$ and a yield strength of 85 ksi at the operating temperature. The wall thickness is 0.75 in., and the operating pressure is 2000 psi.

It is required that the vessel "leak-before-burst." In other words, the crack must be able to grow through the wall thickness before fast fracture occurs. This allows the gas or liquid in the pressure vessel to escape and be detected before an unstable condition develops.

The pressure vessel will be inspected periodically with a technique that can reliabily detect a crack with a surface length larger than 0.5 in. Will the pressure vessel leak before burst when the surface length of the crack is smaller than this size? What is the largest value of the surface crack that can develop and still maintain the leak-before-burst criteria?

Solution The stress intensity for a thumbnail crack in a plate subjected to tension can be calculated from the equations in Fig. 3.4f and a free surface correction of 1.12. The stress intensity for $\theta = \pi/2$ is

$$K_I = \frac{1}{\sqrt{Q}} 1.12\sigma\sqrt{\pi a}$$

where Q is termed the shape factor since it depends on a and c. Figure E3.2 graphically shows this dependence for various ratios of nominal applied stress, σ, to the yield stress, σ_y.

Using this figure, the leak-before-burst problem can be evaluated. The following information is known:

$$K_{Ic} = 60 \text{ ksi} \sqrt{\text{in.}}$$

$$\sigma_y = 85 \text{ ksi}$$

$$p = 2000 \text{ psi}$$

$$r = d/2 = 3 \text{ ft}/2 = 18 \text{ in.}$$

$$t = 0.75 \text{ in.}$$

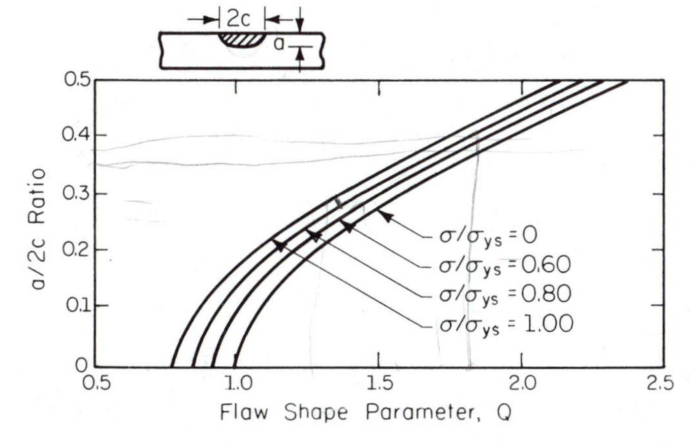

Figure E3.2 Flaw shape parameter as a function of crack aspect ratio. (From Ref. 34.)

The hoop stress in the pressure vessel is

$$\sigma = \frac{pr}{t} = 2000\text{ psi}\left(\frac{18\text{ in.}}{0.75\text{ in.}}\right) = 48.0\text{ ksi}$$

and the ratio of this stress to the yield stress is

$$\frac{\sigma}{\sigma_y} = \frac{48}{85} = 0.56$$

For the leak-before-burst criteria the critical stress intensity factor, K_{Ic}, must be larger than the stress intensity factor due to a 0.75-in. crack in the pressure vessel ($a = 0.75$ in., the wall thickness). From this information, the value for the shape factor, Q, can be determined.

$$K_{Ic} > \frac{\sigma\sqrt{\pi a}}{\sqrt{Q}}(1.12)$$

$$60 > \frac{48\sqrt{\pi(0.75)}}{\sqrt{Q}}(1.12)$$

$$Q > 1.892$$

Using Fig. E3.2, for a value of $Q > 1.892$ and $\sigma/\sigma_y = 0.56$, the value of c, the surface crack length, can now be determined:

$$\frac{a}{2c} > 0.40$$

or

$$\frac{0.75}{2(0.40)} > c$$

Therefore,

$$c < 0.938$$

and

$$2c < 1.875\text{ in.}$$

A surface crack of 1.875 in. length or smaller will ensure that the vessel will leak before break. Thus the vessel will not fail catastrophically when a surface crack of 0.5 in. can be detected.

3.3 FATIGUE CRACK GROWTH

As discussed earlier, the majority of fatigue life may be taken up in the propagation of a crack. By the use of fracture mechanics principles it is possible

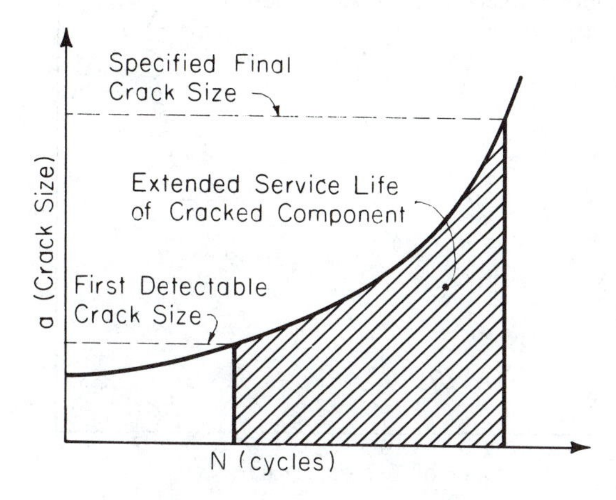

Figure 3.9 Extended service life of a cracked component.

to predict the number of cycles spent in growing a crack to some specified length or to final failure.

The aircraft industry has been instrumental in the effort to understand and predict fatigue crack growth. They have developed the safe-life or fail-safe design approach. In this method, a component is designed such that if a crack forms, it will not grow to a critical size between specified inspection intervals. Thus, by knowing the material growth rate characteristics and with regular inspections, a cracked component may be kept in service for an extended useful life. This concept is shown schematically in Fig. 3.9.

3.3.1 Fatigue Crack Growth Curves

Typical constant amplitude crack propagation data are shown in Fig. 3.10. The crack length, a, is plotted versus the corresponding number of cycles, N, at which the crack was measured. As shown, most of the life of the component is spent while the crack length is relatively small. In addition, the crack growth rate increases with increased applied stress.

The crack growth rate, da/dN, is obtained by taking the derivative of the above crack length, a, versus cycles, N, curve. (Two generally accepted numerical approaches for obtaining this derivative are the spline fitting method and the incremental polynomial method. These methods are explained in detail in many numerical methods textbooks. For example, see Ref. 35.) Values of $\log da/dN$ can then be plotted versus $\log \Delta K$, for a given crack length, using the equation

$$\Delta K = K_{\max} - K_{\min} = f(g)\, \Delta\sigma\sqrt{\pi a} \tag{3.7}$$

where $\Delta\sigma$ is the remote stress applied to the component as shown in Fig. 3.11.

A plot of $\log da/dN$ versus $\log \Delta K$, a sigmoidal curve, is shown in Fig. 3.12. This curve may be divided into three regions. At low stress intensities, Region I, cracking behavior is associated with threshold, ΔK_{th}, effects. In the mid-region,

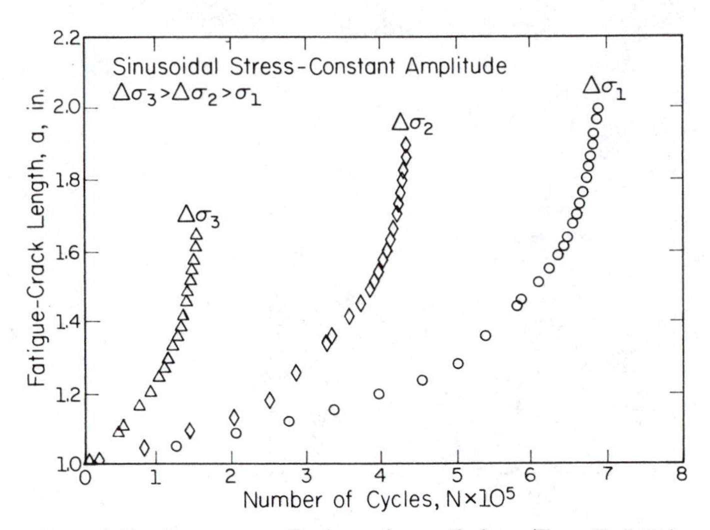

Figure 3.10 Constant amplitude crack growth data. (From Ref. 34.)

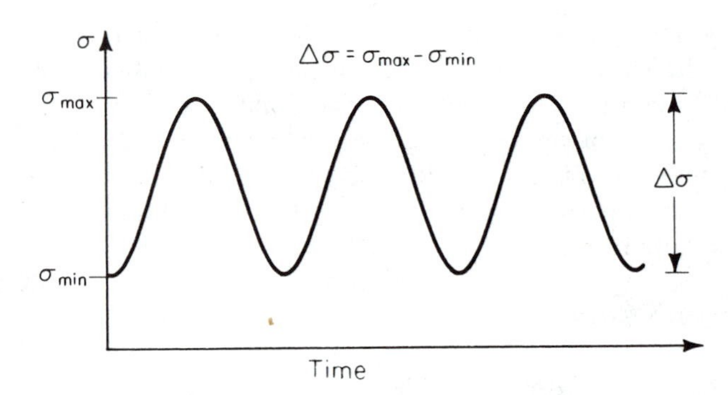

Figure 3.11 Remote stress range.

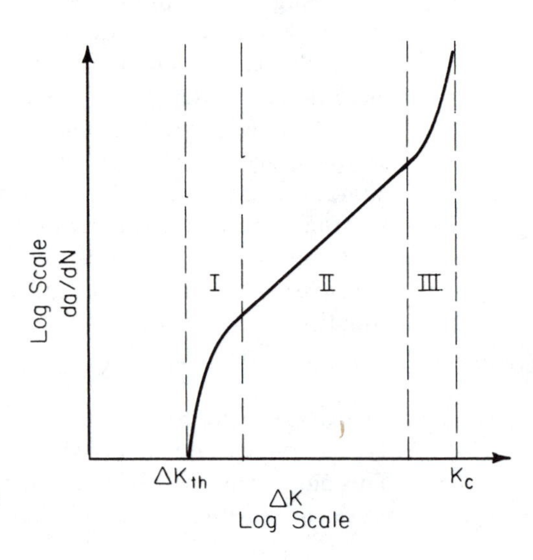

Figure 3.12 Three regions of crack growth rate curve.

Region II, the curve is essentially linear. Many structures operate in this region. Finally, in Region III, at high ΔK values, crack growth rates are extremely high and little fatigue life is involved. These three regions are discussed in detail in the following sections.

3.3.2 Region II

Most of the current applications of LEFM concepts to describe crack growth behavior are associated with Region II. In this region the slope of the log da/dN versus log ΔK curve is approximately linear and lies roughly between 10^{-6} and 10^{-3} in./cycle. Many curve fits to this region have been suggested. The Paris [36] equation, which was proposed in the early 1960s, is the most widely accepted. In this equation

$$\frac{da}{dN} = C(\Delta K)^m \tag{3.8}$$

where C and m are material constants and ΔK is the stress intensity range $K_{max} - K_{min}$.

The material constants, C and m, can be found in the literature and in data books such as Refs. 34 and 37. Values of the exponent, m, are usually between 3 and 4. Reference 38 tabulates values of m for a number of metals. These range from 2.3 to 6.7 with a sample average of $m = 3.5$. In addition, tests may be performed. ASTM E647 sets guidelines for these tests.

The crack growth life, in terms of cycles to failure, may be calculated using Eq. (3.8). The relation may be generally described by

$$\frac{da}{dN} = f(K)$$

Thus, cycles to failure, N_f, may be calculated as

$$N_f = \int_{a_i}^{a_f} \frac{da}{f(K)} \tag{3.9}$$

where a_i is the initial crack length and a_f is the final (critical) crack length. Using the Paris formulation,

$$\begin{gathered} \frac{da}{dN} = C(\Delta K)^m \\ N_f = \int_{a_i}^{a_f} \frac{da}{C(\Delta K)^m} \end{gathered} \tag{3.10}$$

Because ΔK is a function of the crack length and a correction factor that is dependent on crack length [see Eq. (3.7)], the integration above must often be solved numerically. As a first approximation, the correction factor, $f(g)$, can be calculated at the initial crack length and Eq. (3.10) can be evaluated in closed form.

As an example of a closed form integration, fatigue life calculations for a small edge-crack in a large plate are performed below. In this case the correction factor, $f(g)$, does not vary with crack length. The stress intensity factor range is

$$\Delta K = 1.12\Delta\sigma\sqrt{\pi a} \tag{3.11}$$

Substituting into the Paris equation yields

$$\frac{da}{dN} = C(1.12\Delta\sigma\sqrt{\pi a})^m \tag{3.12}$$

Separating variables and integrating (for $m \neq 2$) gives

$$\begin{aligned} N_f &= \int_{a_i}^{a_f} \frac{da}{C(1.12\Delta\sigma\sqrt{\pi a})^m} \\ &= \frac{2}{(m-2)C(1.12\Delta\sigma\sqrt{\pi})^m}\left(\frac{1}{a_i^{(m-2)/2}} - \frac{1}{a_f^{(m-2)/2}}\right) \end{aligned} \tag{3.13}$$

Before this equation may be solved, the final crack size, a_f, must be evaluated. This may be done using Eq. (3.2) as follows:

$$\begin{aligned} K &= f(g)\sigma\sqrt{\pi a} \\ a_f &= \frac{1}{\pi}\left[\frac{K_c}{\sigma f(g)}\right]^2 \end{aligned} \tag{3.14}$$

$$= \frac{1}{\pi}\left(\frac{K_c}{1.12\sigma_{\max}}\right)^2 \tag{3.15}$$

For more complicated formulations of ΔK, where the correction factor varies with the crack length, a, iterative procedures may be required to solve for a_f in Eq. (3.14).

It is important to note that the fatigue-life estimation is strongly dependent on a_i, and generally not sensitive to a_f (when $a_i \ll a_f$). Large changes in a_f result in small changes of N_f, as shown schematically in Fig. 3.13.

An alternative approximate method may be used to predict fatigue crack growth under constant-amplitude loading. This procedure is outlined in Ref. 39 and discussed in Refs. 6 and 16. Briefly, the procedure is as follows:

1. Divide the interval of crack growth from a_i to a_f into a desired number of increments, $n - 1$.
2. In Eq. (3.7), determine $f(g)$ for each of the intermediate crack lengths as well as the initial and final lengths, a_i and a_f, respectively.
3. Calculate a ΔK_n for each crack length, a_n.

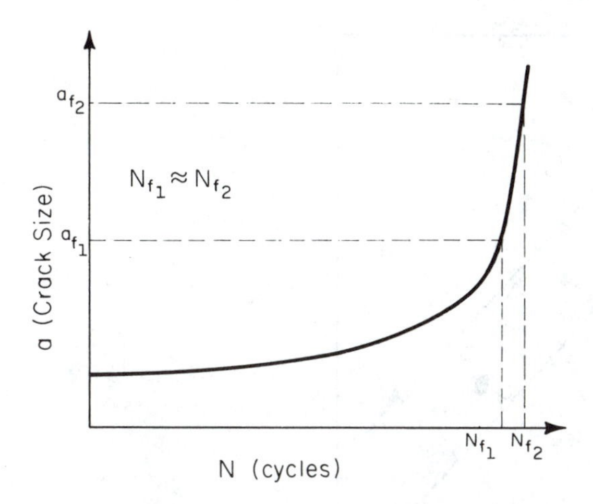

Figure 3.13 Effect of final crack size on life.

4. For each ΔK_n, determine the corresponding da/dN from crack growth rate plots or the Paris equation:

$$\left(\frac{da}{dN}\right)_n = C(\Delta K_n)^m \tag{3.16}$$

5. Average the growth rates for two consecutive crack lengths:

$$\frac{(da/dN)_n + (da/dN)_{n+1}}{2} = \left(\frac{da}{dN}\right)_{\text{average}} \tag{3.17}$$

6. Determine the number of cycles for the growth during the crack increment, a_n to a_{n+1}, by

$$\Delta N = \frac{\Delta a}{(da/dN)_{\text{average}}} = \frac{2(a_{n+1} - a_n)}{(da/dN)_n + (da/dN)_{n+1}} \tag{3.18}$$

Thus an approximate value is obtained for the number of cycles for an increment of crack growth. These values of ΔN for each increment may then be summed for an approximate solution for the number of loading cycles for the growth of the crack between the two lengths, a_i and a_f.

Usually the fatigue life is not sensitive to the fracture toughness of the material. This is a result of the lack of sensitivity of N_f to the final crack size, a_f, as shown in Fig. 3.13. An exception to this would be a case where a very hard material is subjected to large stresses. For instance, the fatigue life of gears is dependent on the fracture toughness of the material because the initial crack size varies little from the final crack length.

3.3.3 Region I

Region I of the sigmoidal crack growth rate curve is associated with threshold effects. Below the value of the threshold stress intensity factor, ΔK_{th}, fatigue

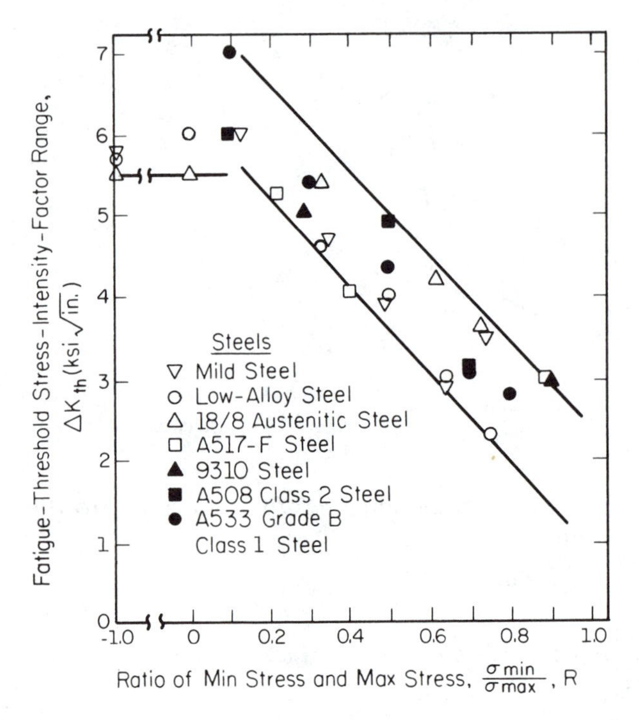

Figure 3.14 Dependence of fatigue-threshold stress-intensity-factor range on stress ratio. (From Ref. 34.)

crack growth does not occur or occurs at a rate too slow to measure. (The smallest measured rates are larger than approximately 10^{-8} in./cycle. This corresponds to the spacing between atoms in most metals.)

The fatigue threshold for steels is usually between 5 and 15 ksi $\sqrt{\text{in.}}$ and between 3 and 6 ksi $\sqrt{\text{in.}}$ for aluminum alloys. The fatigue threshold is dependent on the stress ratio, R ($R = \sigma_{min}/\sigma_{max}$). As seen in Fig. 3.14, the fatigue threshold decreases with increasing stress ratio.

The threshold also depends on frequency of loading and environment (see Section 3.3.5). In addition, many of the published threshold values, ΔK_{th}, were developed for long cracks. The validity of these values for short cracks has recently been questioned. In Ref. 40 an empirical relationship between the ΔK_{th} for short cracks and the ΔK_{th} for long cracks has been proposed. Several methods for measuring ΔK_{th} are reviewed in Ref. 28. In fact, due to the sensitivity of ΔK_{th} to environment and load history, it is felt by many that the best method for determining ΔK_{th} is through testing under conditions that simulate actual service conditions. References 37 and 41 to 43 give more detailed information on fatigue threshold concepts, testing, and results.

Designing a component such that ΔK for service conditions would be below ΔK_{th} would be highly desirable. Although this would ensure a low probability of fatigue failure, this is often impractical due to the low level of operating stress required. Alternatively, ensuring that defects were so small that the ΔK was below the threshold would be equally desirable. Unfortunately, the defect size required is not only impractical but unattainable.

For example, given below are typical values for the endurance limit and fatigue threshold for a common steel. Calculations to determine the maximum defect size in an infinite plate with a center crack are outlined.

$$S_e \approx 50 \text{ ksi}$$

$$\Delta K_{\text{th}} = 5 \text{ ksi} \sqrt{\text{in.}}$$

$$K = \sigma\sqrt{\pi a}$$

$$f(g) = 1 \quad \text{(for an infinite center-cracked plate)}$$

From Eq. (3.14), the critical crack size is calculated:

$$a_c = \frac{1}{\pi}\left[\frac{1}{f(g)}\left(\frac{K_c}{\sigma_{\max}}\right)\right]^2 \tag{3.19}$$

$$= \frac{1}{\pi}\left(\frac{5}{50}\right)^2$$

$$= 0.003 \text{ in.}$$

This defect size is on the order of that obtained due to normal fabrication or machining of a component. Thus, even at the endurance limit, which is a relatively low stress, the defect size is one that would be extremely difficult to detect using nondestructive inspection methods.

The threshold value may be of use when a part is subjected to low stress levels and a very large number of cycles. A good example of this would be power trains that operate at very high speeds.

3.3.4 Region III

In Region III, rapid, unstable crack growth occurs. In many practical engineering situations this region may be ignored because it does not significantly affect the total crack propagation life.

The point of transition from Region II to Region III behavior is dependent on the yield strength of the material, stress intensity factor, and stress ratio. Forman's equation [44] was developed to model Region III behavior, although it is more often used to model mean stress effects. This equation,

$$\frac{da}{dN} = \frac{C\,\Delta K^m}{(1 - R)K_c - \Delta K} \tag{3.20}$$

predicts the sharp upturn in the da/dN versus ΔK curve as fracture toughness is approached. (This equation is discussed further in Section 3.3.5).

Region III is of most interest when the crack propagation life is on the order of 10^3 cycles or less. At high stress intensities, though, the effects of plasticity start to influence the crack growth rate because the plastic zone size becomes large compared to the dimensions of the crack. In this case, the problem should be analyzed by some elastic–plastic fracture approach such as the J-integral or the crack-tip opening displacement (COD) methods.

3.3.5 Factors Influencing Fatigue Crack Growth

Stress ratio effects. The applied stress ratio, R, can have a significant effect on the crack growth rate. Recall, as defined in Section 3.3.3, $R = \sigma_{min}/\sigma_{max} = K_{min}/K_{max}$. In general, for a constant ΔK, the more positive the stress ratio, R, the higher the crack growth rates, as shown in Fig. 3.15. The stress ratio sensitivity, though, is strongly dependent on material as shown in Figs. 3.15 and 3.16. (Note the difference of the scales on the two figures.)

Forman's equation [Eq. (3.20)] is often used to predict stress ratio effects. As R increases, the crack growth rate, da/dN, increases. This is consistent with test observations. Forman's equation is valid only when $R > 0$. Generally, it is believed that when $R < 0$, no significant change in growth rate occurs compared to the $R = 0$ growth rate. Again this is material dependent, as some researchers have obtained data for certain materials which show higher growth rates for $R < 0$ loading [45, 46].

Another method used to compensate for stress ratio effects is *Walker's equation* [47]:

$$\frac{da}{dN} = C[(1 - R)^m K_{max}]^n \tag{3.21}$$

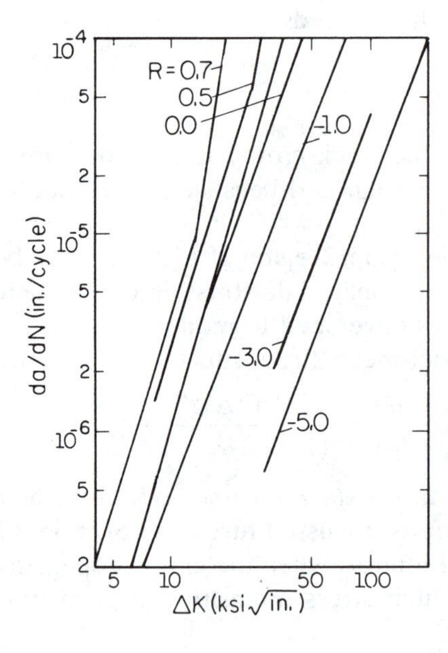

Figure 3.15 Influence of R on fatigue crack growth in Ti–6Al–4V. (From Ref. 48.)

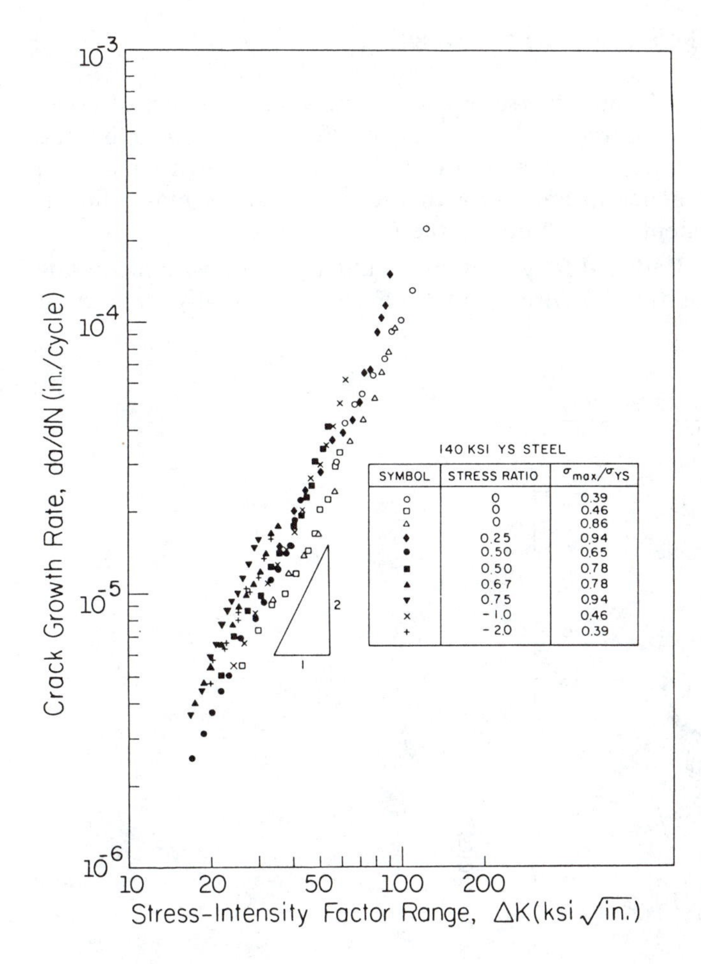

Figure 3.16 No major influence of R on fatigue crack growth in 140-ksi yield strength steel. (From Ref. 49.)

Use of this equation requires that stress ratio data be available to fit the exponents m and n for a particular material.

Crack closure arguments, as well as arguments based on environmental effects, have been used to explain the stress ratio effect on crack growth rates. Both of these topics are discussed in further detail in later sections.

Environmental effects. The fatigue crack growth rate can be greatly influenced by environmental effects. These effects are extremely complicated due to the large number of mechanical, metallurgical, and chemical variables and the interaction between them. Because of this complexity, only an overview is presented here. A more detailed discussion is found in Ref. 50 with a comprehensive literature review presented in Ref. 28.

The environmental effect on fatigue crack growth rate is strongly dependent on the material–environment combination. Several additional factors that influence the environmental effect are the following:

Frequency of loading. In an adverse environment, a strong effect of cyclic loading frequency is observed. No frequency effect is observed on the fatigue crack growth rate for a material tested in an inert environment. In general, at low frequencies, crack growth rates increase as more time is allowed for environmental attack during the fatigue process.

Temperature effects. Reduced fatigue life is usually observed with increasing temperature. In addition, environmental effects are usually greater at

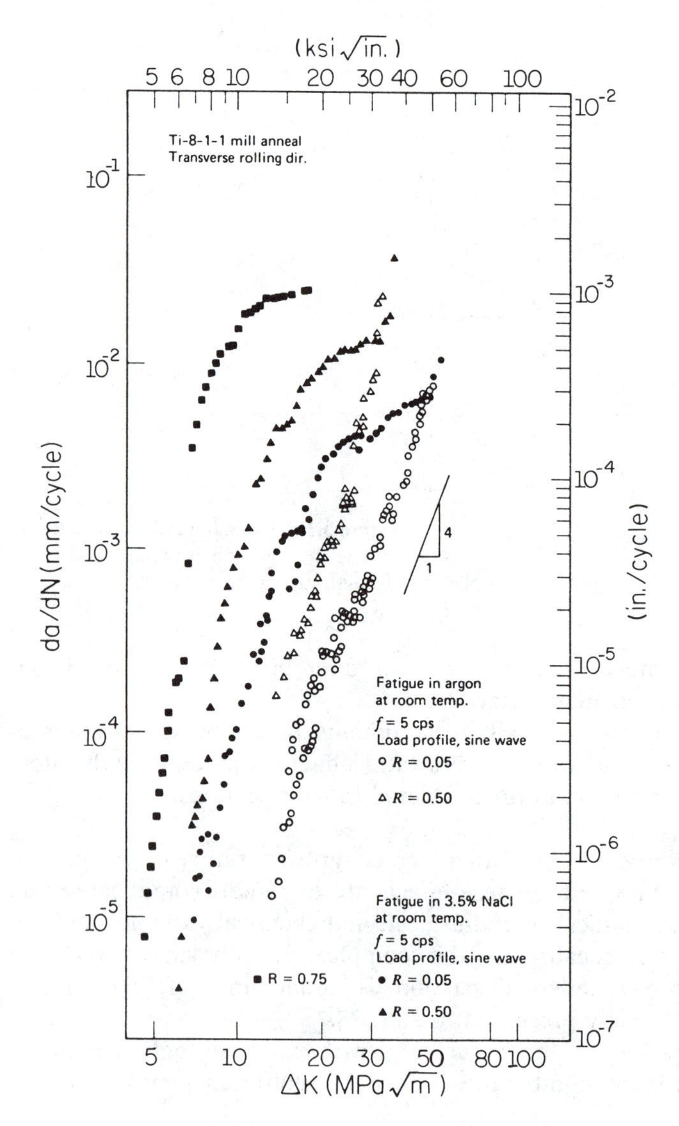

Figure 3.17 Effect of load ratio R on fatigue crack propagation in Ti–8Al–1Mo–1V alloy. Tests conducted in 3.5% NaCl solution and in argon. (From Ref. 51.)

elevated temperatures. This is due in part to oxide growth, which both promotes intergranular cracking and accelerates transgranular cracking.

Waveform of loading cycle. Higher fatigue crack growth rates generally occur if the increasing (tensile) portion of the loading cycle occurs more slowly. In other words, when the load rise time is small, the environmental influence is minimized. For example, a positive sawtooth waveform, /|/|/|, results in a higher environmental effect and consequently, increased crack growth rate than a negative sawtooth waveform, |\|\|\. No effect of waveform profile is usually observed in air.

Stress ratio effects. As discussed previously, some researchers feel that environmental effects may cause fatigue crack growth rate sensitivity to stress ratio, R, effects. At high R ratios, enhanced corrosion occurs, as demonstrated in Fig. 3.17.

Finally, environmental effects have been observed to cause either an increase or a decrease in ΔK_{th}, depending on material and environment. The increase in ΔK_{th} may be explained in some situations by local corrosion or oxides on the crack surfaces. These oxides increase the volume of material, contributing to the crack closure effect. The principles of crack closure are discussed below.

3.3.6 Crack Closure

Crack closure arguments are often used to explain the stress ratio effect of crack growth rates as well as environmental effects on ΔK_{th}. In addition, crack closure theories are very important in variable amplitude fatigue crack growth predictions, which are discussed in Chapter 5.

In the early 1970s, Elber [52] observed that the surfaces of fatigue cracks close (contact each other) when the remotely applied load is still tensile and do not open again until a sufficiently high tensile load is obtained on the next loading cycle. He developed the theory of crack closure to explain this phenomenon.

Elber proposed that crack closure occurs as a result of crack-tip plasticity. Recall from Section 3.2.5 that a plastic zone develops around the crack tip as the yield stress of the material is exceeded. As shown in Fig. 3.18, as the crack grows, a wake of plastically deformed material is developed while the surrounding body

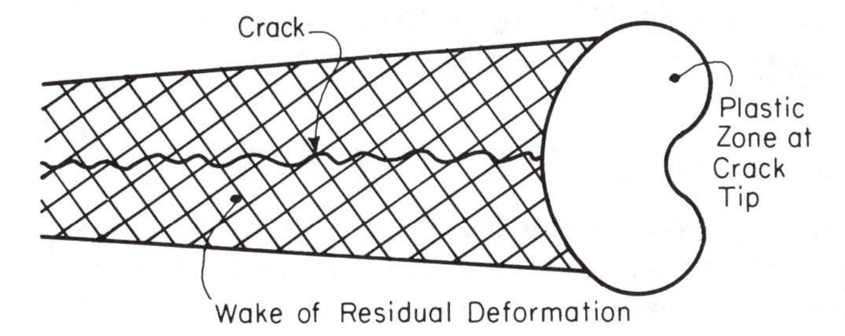

Figure 3.18 Wake of plastically deformed material.

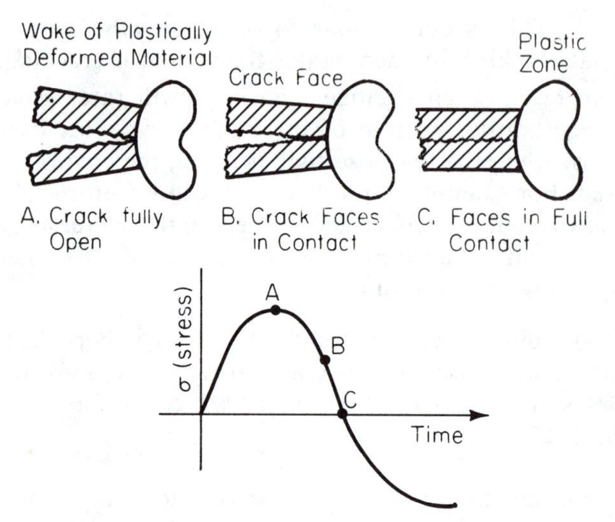

Figure 3.19 Crack closure phenomenon.

remains elastic. (Figure 3.18 shows the case of a gradually increasing ΔK and consequently, gradually increasing plastic zone size.) Elber proposed that as the component is unloaded, the plastically "stretched" material causes the crack surfaces to contact each other before zero load is reached (see Fig. 3.19).

Elber further introduced the idea of a crack-opening stress. This is the value of applied stress at which the crack is just fully open, σ_{op}. He suggested that for fatigue crack growth to occur, the crack must be fully open:

$$\begin{aligned} \Delta K_{eff} &= K_{max} - K_{open} \\ \Delta K &= K_{max} - K_{min} \end{aligned} \tag{3.22}$$

since

$$K_{open} > K_{min}$$

Consequently,

$$\Delta K > \Delta K_{eff}$$

Therefore, an effective stress intensity factor range, ΔK_{eff}, which is smaller than ΔK, should be used in fatigue crack growth predictions.

$$\frac{da}{dN} = f(\Delta K_{eff}) \tag{3.23}$$

Elber proposed that ΔK_{eff} accounts for the R effect on growth rates. At higher values of R, less crack closure results and ΔK_{eff} becomes closer to ΔK because K_{open} approaches K_{min}. This results in the crack being subjected to a greater range of loading. He obtained the empirical relationship

$$\Delta K_{\text{eff}} = U\,\Delta K$$
$$U = \frac{\Delta K_{\text{eff}}}{\Delta K} = 0.5 + 0.4R \quad (3.24)$$

Note that Eq. (3.24) is valid only when $R > 0$. Other researchers have subsequently developed expressions for U [53] and extended these for ratios of $R < 0$.

Crack closure arguments are further discussed in connection with variable amplitude loading and crack growth retardation in Chapter 5.

3.4 IMPORTANT CONCEPTS

- Fracture mechanics approaches provide an estimate of the crack propagation fatigue life.
- In the fracture mechanics approach, the local stresses and strains are related to the remote (applied) stresses and strains by the stress intensity factor, K.
- The linear elastic fracture mechanics approach is based on the assumption that the plastic zone at the crack tip is small compared to the crack length and the size of the cracked component.
- The fatigue crack growth rate can be related to the stress intensity factor range. From this, cycles to failure may be calculated.
- The fatigue life estimate is strongly dependent on the initial crack size, a_i. Large changes in the estimate of final crack size, a_f, result in only small changes in the life estimate.

3.5 IMPORTANT EQUATIONS

Stress Intensity Factor

$$K = f(g)\sigma\sqrt{\pi a} \quad (3.2)$$

Stress Intensity Factor Range

$$\Delta K = K_{\max} - K_{\min} = f(g)\,\Delta\sigma\sqrt{\pi a} \quad (3.7)$$

Crack Growth Rate (Paris Law)

$$\frac{da}{dN} = C(\Delta K)^m \quad (3.8)$$

Cycles to Failure

$$N_f = \int_{a_i}^{a_f} \frac{da}{C(\Delta K)^m} \tag{3.10}$$

Critical (Final) Crack Size

$$a_f = \frac{1}{\pi}\left[\frac{K}{\sigma f(g)}\right]^2 \tag{3.14}$$

Effective Stress Intensity Factor Range

$$\Delta K_{\text{eff}} = K_{\max} - K_{\text{open}} \tag{3.22}$$

REFERENCES

1. A. A. Griffith, *Philos. Trans. R. Soc. London,* Vol. A221, 1920, p. 163. (This article has been republished with additional commentary in *Trans. Am. Soc. Metals,* Vol. 61, 1968, p. 871.)
2. G. R. Irwin, *Fracturing of Metals,* American Society for Metals, Cleveland, Ohio, 1949, p. 147.
3. G. R. Irwin, "Analysis of Stresses and Strains near the End of a Crack Traversing a Plate," *Trans. ASME, J. Appl. Mech.,* Vol. E24, 1957, p. 361.
4. J. F. Knott, *Fundamentals of Fracture Mechanics,* Butterworth, London, 1973.
5. D. Broek, *Elementary Engineering Fracture Mechanics,* Martinus Nijhoff, The Hague, 1982.
6. H. L. Ewalds and R. J. H. Wanhill, *Fracture Mechanics,* Edward Arnold, London, 1985.
7. H. Tada, P. C. Paris and G. R. Irwin, *The Stress Analysis of Cracks Handbook,* Del Research Corporation, Hellertown, Pa., 1973.
8. D. P. Rooke, and D. J. Cartwright, *Compendium of Stress Intensity Factors,* H.M. Stationery Office, London, 1975.
9. G. C. M. Sih, *Handbook of Stress Intensity Factors,* Lehigh University, Bethlehem, Pa., 1973.
10. C. E. Fedderson, Discussion to: "Plane Strain Crack Toughness Testing," in ASTM STP 410, American Society for Testing and Materials, Philadelphia, 1966, p. 77.
11. B. Gross and J. E. Srawley, "Stress Intensity Factors for a Single Notch Tension Specimen by Boundary Collocation of a Stress Function," NASA TN D-2395, 1964.
12. W. F. Brown and J. E. Srawley, "Plane Strain Crack Toughness Testing of High Strength Metallic Materials," in ASTM STP 410, American Society for Testing and Materials, Philadelphia, 1966.
13. I. N. Sneddon, "The Distribution of Stress in the Neighborhood of a Crack in an Elastic Solid," in *Proceedings of the Royal Society of London,* Series A, Vol. A187, 1946, p. 229.

14. M. A. Sadowsky and E. G. Sternberg, "Stress Concentration around a Triaxial Ellipsoidal Cavity," *Trans. ASME, J. Appl. Mech.*, Vol. E16, 1949, pp. 149–157.
15. A. E. Green and I. N. Sneddon, "The Distribution of Stress in the Neighborhood of a Flat Elliptical Crack in an Elastic Solid," *Proc. Cambridge Philos. Soc.*, Vol. 46, 1950, p. 159.
16. A. P. Parker, *The Mechanics of Fracture and Fatigue: An Introduction*, E. and F. N. Spon, London, 1981.
17. D. J. Cartwright and D. P. Rooke, "Approximate Stress Intensity Factors Compounded from Known Solutions," *Eng. Fract. Mech.*, Vol. 6, 1974, pp. 563–571.
18. D. P. Rooke, *Stress Intensity Factors for Cracks at the Edges of Holes*, RAE TR 76087, Royal Aircraft Establishment, Farnborough, England, 1976.
19. D. P. Rooke, "Stress Intensity Factors for Cracked Holes in the Presence of Other Boundaries," in *Fracture Mechanics in Engineering Practice*, Applied Science Publishers, Barking, Essex, England, 1977, pp. 149–163.
20. D. P. Rooke and D. J. Cartwright, "The Compounding Method Applied to Cracks in Stiffened Sheets," *Eng. Fract. Mech.*, Vol. 8, 1976, pp. 567–573.
21. *The Compounding Method of Estimating Stress Intensity Factors for Cracks in Complex Configurations Using Solutions for Simple Configurations*, Engineering Sciences Data Unit, Item No. 78036, London Royal Aeronautical Society, Nov. 1978.
22. J. R. Rice, "Some Remarks on Elastic Crack-Tip Stress Fields," *Int. J. Solids Struct.*, Vol. 8, 1972, pp. 751–758.
23. H. F. Bueckner, "A Novel Principle for the Computation of Stress Intensity Factors," *Z. Angew. Math. Mech.*, Vol. 50, 1970, p. 529.
24. P. C. Paris, R. M. McMeeking, and H. Tada, "The Weight Function Method for Determining Stress Intensity Factors," in *Cracks and Fracture*, ASTM STP 601, American Society for Testing and Materials, Philadelphia, 1976, pp. 471–489.
25. T. A. Cruse, and P. M. Besuner, "Residual Life Prediction for Surface Cracks in Complex Structural Details," *J. Aircr.*, Vol. 12, No. 4, 1975, pp. 369–375.
26. R. J. Hartranft and G. C. Sih, "Alternating Method Applied to Edge and Surface Crack Problems," in *Mechanics of Fracture I*, G. C. Sih (ed.), Sÿthoffen Noordhoff, Alphen ann den Rijn, the Netherlands, 1973, Chap. 4, pp. 179–238.
27. T. K. Hellen, "Numerical Methods in Fracture Mechanics," in *Developments in Fracture Mechanics*, Vol. 1, G. G. Chell (ed.), Applied Science Publishers, Barking, Essex, England, 1979, pp. 145–181.
28. P. M. Besuner, (ed.), *A Review of Fracture Mechanics Life Technology*, Final Report Contract NAS8-34746, prepared by Failure Analysis Associates, Palo Alto, Ca, Sept. 1983.
29. V. E. Saouma and D. Schwemmer, "Numerical Evaluation of the Quarter-Point Crack-Tip Element," *Int. J. Numer. Methods Eng.*, Vol. 20, 1984, pp. 1629–1641.
30. R. D. Henshell and K. G. Shaw, "Crack Tip Elements Are Unnecessary," *Int. J. Numer. Methods Eng.*, Vol. 9, 1975, pp. 495–507.
31. R. S. Barsoum, "On the Use of Isoparametric Elements in Linear Elastic Fracture Mechanics," *Int. J. Numer. Methods Eng.*, Vol. 10, 1976, pp. 25–37.
32. J. J. Oglesby and O. Lomacky, "An Evaluation of Finite Element Methods for the

Computation of Elastic Stress Intensity Factors," *J. Eng. Ind.*, Feb. 1973, pp. 177–184.

33. D. M. Parks, "A Stiffness Derivative Finite Element Technique for Determination of Crack Tip Stress Intensity Factors," *Int. J. Fract.*, Vol. 10, 1974, pp. 487–502.
34. S. T. Rolfe and J. M. Barsom, *Fracture and Fatigue Control in Structures,* Prentice-Hall, Englewood Cliffs, N.J., 1977.
35. S. C. Chapra and R. P. Canale, *Numerical Methods for Engineers: With Personal Computer Applications,* McGraw-Hill, New York, 1985.
36. P. C. Paris and F. Erdogan, "A Critical Analysis of Crack Propagation Laws," *Trans. ASME, J. Basic Eng.*, Vol. D85, 1963, pp. 528–534.
37. J. M. Barsom, "Fatigue-Crack Propagation in Steels of Various Yield Strengths," *Trans. ASME, J. Eng. Ind.*, Vol. B73, No. 4, Nov. 1971, p. 1190.
38. J. F. Throop, and G. A. Miller, "Optimum Fatigue Crack Resistance," in *Achievement of High Fatigue Resistance in Metals and Alloys,* ASTM STP 467, American Society for Testing and Materials, Philadelphia, 1970, p. 154.
39. Engineering Sciences Data Unit, *Examples of the Use of Data Items on Fatigue Crack Propagation Rates,* Item No. 74017, Engineering Sciences Data Unit, London Royal Aeronautical Society, 1977.
40. O. N. Romaniv, V. N. Siminkovich and A. N. Tkach, "Near Threshold Short Fatigue Crack Growth," *in Fatigue Thresholds, Fundamentals and Engineering Applications,* Vol. 2, J. Backlund, A. F. Blom, and J. C. Beevers (eds.), Engineering Materials Advisory Services Ltd., West Midlands, England, 1982, pp. 799–807.
41. J. K. Musuva and J. C. Radon, "The Effect of Stress Ratio and Frequency on Fatigue Crack Growth," *Fatigue Eng. Mater. Struct.*, Vol. 1, 1979, pp. 457–470.
42. J. Backlund, A. F. Blom and J. C. Beevers (eds.), *Fatigue Thresholds, Fundamentals and Engineering Applications,* Vols. 1 and 2, Engineering Materials Advisory Services Ltd., West Midlands, England, 1982.
43. R. A. Smith, "Fatigue Thresholds—A Design Engineer's Guide through the Jungle," in *Fatigue Thresholds, Fundamentals and Engineering Applications,* Vol. 1, J. Backlund, A. F. Blom, and J. C. Beevers (eds.), Engineering Materials Advisory Services Ltd., West Midlands, England, 1982, pp. 33–44.
44. R. G. Forman, V. E. Kearney, and R. M. Engle, "Numerical Analysis of Crack Propagation in a Cyclic-Loaded Structure," *Trans. ASME, J. Basic Eng.*, Vol. D89, No. 3, 1967, pp. 459–464.
45. C. M. Hudson, "Effect of Stress Ratio on Fatigue Crack Growth in 7075–T6 and 2024–T3 Aluminum Alloy Specimens," NASA TN–D 5390, National Aeronautics and Space Administration, Aug. 1969.
46. W. Illg, and A. J. McEvily, "The Rate of Fatigue Crack Propagation for Two Aluminum Alloys under Completely Reversed Loading," NASA TN-D-52, National Aeronautics and Space Administration, Oct. 1959.
47. K. Walker, "The Effect of Stress Ratio during Crack Propagation and Fatigue for 2024–T3 and 7075–T6 Aluminum," in *Effects of Environment and Complex Load History on Fatigue Life,* ASTM STP 462, American Society for Testing and Materials, Philadelphia, 1970, p. 1.
48. A. Yuen, S. W. Hopkins, G. R. Leverant and C. A. Rau, "Correlations between

Fracture Surface Appearance and Fracture Mechanics Parameters for Stage II Fatigue Crack Propagation in Ti-6Al-4V," *Metall. Trans.*, Vol. 5, Aug. 1974, pp. 1833–1842.
49. T. W. Crooker and D. J. Krause, "The Influence of Stress Ratio and Stress Level on Fatigue Crack Growth Rates in 140 ksi YS Steel," Report of NRL Progress, Naval Research Laboratory, Washington D.C., Dec. 1972, pp. 33–35.
50. R. W. Hertzberg, *Deformation and Fracture Mechanics of Engineering Materials,* 2nd ed., Wiley, New York, 1983.
51. R. J. Bucci, PhD. Dissertation, Lehigh University, 1970.
52. W. Elber, "The Significance of Fatigue Crack Closure," in *Damage Tolerance in Aircraft Structures,* ASTM STP 486, American Society for Testing and Materials, Philadelphia, 1971, pp. 230–242.
53. J. Schijve, *The Stress Ratio Effect on Fatigue Crack Growth in* 2024-*T*3 *Alclad and the Relation to Crack Closure,* Memorandum M-336, Delft University of Technology Department of Aerospace Engineering, Delft, the Netherlands, Aug. 1979.
54. E. M. Caufield, "Evaluation of Fracture Mechanics Parameters for A27 Cast Steel," Fracture Control Program Report No. 28, University of Illinois at Urbana–Champaign, 1977.

PROBLEMS

SECTION 3.2

3.1. A large plate made of AISI 4340 steel contains an edge crack and is subjected to a tensile stress of 40 ksi. The material has an ultimate strength of 260 ksi and a K_c value of 45 ksi $\sqrt{\text{in}}$. Assume that the crack is much smaller than the width of the plate. Determine the critical crack size.

3.2. If the plate in Problem 3.1 is now of a finite width such that the ratio of the crack length to the plate width is 0.1, determine the critical crack size. Determine the critical crack size for a crack length to plate width ratio of 0.2.

3.3. Determine the stress intensity factor for the edge-cracked beam shown below when subjected to a moment of 400 ft-kips. If the beam was made from an extremely tough steel that has a yield strength of 195 ksi and a K_c of 160 ksi $\sqrt{\text{in.}}$ and the moment applied to the beam was increased to 1600 ft-kips, would this beam fail?

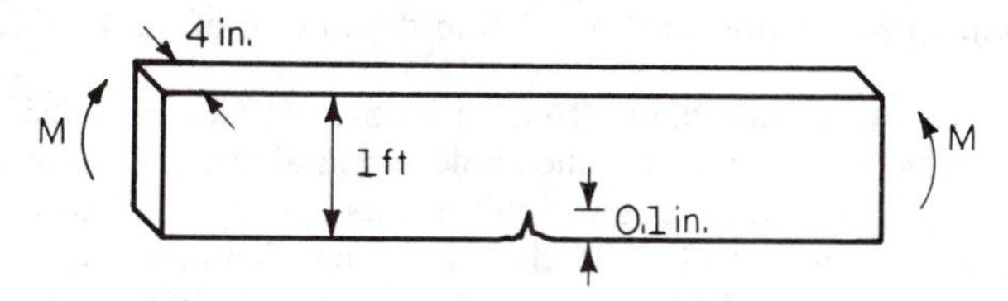

3.4. The fracture toughness of a material decreases, often dramatically, as the yield strength of the material increases. For example, for the titanium–aluminum alloy, Ti–6Al–4V, with a yield stress of 130 ksi, the fracture toughness is 105 ksi $\sqrt{\text{in.}}$ If the yield stress is increased to 150 ksi, the fracture toughness decreases to 50 ksi $\sqrt{\text{in.}}$

An engineer is faced with the following problem. His company has been manufacturing a component in the shape of a large sheet or plate using the alloy above in the 130-ksi yield strength condition. It has been suggested to him that a weight reduction could be obtained by using the alloy in the 150-ksi yield strength condition. Nondestructive testing of this component can reliably detect an edge crack of 0.2 in. Thus design requirements specify that the critical edge crack size be larger than this value (0.2 in). In addition, a factor of safety of 2 is specified for the design stress. (The design stress must be less than or equal to one-half the yield stress.) He has been asked to evaluate the proposed change in material. Should he approve the proposed change? Verify with calculations and comments. What is the maximum design stress that could be used with the higher strength material? Would use of the higher strength material result in a weight reduction?

3.5. A large cylindrical bar made of 4140 steel ($\sigma_y = 90$ ksi) contains an embedded circular (penny shaped) crack with a 0.1 in. diameter. Assume that the crack radius, a, is much smaller than the radius of the bar, R, so that the bar may be considered infinitely large compared to the crack. The bar is subjected to a tensile stress of 50 ksi. Determine the plastic zone size at the crack tip. Are the basic LEFM assumptions violated?

3.6. A large plate made of 4140 steel ($\sigma_y = 90$ ksi) containing a 0.2 in. center crack is subjected to a tensile stress of 30 ksi. Determine the plastic zone size. Are LEFM assumptions violated? If the yield strength of the material is reduced by a factor of 2, calculate the plastic zone size. Are LEFM assumptions violated? Discuss the relationship between yield strength and plastic zone size. What effect does the thickness of the plate have on the plastic zone size?

3.7. If the plate in Problem 3.6 was made from the material with the lower yield strength and subjected to a reversed stress of 30 ksi, calculated the reversed plastic zone size. Are LEFM assumptions violated? Discuss.

3.8. Fracture toughness often decreases significantly with decreasing temperature. As an example of this, the fracture toughness of A27 cast steel is plotted versus temperature (°F) below. (Data taken from Ref. 54.)

A component, which can be modeled as a beam subjected to a bending moment, is made from this material and experiences temperatures ranging from −150 to +150°F. Quality control procedures can only ensure that the component will have no cracks larger than 0.4 in. Assume that the crack length is much smaller than the beam depth. Determine the maximum bending moment that this component may withstand for a beam depth of 6 in. and a thickness of 3 in.

3.9. A very wide plate made from Al 7075-T651 ($K_{Ic} = 27 \text{ ksi}\sqrt{\text{in.}}$, $\sigma_y = 80$ ksi) contains an edge crack. Plot the allowable nominal stress (ksi) as a function of crack size, a (in inches), if the design requirements specify a factor of safety of 2 on the critical stress intensity factor. If the plate specifications were changed so that Al 7050-T73651 was used ($K_{Ic} = 35 \text{ ksi}\sqrt{\text{in.}}$, $\sigma_y = 70$ ksi), replot the curve. For a nominal stress of one-half the yield stress, determine the increase in allowable flaw size by changing from the Al 7075 alloy to the Al 7050 alloy.

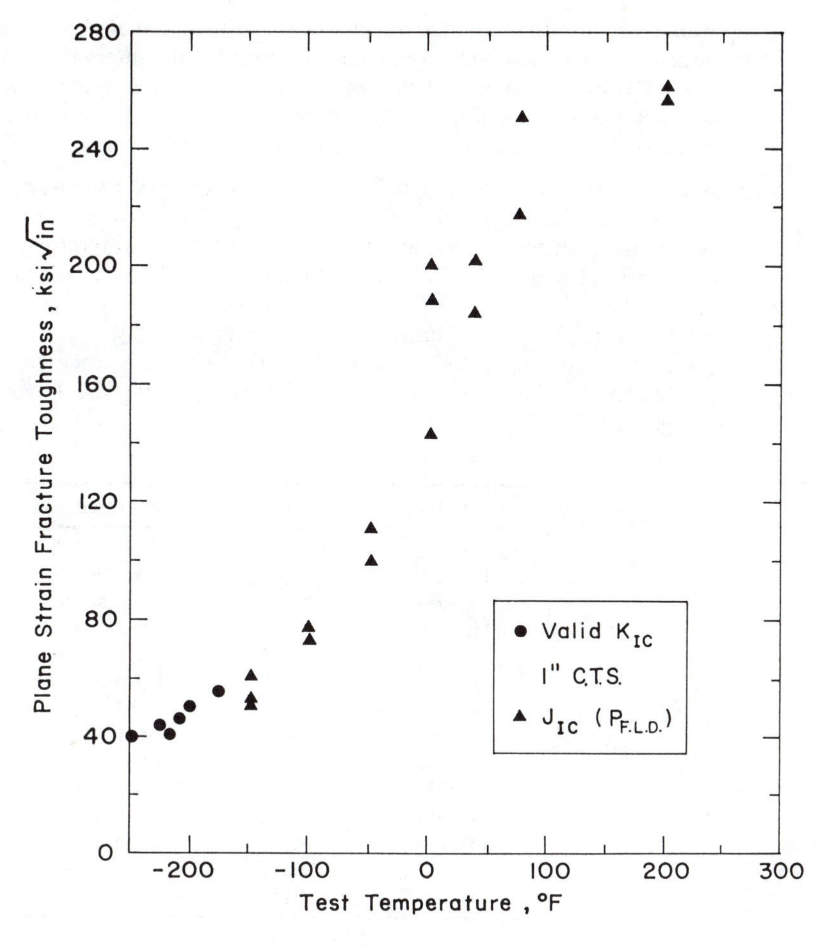

Figure P3.8

3.10. Design a pressure vessel that is capable of withstanding a static pressure of 1000 psi and that will "leak-before-burst." The required material has a fracture toughness of 60 ksi $\sqrt{\text{in.}}$ and a yield strength of 85 ksi. The diameter of the vessel is specified to be 4 ft. A crack with surface length of 1 in. can reliably be detected. Since the cost of the vessel is related directly to the amount of material used, optimize the design so that the cost is minimized.

SECTION 3.3

3.11. The crack in an edge-cracked plate extends due to repeated loading. The crack is initially 0.01 in. and the plate width is 5 in. Calculate the geometry correction factor, $f(g)$, for crack lengths of 0.01, 0.05, 0.1, and 0.2 in. Discuss the amount of error that would be introduced in the fatigue crack growth calculation in assuming that the geometry correction factor remained constant for the interval $a = 0.01$ to 0.2 in.

Discuss the impact of this on the method of crack growth predictions outlined in Section 3.3.2. Discuss the amount of error that would be introduced if the initial crack length were assumed to be 0.2 in. instead of 0.01 in.

3.12. A component made from 7005-T53 aluminum contains a semi-circular surface crack ($a/c = 1$) and is subjected to $R = 0.1$ loading with a stress range, $\Delta\sigma$, of 250 MPa. (Refer to Example 3.1 for an expression for the stress intensity range, ΔK.) The following crack growth data were obtained in laboratory air environment. Using these data:

(a) Plot crack length, a (mm), versus cycles, N.

(b) Plot da/dN versus ΔK. Identify the three regions of crack growth.

(c) Determine the Paris law constants, C and m, for the linear region of crack growth.

N (cycles)	a (mm)	da/dN (mm)
95,000	0.244	
100,000	0.246	7.00×10^{-7}
105,000	0.251	3.920×10^{-6}
110,000	0.285	9.665×10^{-6}
115,000	0.347	1.053×10^{-5}
125,000	0.414	1.230×10^{-5}
130,000	0.490	2.063×10^{-5}
135,000	0.621	4.661×10^{-5}
140,000	0.956	9.565×10^{-5}
145,000	1.577	3.964×10^{-4}
147,000	2.588	1.105×10^{-3}
147,400	3.078	1.554×10^{-3}
147,500	3.241	8.758×10^{-3}
147,500	3.445	

3.13. The following crack growth data were obtained from a center-cracked panel subjected to a stress range of 50 ksi. Plot the crack size, a, as a function of life, N. Determine the crack growth rate da/dN. Plot da/dN as a function of ΔK and determine the Paris constants. Compare the da/dN values calculated to the following values given. How sensitive are the Paris constants to the da/dN values?

3.14. An aluminum alloy has the following fatigue crack propagation relationship for $R = 0$ loading:

$$\frac{da}{dN} = 10^{-8}(\Delta K)^4 \quad \text{in./cycle}$$

A component that is made from this material is subjected to 0.1 Hz, constant amplitude, zero to maximum loading in service. The component is inspected every

Cycles	Crack Length (in.)	da/dN (in./cycle)
10,000	0.009	
20,000	0.010	1.56×10^{-7}
30,000	0.012	2.46×10^{-7}
40,000	0.015	5.10×10^{-7}
50,000	0.018	4.50×10^{-7}
60,000	0.025	1.05×10^{-6}
70,000	0.040	1.74×10^{-6}
80,000	0.060	2.94×10^{-6}
90,000	0.100	6.81×10^{-6}
100,000	0.200	

1000 hours using a facility capable of detecting a crack of 0.2 in. length on the surface. In rare cases where the failure occurred in service, the crack was found to be semi-circular with a depth of 1 in. ($a = 1$ in.). Assuming that the aspect ratio remains equal to 1 ($a/c = 1$) throughout the life, determine the limit of ΔS_{max} for the inspection program to be successful. Assume that the crack depth is much smaller than the thickness of the part so that the backwall effect may be ignored. (*Hint*: See Example 3.1 for the stress intensity factor.)

A new inspection technique promises to reduce the minimum detectable crack by a factor of 10. Estimate how much the inspection interval can be increased and still assure safe operation in service without changing the stress. How much could ΔS_{max} be increased if the new facility were used with the original inspection interval?

3.15. A component made from A27 cast steel was inspected and found to have a circular corner crack with a radius of 0.1 in. The fracture toughness for this material at the operating temperature (75°F) is about 220 ksi $\sqrt{\text{in}}$. Using the da/dN versus ΔK curve shown below [54], determine the number of constant amplitude cycles of 50 ksi (zero-to-maximum loading) that the component may experience before fast fracture. Assume that the crack size is negligible compared to the thickness throughout the life of the component. (*Hint*: See Example 3.1 for the stress intensity.)

If the operating temperature were decreased to 0°F, so that the fracture toughness was 120 ksi $\sqrt{\text{in}}$., determine the number of cycles before failure. (Assume that the crack growth rate remains constant with temperature.) How does the final crack size, a_f, affect the number of cycles that this component may experience?

3.16. A very wide plate containing a central crack of length $2a$ is made of a material with a yield strength of 70 ksi and a fracture toughness of 100 ksi $\sqrt{\text{in}}$. The plate is subjected to a zero-to-maximum constant nominal stress range. Assuming that the plate fails catastrophically when $K_{max} = K_{Ic}$, determine the number of cycles to failure for $S_{max} = 20$, 30, 40, 50, and 60 ksi for initial crack lengths of $a_0 = 0.005$, 0.01, 0.05, and 0.1 in. The crack growth constants for the Paris law equations are

$$C = 10^{-8} \qquad m = 3$$

Plot the results as a S–N curve with a_0 as the parameter.

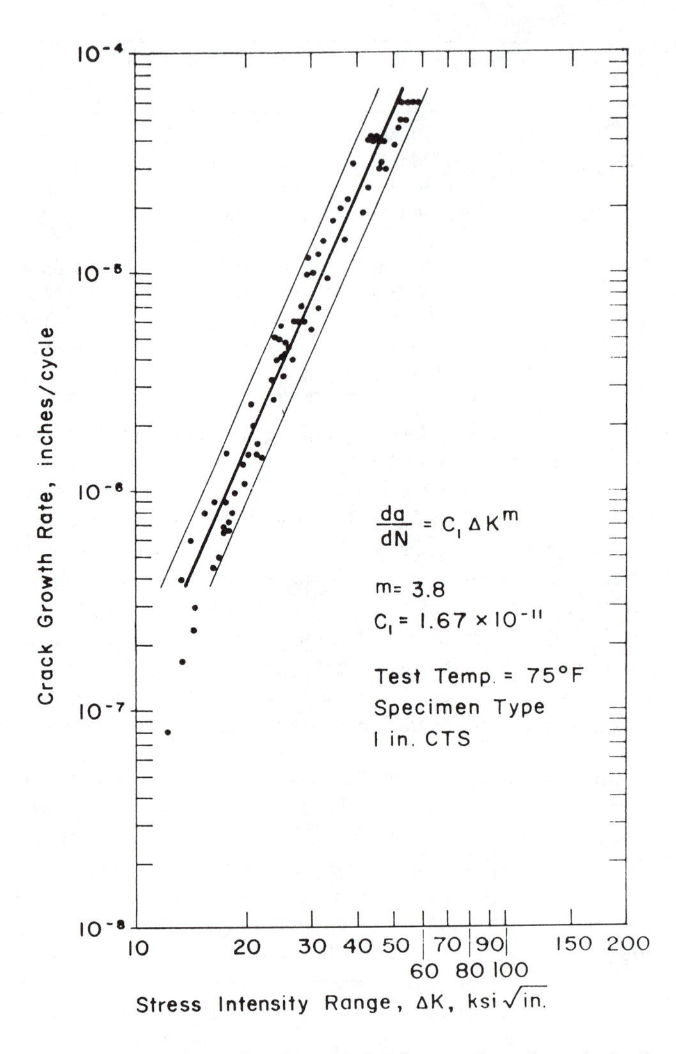

Figure P3.15

3.17. A surface crack of 0.1 in. depth and 0.2 in. surface length is found in a thick component. The component is scheduled to be repaired in 6 months. From loading and material information, it has been determined that catastrophic failure will occur when the crack size reaches 0.5 in. in depth. The component is subjected to zero-to-maximum loading 10 times per hour, with the maximum stress equal to 50 ksi. Assuming that the crack ratio remains constant, will the component fail before repair? (Assume that the crack growth rate, da/dN, calculated for $a = 0.1$ in. remains constant until the crack length reaches $a = 0.2$. Then assume that the crack growth remains constant until the crack increases by another 0.1 in. Continue to use this assumption to determine the number of cycles that the component may withstand before failure.) $C = 6 \times 10^{-6}$, $m = 3$.

3.18. If the beam in Problem 3.3 was subjected to the loading histories shown below, determine the stress intensity factor ranges, ΔK, for each history and discuss which history would be more damaging (neglect crack closure effects). Determine the stress ratio, R, for these two histories. Discuss the effect that crack closure would have on the damage produced by the two histories.

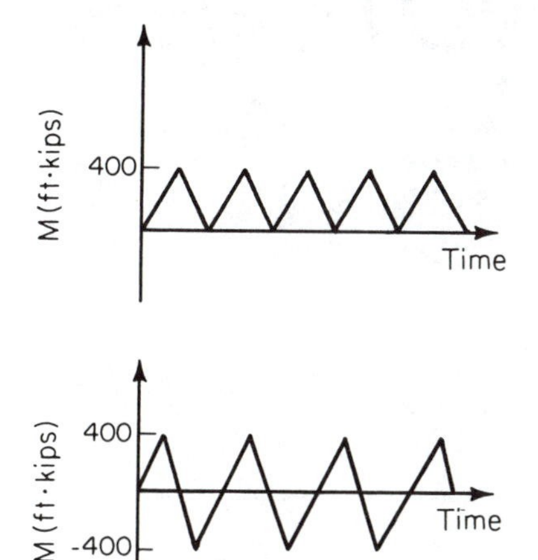

Moment vs Time Loading Histories

0.1 → 0.5 depth
0.2 → 1.0 depth
DK = 0-50 = 50 ksi

N= 6ro x

c = 6×10
n=3

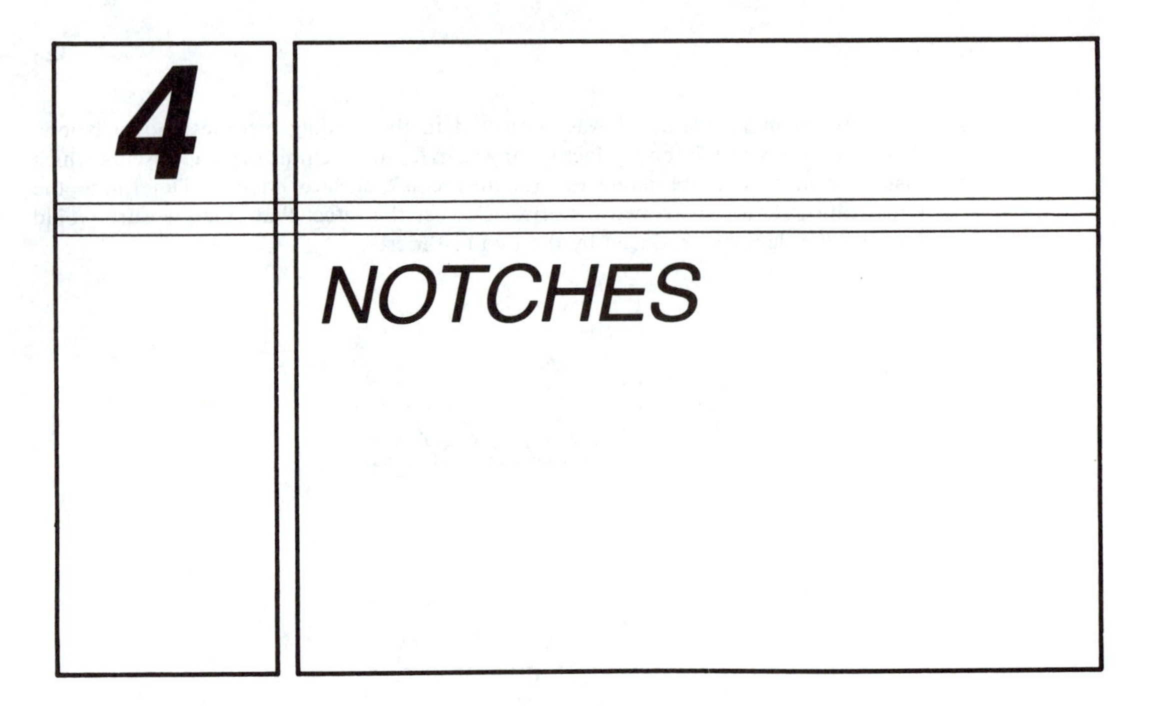

4.1 INTRODUCTION

In Chapters 1, 2, and 3, the three basic methods of fatigue analysis were presented. In practice, fatigue failures usually occur at notches or stress concentrations. In this chapter we review the application of the three approaches to analyses of notched members.

In Chapter 1 the stress–life approach was presented as a means of determining the fatigue life of a smooth specimen subjected to an applied alternating stress. This method is best suited for high cycle fatigue (HCF), where the notch strains are predominantly elastic. This approach does not account for inelastic behavior at the notch and cannot account for load sequence events.

The basic concepts and background of the strain–life method were presented in Chapter 2. This method was developed to account for notch root plasticity and the influence of load sequence effects on local mean and residual stresses.

Fracture mechanics principles were discussed in Chapter 3. The basic concepts can be extended to account for fatigue crack growth at a notch. Fracture mechanics approaches, used alone or in combination with the strain–life method, allow estimates to be made that include the propagation portion of fatigue life.

4.2 STRESS–LIFE APPROACH

Almost all machine components and structural members contain some form of geometrical or microstructural discontinuities. These discontinuities, or stress

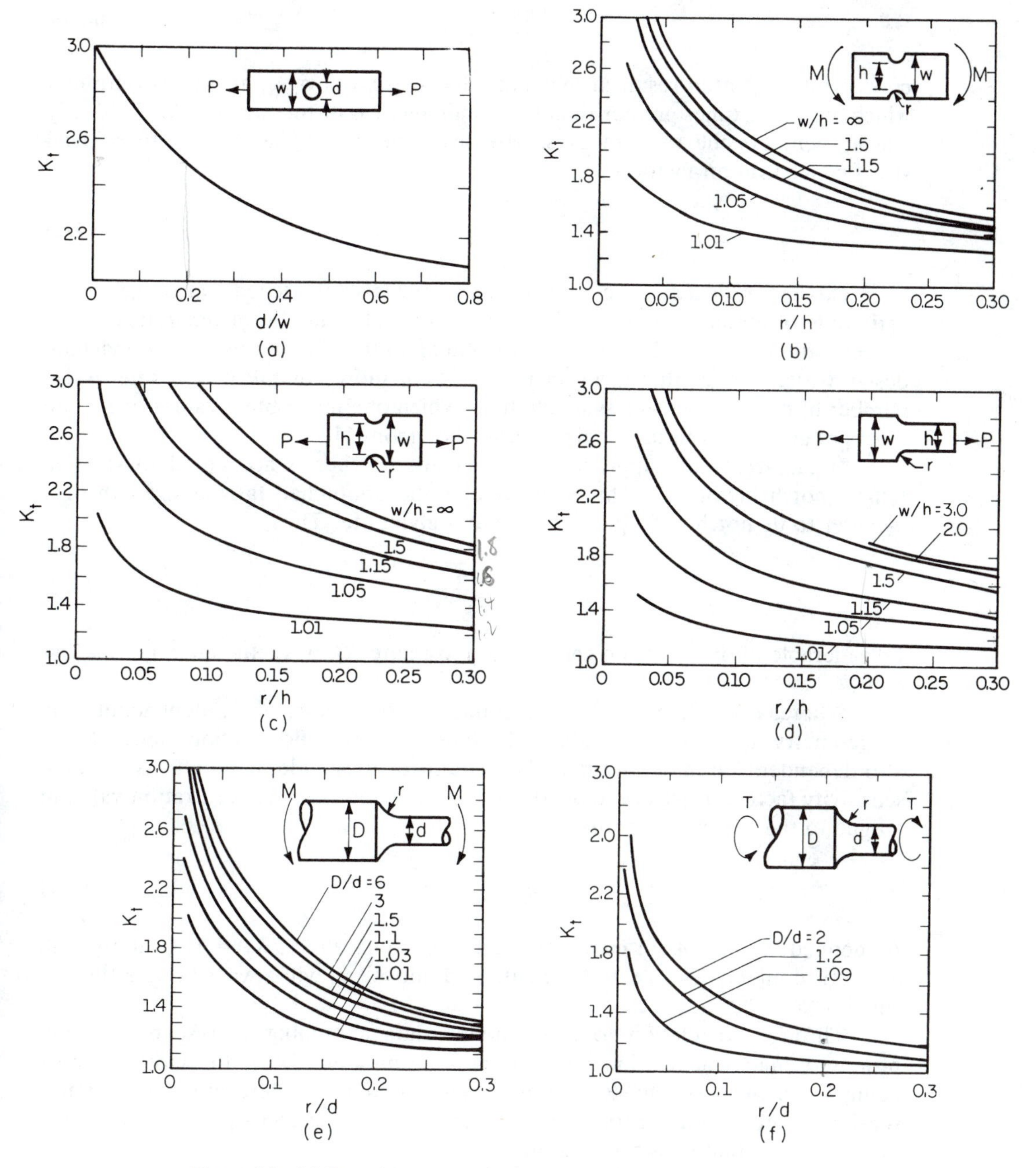

Figure 4.1 (a) Bar with a transverse hole in tension or simple compression: $S_{nom} = P/A$; $A = (w - d)t$; t = thickness. (b) Notched bar in bending: $S_{nom} = Mc/I = 6M/th^2$; t = thickness. (c) Notched bar in tension or simple compression: $S_{nom} = P/A = P/th$; t = thickness. (d) Bar with a shoulder fillet in tension or simple compression: $S_{nom} = P/A = P/th$; t = thickness. (e) Shaft with fillet in bending: $S_{nom} = Mc/I = 32M/(\pi d^3)$. (f) Shaft with fillet in torsion: $\tau_{nom} = Tc/J = 16T/(\pi d^3)$. (R. E. Peterson, *Stress Concentration Factors*, John Wiley and Sons, New York, 1974. Reprinted with permission.)

concentrations, often result in maximum local stresses, σ_{max}, at the discontinuity which are many times greater than the nominal stress of the member, S. In ideally elastic members the ratio of these stresses is designated as K_t, the theoretical stress concentration factor.

$$K_t = \frac{\sigma_{max}}{S} \tag{4.1}$$

This theoretical stress concentration factor is solely dependent on geometry and mode of loading. Shown in Fig. 4.1 are typical examples of theoretical stress concentration curves. These curves are similar to those found in general machine design texts, such as that given in Ref. [2]. For more complicated geometries, a number of publications are available from which one may obtain values of K_t, the most popular and well used being that of Peterson [1].

In the stress–life approach the effect of notches is accounted for by the fatigue notch factor, K_f. This value relates the unnotched fatigue strength of a member to its notched fatigue strength at a given life. Thus

$$K_f = \frac{S_e^{(\text{unnotched})}}{S_e^{(\text{notched})}} \tag{4.2}$$

Considerable effort has been spent trying to relate K_f to K_t. In general, K_f is less than K_t.

Whereas the theoretical stress concentration factor, K_t, is dependent only on geometry and mode of loading, the fatigue stress concentration factor, K_f, is also dependent on material type. To account for these additional effects, a notch sensitivity factor, q, was developed which relates the stress concentration value in fatigue to the theoretical value.

$$q = \frac{K_f - 1}{K_t - 1} \tag{4.3}$$

As defined, values of q range from zero (no notch effect, $K_f = 1$) to unity (full theoretical effect, $K_f = K_t$). As mentioned above, K_f will always be less than or equal to K_t.

Through correlation to experimental data a number of researchers have proposed analytical relationships for the determination of q, the most common being those proposed by Peterson [3] and Neuber [4]. Peterson plotted notch sensitivity versus notch root radius for a variety of materials [5] and obtained a plot similar to that shown in Fig. 4.2.

From these tests Peterson proposed the following empirical relationship:

$$q = \frac{1}{\left(1 + \frac{a}{r}\right)} \tag{4.4}$$

where r is the notch root radius and a is a material constant. The constant a

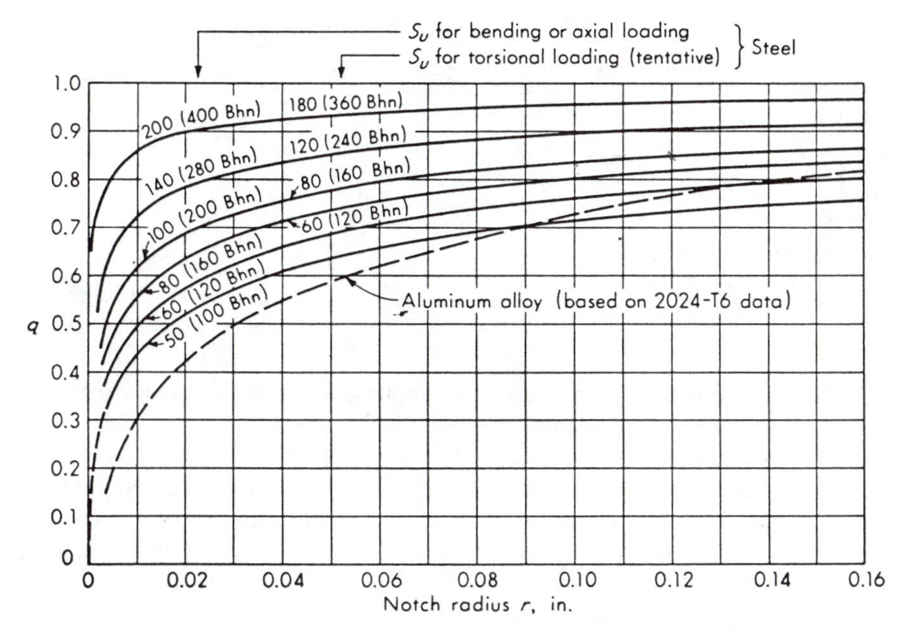

Figure 4.2 Notch sensitivity curves. (From Ref. 6.)

depends on material strength and ductility and is obtained from long life fatigue tests conducted on notched and unnotched specimens. Combining Eqs. (4.3) and (4.4) yields K_f as a function of K_t, a, and r.

$$K_f = 1 + \frac{K_t - 1}{\left(1 + \dfrac{a}{r}\right)} \tag{4.5}$$

For ferrous-based wrought metals, a is approximately given by [7]

$$a = \left[\frac{300}{S_u(\text{ksi})}\right]^{1.8} \times 10^{-3}\text{ in.} \tag{4.6}$$

or, using the approximation

$$S_u \approx 0.5\text{BHN} \tag{4.7}$$

$$a = \left(\frac{300}{0.5\text{BHN}}\right)^{1.8} \times 10^{-3}\text{ in.} \tag{4.8}$$

Then a can easily be estimated for steel in a variety of conditions, as given below.

a for normalized or annealed steels $\approx$ 0.010 in. (BHN $\approx$ 170)
a for quench and tempered steels $\approx$ 0.0025 in. (BHN $\approx$ 360)
a for highly hardened steels $\approx$ 0.001 in. (BHN $\approx$ 600)

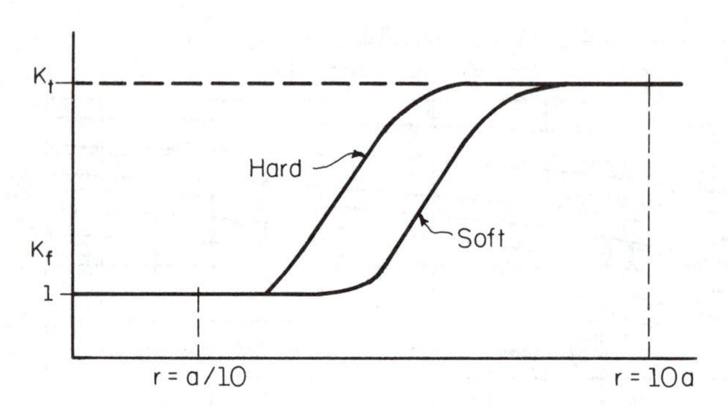

Figure 4.3 Effect of notch root radius on notch sensitivity factor.

From the values above, the trend is apparent that for a sharp notch, hard steels tend to be notch sensitive.

By combining Peterson's relation with Eq. (4.3), the effect of notch root radius, r, on K_f for hard and soft materials may be illustrated. Figure 4.3 contains a qualitative representation of this effect. From this figure it is recognized that the effect of changing r on K_f is greatest when $r = a$. On the other hand, K_f is affected very little when r is less than $a/10$ or greater than $10a$.

Alternative testing conducted by Neuber [4] led him to propose that

$$K_f = 1 + \frac{K_t - 1}{1 + \sqrt{\rho/r}} \tag{4.9}$$

where ρ is a material constant. When related to the definition of notch sensitivity presented earlier, this gives

$$q = \frac{1}{1 + \sqrt{\rho/r}} \tag{4.10}$$

In this relation r once again represents the radius at the root of the notch, but ρ, a material constant, is related to the grain size of the material. Typical values of ρ are given in Fig. 4.4.

Both the Peterson and Neuber relations are empirical fits to data. When used for analysis there is little difference in the approaches. The important point is that both methods show that q is related to material, notch geometry, and notch size.

The notch sensitivity relations presented above are not only a function of material properties (strength and ductility for a, and grain size for ρ) but also a function of the notch root radius, r. In this manner the sensitivity factor, q, incorporates notch size effects into the analysis of a notch subjected to fatigue loading. Two geometrically similar notches, such as those shown in Fig. 4.5, have identical theoretical stress concentration factor values. Inspection of either Peterson's or Neuber's relations [Eqs. (4.4) and (4.9), respectively] reveals that these two notches will have different fatigue stress concentration values ($K_{f_1} >$

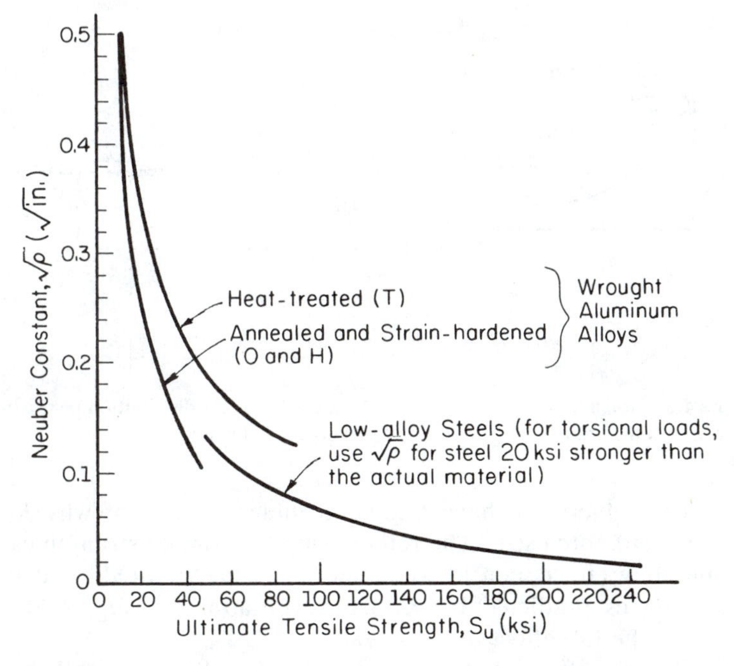

Figure 4.4 Neuber constants for steel and aluminum. (From Ref. 6.)

K_{f_2}). This result shows that for geometrically similar notches, notch sensitivity increases with increasing notch root radius.

Plate 1:	Plate 2:
$W_1 = 5.0$ in.	$W_2 = 0.5$ in.
$d_1 = 0.5$ in.	$d_2 = 0.05$ in.
$S_u = 100$ ksi	$S_u = 100$ ksi
$K_t = 2.7$ (from Fig. 4.1a)	$K_t = 2.7$ (from Fig. 4.1a)
$q = 0.97$ [from Eqs. (4.4) and (4.6)]	$q = 0.78$ [from Eqs. (4.4) and (4.6)]
$K_{f1} = 2.65$ [from Eq. (4.5)]	$K_{f2} = 2.32$ [from Eq. (4.5)]

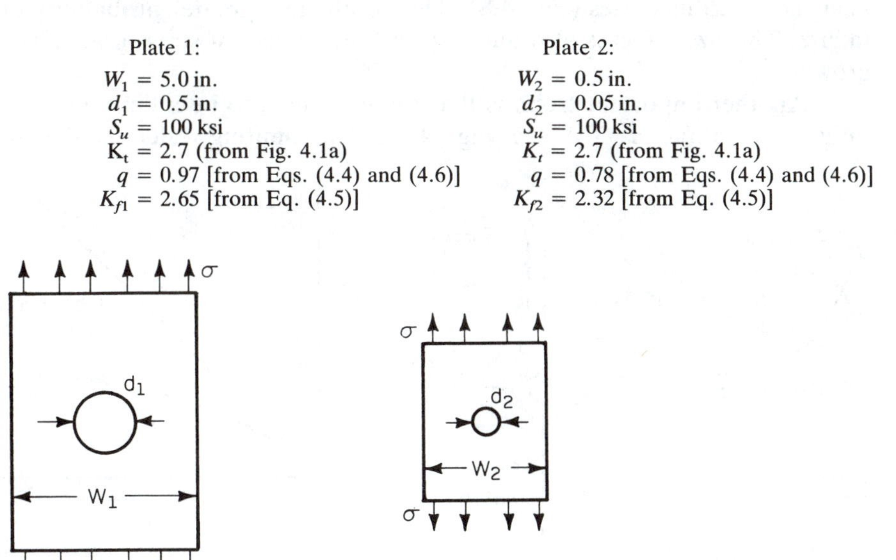

Figure 4.5 Notch sensitivity factor for two geometrically similar plates.

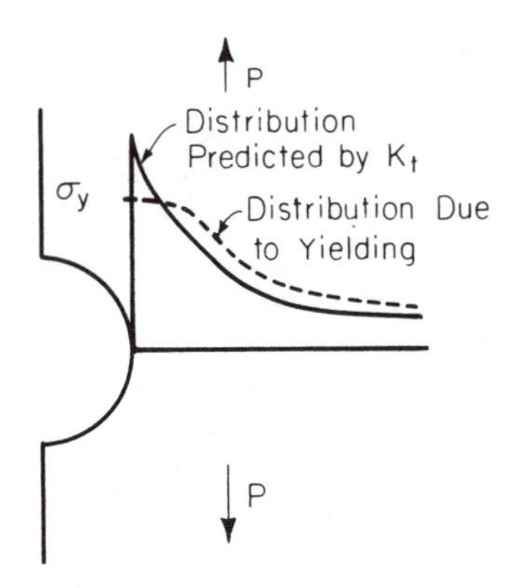

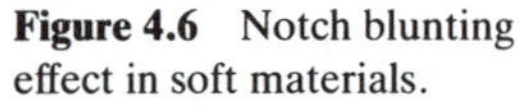
Figure 4.6 Notch blunting effect in soft materials.

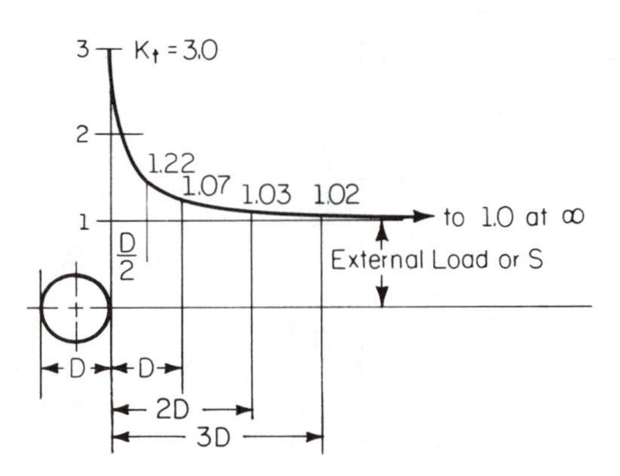

Figure 4.7 Stress distribution near a hole in an axially loaded plate.

It is valuable to have a general understanding of why K_f is dependent on material and notch size. The relationship to ultimate strength can be explained by the blunting effect seen in soft materials. Due to yielding at the notch root the peak stresses predicted by K_t are never attained (Fig. 4.6). In high strength materials the full effect of K_t is realized.

The size effect of notch root radius can be attributed to the volume effect discussed in Chapter 1. For a given notch geometry the shape of the stress gradient is constant when normalized by the dimension of the notch radius (Fig. 4.7). As the size of the notch increases, the volume of highly stressed material near the notch increases (Fig. 4.8). This results in a greater probability of fatigue failure. The size effect is also due in part to the effect of stress gradients on crack growth.

Another important trend is that there appears to be a limiting value on the fatigue notch factor, K_f (see Fig. 4.9). The limiting value is dependent on

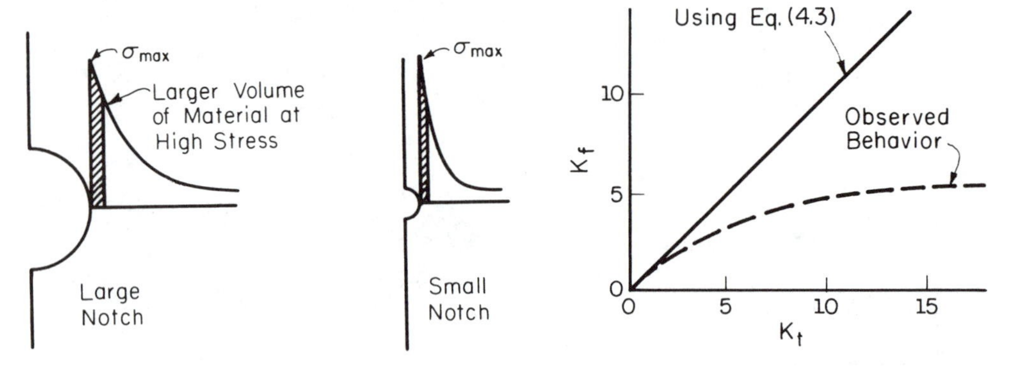

Figure 4.8 Volume of critically stressed material at large and small notches.

Figure 4.9 General relationship between K_t and K_f.

material, but in most cases it is around 5 or 6. There are two possible explanations for this. First, it may be due to the blunting effect caused by yielding at the notch root. This blunting does not allow the maximum stress predicted by K_t to occur. Second, the phenomenon may be due to initiation–propagation effects. In very sharp notches the initiation life is small and the total life is dependent on propagation. The limit on K_f may be an empirical method to account for this loss of initiation life in sharp notches.

The fatigue notch factor, K_f, and the notch sensitivity factor, q, are usually developed to correct the endurance limit, S_e, for notched members. Although this value of K_f can be used to correct the entire S–N curve for notched members, results tend to be conservative. A general trend is that the value of fatigue notch factor decreases with increasing stress level. This trend is material dependent and can probably be explained by the blunting effect near the notch. In most design cases the fatigue notch factor needs to be corrected for shorter lives.

The fatigue notch factor for stresses corresponding to lives of 1000 cycles has been defined as K_f'. Figure 4.10 shows an empirical relationship between the correction for K_f' and ultimate strength of different materials. As can be seen, the notch effect at short lives is greatly reduced for low strength or soft materials, whereas the notch effect remains almost constant with life for high strength or hard materials. The resulting corrected S–N curve for a notched member is shown in Fig. 4.11. This method is referred to as the modified Juvinall approach.

An alternative method for predicting the S–N curve for notched components is shown in Fig. 4.12. This method uses a straight line to connect the corrected endurance limit (S_e/K_f) to the true fracture stress, σ_f, at one cycle. Although this method incorporates the trend that the notch effect decreases with decreasing fatigue life, the method tends to be conservative. Another point is that

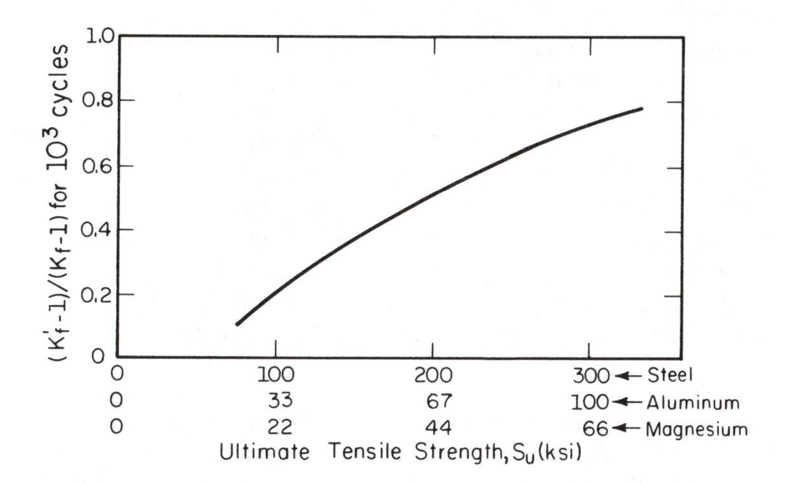

Figure 4.10 Relationship between K_f' and K_f as a function of ultimate strength. (From Ref. 6.)

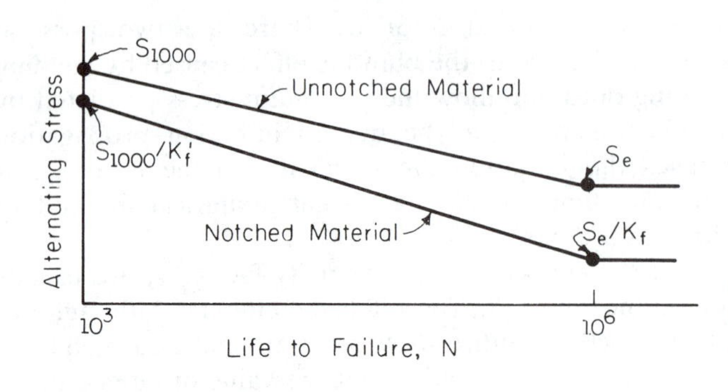

Figure 4.11 Modification of *S–N* curve for notched components: Juvinall approach.

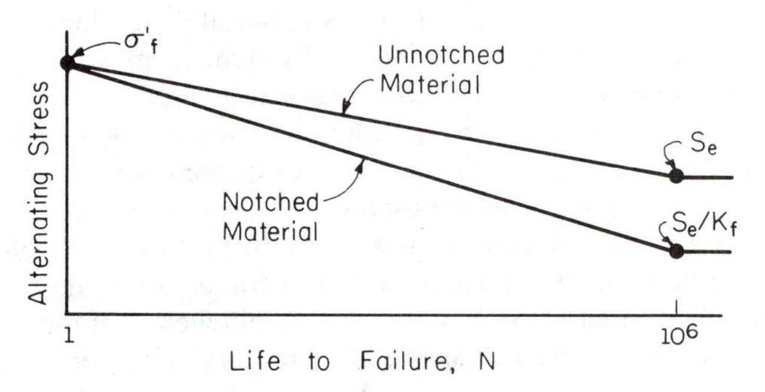

Figure 4.12 Alternative modification of *S–N* curve for notched components.

it defines the *S–N* curve in the low cycle region (<1000 cycles), where the approach is inappropriate.

The assumptions and modification factors discussed in Chapter 1 (size, surface finish, etc.) should be incorporated when using the stress–life method to analyze notched members. The stress–life method is best suited for high cycle fatigue (HCF), where the notch strains are predominantly elastic and loading is essentially constant amplitude. There are limitations to the stress–life method that make this method inappropriate in certain situations. This approach does not account for inelastic behavior at the notch and cannot properly account for changes in notch mean or residual stresses. It cannot account for load sequence events in anything other than an empirical manner. One example of this is the increase in crack initiation life due to a compressive residual stress developed at the notch following a high tensile load. Nevertheless, this method is still widely used in fatigue analyses. It is especially useful for long lives, where surface finish and other processing variables have a large effect.

Example 4.1

A notched steel component consists of a bar 1.0 in. wide and 0.25 in. thick with two semi-circular edge notches with radii of 0.1 in. This gives the plate a width at the reduced section of 0.8 in. Determine the life of the component subjected to a fully reversed ($R = -1$) load with an amplitude of 10 kips. Use the Juvinall approach shown in Fig. 4.11. The steel has an ultimate strength, S_u, of 114 ksi.

Solution Determine the theoretical stress concentration factor, K_t, using component geometry and K_t charts (Fig. 4.1). Using the geometry of the specimen, K_t (for net section) can be found using Fig. 4.1c.

$$K_t = 2.42$$

Determine the notch sensitivity factor, q, using Eq. (4.4) or Fig. 4.2.

$$q = \frac{1}{\left(1 + \frac{a}{r}\right)}$$

where r is the notch root radius and a is a material constant. For ferrous-based wrought steels [Eq. (4.6)], this is

$$a = \left[\frac{300}{S_u(\text{ksi})}\right]^{1.8} \times 10^{-3}\text{ in.}$$

For this case,

$$a = 10^{-3}\left(\frac{300\text{ ksi}}{114\text{ ksi}}\right)^{1.8} = 0.00571\text{ in.}$$

$$q = \frac{1}{1 + \frac{a}{r}} = \frac{1}{1 + \frac{0.00571}{0.10}} = 0.946$$

Determine the fatigue notch factor, K_f, using Eq. (4.3).

$$q = \frac{K_f - 1}{K_t - 1}$$

$$\begin{aligned} K_f &= (K_t - 1)q + 1 \\ &= (2.42 - 1)(0.946) + 1 \\ &= 2.34 \end{aligned}$$

Determine the fatigue stress concentration factor at 1000 cycles, K_f', using Fig. 4.10.

From Fig. 4.10 at $S_u = 114$ ksi,

$$\frac{K_f' - 1}{K_f - 1} = 0.28$$

$$\begin{aligned} K_f' &= (2.34 - 1)0.28 + 1 \\ &= 1.37 \end{aligned}$$

Determine the endurance limit, S_e, and the alternating stress level corresponding to 1000 cycles, S_{1000}. Since these values are not available, they may be estimated using Eqs. (1.2) and (1.6).

$$\begin{aligned} S_e &\approx 0.5S_u \\ &\approx 0.5(114\text{ ksi}) \\ &\approx 57\text{ ksi} \\ S_{1000} &\approx 0.9S_u \\ &\approx 0.9(114\text{ ksi}) \\ &\approx 102.6\text{ ksi} \end{aligned}$$

Next, the design S–N curve is constructed by locating the following points on log–log coordinates. (*Note*: In these calculations, K_f and K'_f are used as strength reduction factors.)

Life (cycles)	Stress (ksi)
10^6	S_e/K_f
10^3	S_{1000}/K'_f
1	S_u + 50 ksi

(*Note*: In the solution of this problem the S–N curve is extended into the range 1 to 1000 cycles. This is done for comparison purposes only and is not recommended for general design purposes. The estimate at one cycle is the true fracture strength, σ_f. For steels this is approximately the ultimate strength plus 50 ksi.)

These points are plotted on log–log coordinates as shown in Fig. E4.1.

$$\text{At } 10^6 \text{ cycles:} \quad \frac{S_e}{K_f} = \frac{57\text{ ksi}}{2.34} = 24.4\text{ ksi}$$

$$\text{At } 10^3 \text{ cycles:} \quad \frac{S_{1000}}{K'_f} = \frac{102.6\text{ ksi}}{1.37} = 75\text{ ksi}$$

$$\text{At 1 cycle:} \quad S_u + 50 = 164\text{ ksi}$$

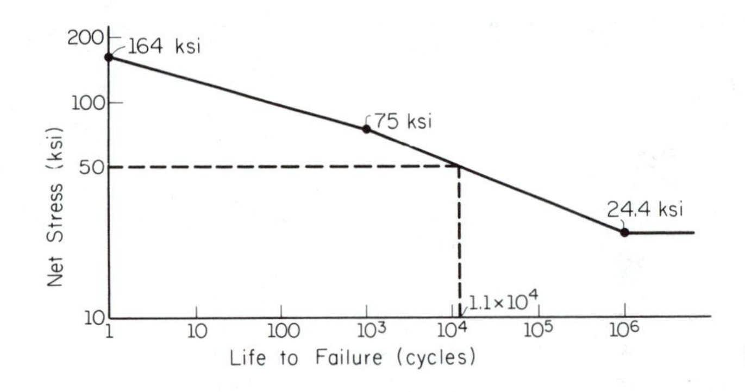

Figure E4.1 S–N design curve.

Determine the net or gross section stress in the component. (The use of net or gross section stress must correspond to the K_t value used in the solution.)

$$S_{\text{net}} = \frac{P}{A_{\text{net}}} = \frac{10 \text{ kips}}{(0.25 \text{ in.})(0.8 \text{ in.})} = 50 \text{ ksi}$$

Determine the life to failure, N_f, by locating the life corresponding to the net or gross section stress on the design S–N curve. The life corresponding to 50 ksi on the S–N curve is

$$N_f = \underline{\underline{1.1 \times 10^4 \text{ cycles}}}$$

4.3 STRAIN–LIFE APPROACH

The strain–life method accounts for notch-root plasticity. By knowing the notch root strain history and smooth specimen strain–life data or fatigue properties, fatigue-life evaluations may be performed for notched members. One of the advantages of this method is that it accounts for changes in local mean and residual stresses.

The basic concepts and background of the strain–life method were presented in Chapter 2. In the following section we discuss how those concepts may be applied in a fatigue analysis of a notched component for constant amplitude loading. Variable amplitude loading, sequence, and mean stress effects are discussed in Chapter 5.

4.3.1 Notch Root Stresses and Strains

The strain–life method requires that notch root stresses and strains be known. These may be determined by the following methods:

1. Strain gage measurements
2. Finite element analyses
3. Methods that relate local stresses and strains to nominal values

Often the least time-consuming and least expensive method of determining these values is by relating the local stresses and strains to nominal values (item 3 above). The following is a brief description of this procedure.

As discussed in Section 4.2, the theoretical stress concentration factor, K_t, is often used to relate the nominal stresses, S, or strains, e, to the local values, σ and ϵ. For increasing nominal stress, S, K_t remains constant until yielding begins. Upon yielding, the local stress, σ, and local strain, ϵ, are no longer linearly related and the local values are no longer related to the nominal values by K_t. Instead, the nominal and local values are related in terms of stress and strain

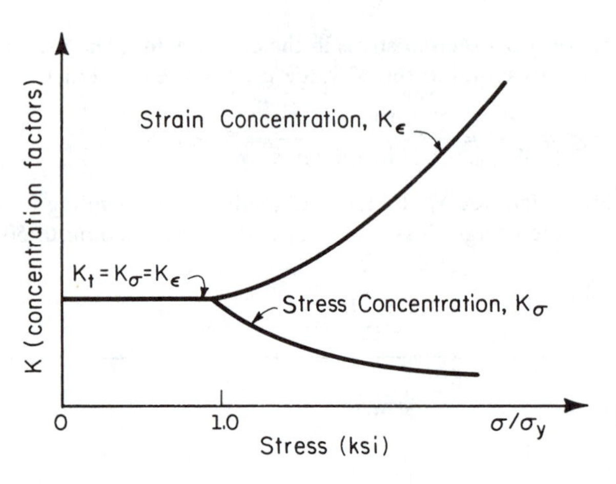

Figure 4.13 Effect of yielding on K_σ and K_ε.

concentration factors:

$$K_\sigma = \frac{\sigma}{S} \qquad \text{for } \sigma > \sigma_y \qquad (4.11)$$
$$K_\epsilon = \frac{\epsilon}{e}$$

After yielding occurs, the local stress concentration, K_σ, decreases with respect to K_t, and K_ϵ increases with respect to K_t (see Fig. 4.13). In other words, after yielding, the actual local stress is less than that predicted using K_t, while the actual local strain is greater than that predicted using K_t, as shown in Fig. 4.14.

The local stress–strain response at the notch root may vary from the nominally applied loading. An extreme example of this is shown in Fig. 4.15. In this figure, the nominally applied stress is from 0 to σ_{max}, while the local stress–strain response is completely reversed. This local response is due to residual stresses developed as a result of local yielding at the notch root.

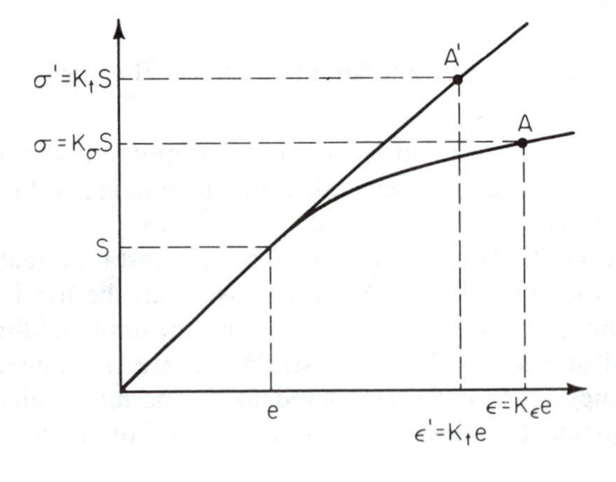

Figure 4.14 Differences between local stress and strain predictions using K_t and K_σ, K_ε values.

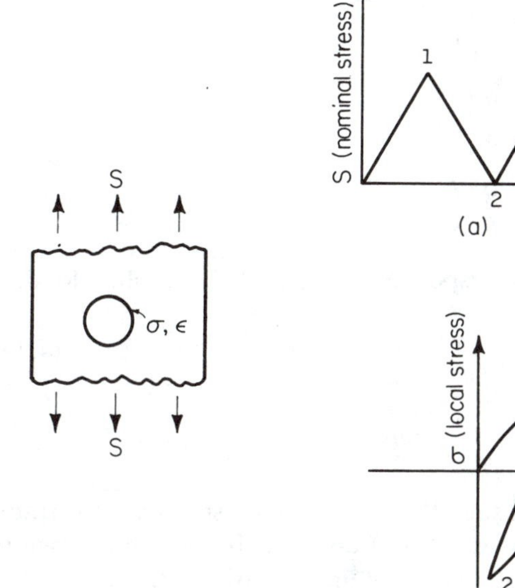

Figure 4.15 (a) Nominal stress history; (b) local stress–strain response.

Neuber [8] analyzed a specific notch geometry and derived the following relationship:

$$K_t = \sqrt{K_\sigma K_\epsilon} \tag{4.12}$$

This may be rearranged using Eq. (4.11) as

$$\begin{aligned} K_t^2 &= \frac{\sigma}{S}\frac{\epsilon}{e} \\ K_t^2 S e &= \sigma\epsilon \end{aligned} \tag{4.13}$$

Neuber's rule [Eq. (4.12)] states that the theoretical stress concentration is the geometric mean of the stress and strain concentration or the square root of the product of K_σ and K_ϵ. This seems intuitively reasonable since after yielding occurs, K_σ decreases while K_ϵ increases, as shown in Fig. 4.13. Although this method was proven only for one notch geometry, it is assumed that this relationship holds true for most notch geometries.

Three versions of Eq. (4.12), termed Neuber's rule, are often used in the local strain approach to relate nominal stresses and strains to local values. The three versions are described below.

Nominally Elastic Behavior. For nominally elastic behavior, the remote strain, e, can be related to the remote stress, S, using Hooke's law. Neuber's relation then takes the form

$$K_t^2 = K_\sigma K_\epsilon$$

$$K_t^2 = \frac{\sigma}{S}\frac{\epsilon}{e}$$

$$K_t^2 = \frac{\sigma}{S}\frac{\epsilon E}{S}$$

This can be restated by describing notch response in terms of the applied load.

$$\underbrace{\frac{(K_tS)^2}{E}}_{\text{applied load}} = \underbrace{\sigma\epsilon}_{\text{notch response}} \tag{4.14}$$

Since the component is usually designed such that the nominal stresses and strains remain elastic, this form of Neuber's rule is most often used. It has been shown to work fairly well as long as there is not an excessive amount of plasticity.

Limited Yielding. Neuber's rule, in the form of Eq. (4.13), is sometimes used when yielding occurs in the nominal stresses or strains. In this case, Hooke's law can no longer be used to relate the two ($S \neq Ee$). Restating Eq. (4.13) gives

$$K_t^2 Se = \sigma\epsilon \tag{4.13}$$

where e is nominal strain such that S and e are coordinates of a point on the cyclic stress–strain curve.

Seeger's Version. Finally, a third form of Neuber's rule is used when very high strains are present such that general yielding occurs in the component (nominal stresses and strains are well beyond yielding). For this case Seeger and Heuler [9] proposed the following version of Neuber's rule:

$$K_p^2 S^* e^* = \sigma\epsilon \tag{4.15}$$

where

$$K_p = \frac{S \text{ at onset of general yielding}}{S \text{ at first notch yielding}} = \frac{S_L}{\sigma_y/K_t}$$

(For an elastic–perfectly plastic material, with the same yield strength as the actual material, S_L is the limit load.)

$$S^* = \frac{K_t}{K_p} S \tag{4.16}$$

and the values of S^* and e^* must lie on the cyclic stress–strain curve. At these

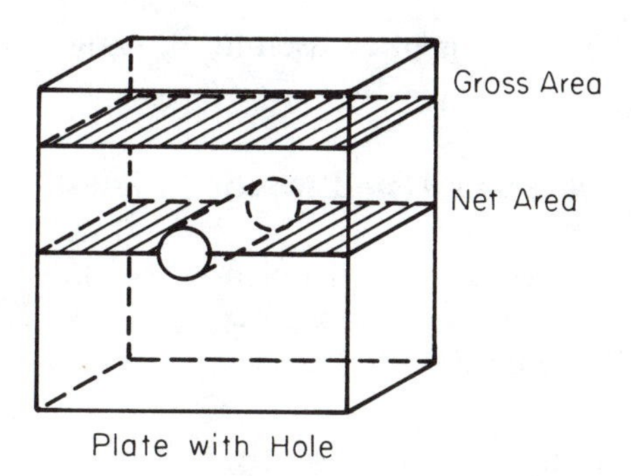

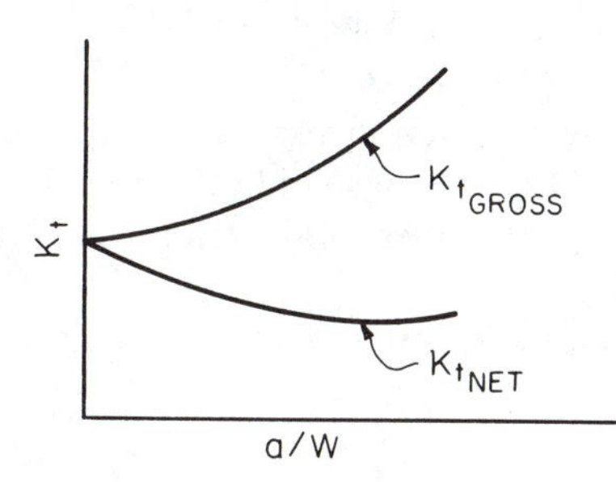

Figure 4.16 Gross and net section areas (a = hole radius, W = plate width).

load levels, failure criteria of general yielding and buckling must be considered as well as fatigue.

In all three versions of Neuber's rule, K_t and S must be defined consistently. They must both be defined using the same cross-sectional area, either net section or gross section ($K_{t_{net}}$ and S_{net} or $K_{t_{gross}}$ and S_{gross}). For example, in Fig. 4.16 if using $K_{t_{net}}$, S must be calculated based on the net section area as shown.

Dowling [10] compared predicted lives based on the three versions of Neuber's rule to test data. For lives greater than 10^3 cycles, predictions made using all three versions were close to actual lives. At lives below 10^3 cycles, Seeger's version [Eq. (4.15)] resulted in predictions that were much closer than those obtained using the other two versions of Neuber's rule. Dowling attributed this lack of correlation between test data and predictions made using Eqs. (4.13) and (4.14) for lives below 10^3 cycles to the large errors that resulted between the predicted and actual values of strain. Fortunately, he found that in this life range, "estimated fatigue lives are not highly sensitive to estimated strains. Significant errors in life only occur where gross errors are made in estimating strain."

In fatigue, since small notches have less effect than is indicated by K_t, Topper et al. [11] proposed replacing K_t in Neuber's rule by K_f. For example, Eq. (4.13) is rewritten as

$$\frac{(K_f S)^2}{E} = \sigma\epsilon \tag{4.17}$$

This modification of Neuber's rule is often used in the fatigue analysis of notched components.

4.3.2 Example of Notch Analysis Using Neuber's Rule

In the following section we describe in detail how Neuber's rule is applied in constant amplitude fatigue analyses of notched components. In the example, the following assumptions and values are used:

1. Assume nominally elastic behavior [use Eq. (4.14)].
2. Use K_f instead of K_t [Eq. (4.17)].
3. Use net section properties $K_{t_{net}}$ and S based on A_{net}.

Step 1: Initial Loading. A nominal stress, S_1, is applied to a previously unstressed component (see Fig. 4.17). Using Neuber's rule, we must satisfy Eq. (4.14). At the same time the local stress and strain must lie on the cyclic stress–strain curve of the material.

Using Eq. (4.17), the nominally elastic version of Neuber's rule,

$$\underbrace{\sigma\epsilon}_{\text{notch response}} = \underbrace{\frac{(K_f S)^2}{E}}_{\text{applied load}} \tag{4.18}$$

The values on the right-hand side are known while σ and ϵ are to be determined. For the loading, S_1, the value on the right-hand side is a constant. Thus

$$\sigma\epsilon = \text{constant}$$

From elementary mathematics, $(x)(y) = \text{constant}$ is an equation of a hyperbola.

Since the notch stress–strain response must lie on the stress–strain curve of the material, the intersection of these two curves [the cyclic stress–strain curve

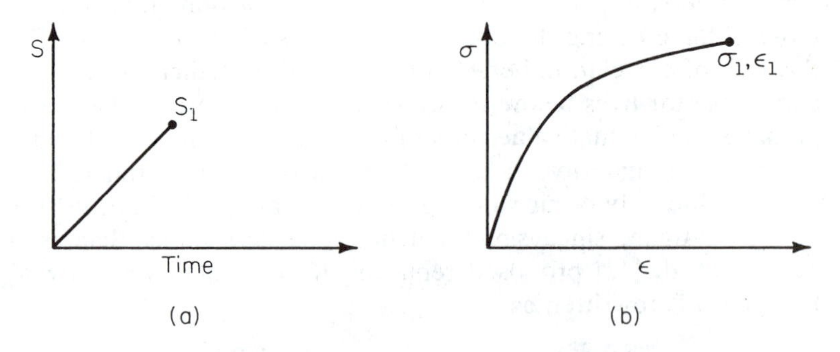

Figure 4.17 (a) Nominal stress versus time; (b) local (notch) stress–strain response on initial loading.

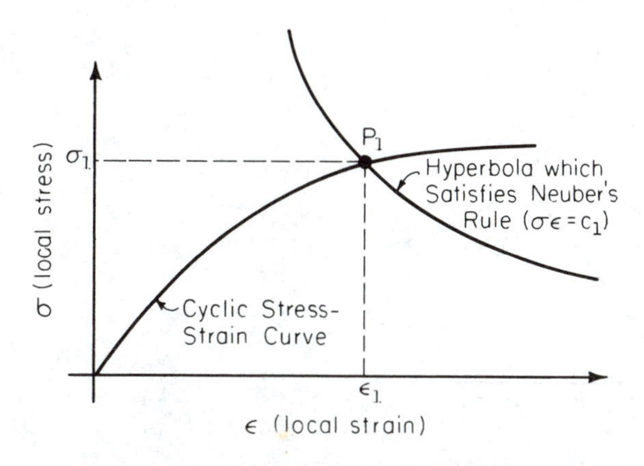

Figure 4.18 Intersection of cyclic stress–strain curve and Neuber's hyperbola.

and Neuber's hyperbola from Eq. (4.18)] provides the correct values of σ_1 and ϵ_1 for the initial loading to S_1 (see Fig. 4.18).

Analytically, the stress and strain coordinates of this point, P_1, may be found using the equation of the cyclic stress–strain curve (see Chapter 2).

$$\epsilon = \frac{\sigma}{E} + \left(\frac{\sigma}{K'}\right)^{1/n'} \tag{4.19}$$

Substituting into Eq. (4.18) yields

$$\left[\frac{\sigma}{E} + \left(\frac{\sigma}{K'}\right)^{1/n'}\right]\sigma = \frac{(K_f S)^2}{E} \tag{4.20}$$

This may be solved using numerical methods such as an iteration technique.

Step 2: Nominal Stress Reversal. The nominal stress is then reversed to some stress value, S_2, resulting in a change of stress $\Delta S = S_1 - S_2$ (see Fig. 4.19). This results in a change of the local or notch root stress, $\Delta\sigma$, and strain, $\Delta\epsilon$.

The general procedure in step 1 is repeated except for two modifications. The origin of the rectangular coordinate system is now located at point P_1, and the hysteresis loop is used instead of the cyclic stress–strain curve as discussed below.

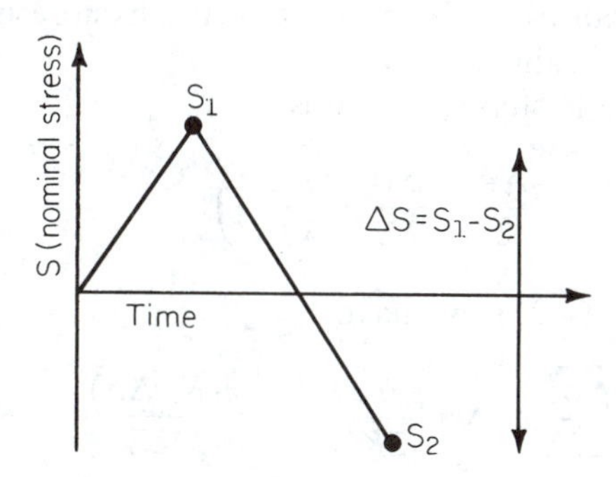

Figure 4.19 Nominal stress reversal.

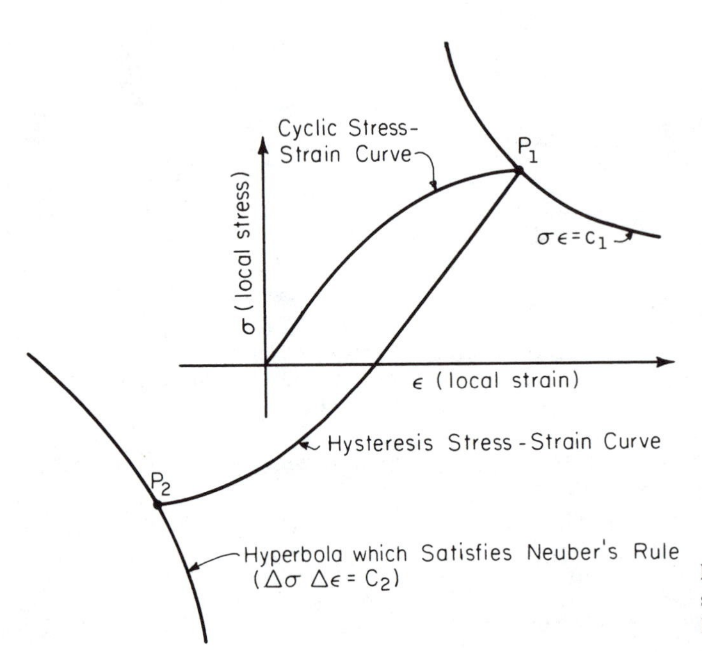

Figure 4.20 Intersection of hysteresis stress–strain curve and Neuber's hyperbola ($\Delta\sigma\,\Delta\varepsilon$).

Neuber's rule must again be satisfied, for this case in the form

$$\Delta\sigma\,\Delta\epsilon = \frac{(K_f\,\Delta S)^2}{E} \tag{4.21}$$

Again, the value on the right-hand side of the equation is constant, although not equal to the constant in Eq. (4.18). Another hyperbola is formed as shown in Fig. 4.20. The origin of the axis is now at point P_1.

The values of $\Delta\sigma$ and $\Delta\epsilon$ can then be determined. Since the change in stress and strain is being considered, the point P_2 lies at the intersection of the hyperbola defined by Eq. (4.21) and the *hysteresis* stress–strain curve. The hysteresis stress–strain is now used instead of the *cyclic* stress–strain curve which was used in step 1. (Recall from Chapter 2 that the hysteresis stress–strain curve is twice the cyclic stress–strain curve.)

The equation of the hysteresis curve is

$$\frac{\Delta\epsilon}{2} = \frac{\Delta\sigma}{2E} + \left(\frac{\Delta\sigma}{2K'}\right)^{1/n'} \tag{4.22}$$

Thus combined with Eq. (4.21), we have

$$\frac{\Delta\sigma^2}{2E} + \Delta\sigma\left(\frac{\Delta\sigma}{2K'}\right)^{1/n'} = \frac{(K_f\,\Delta S)^2}{2E} \tag{4.23}$$

The values of $\Delta\sigma$ and $\Delta\epsilon$ can be determined by solving Eq. (4.23) using iterative techniques. Once these are determined, the values of σ and ϵ corresponding to P_2 can be determined by subtracting $\Delta\sigma$ and $\Delta\epsilon$ from the values of σ_1 and ϵ_1 obtained previously. Note that $\Delta\sigma$ and $\Delta\epsilon$ are added or subtracted to the previous values of σ and ϵ, depending on the sign of the change in nominal load.

All successive load steps use Eqs. (4.21), (4.22), and (4.23), since further cycling causes material response to follow the hysteresis loop. This approach is easily adapted for computer applications. A Neuber analysis may be used to evaluate the fatigue life of a notched component under variable amplitude loading. This is described in Chapter 5.

This strain–life method is more complicated than the stress–life approach, yet provides more physical insight. Although it requires more material data and use of simple iterative techniques, it allows the stress–strain behavior at the notch root to be followed. Because of this, load sequence effects may be accounted for.

Example 4.2

Solve the problem described in Example 4.1 using a Neuber analysis. The stress–strain and strain–life properties of the steel are

$$E = 30 \times 10^3 \text{ ksi} \qquad K' = 154 \text{ ksi} \qquad n' = 0.123$$

$$\sigma_f' = 169 \text{ ksi} \qquad b = -0.081$$

$$\epsilon_f' = 1.14 \qquad c = -0.67$$

Solution The first step is to determine the notch root stress–strain behavior using a Neuber analysis. The elastic–plastic form of Neuber's rule is given by Eq. (4.13):

$$K_t^2 Se = \sigma\epsilon$$

or in terms of stress and strain ranges,

$$K_t^2 \frac{\Delta S\,\Delta e}{4} = \frac{\Delta\sigma\,\Delta\epsilon}{4}$$

The nominal and local stress–strain relationships are obtained using Eq. (2.36). These are

$$\frac{\Delta\epsilon}{2} = \frac{\Delta\sigma}{2E} + \left(\frac{\Delta\sigma}{2K'}\right)^{1/n'}$$

and

$$\frac{\Delta e}{2} = \frac{\Delta S}{2E} + \left(\frac{\Delta S}{2K'}\right)^{1/n'}$$

Combining these three equations, the following expression may be developed:

$$K_t^2 \frac{\Delta S}{2}\left[\frac{\Delta S}{2E} + \left(\frac{\Delta S}{2K'}\right)^{1/n'}\right] = \frac{\Delta\sigma}{2}\left[\frac{\Delta\sigma}{2E} + \left(\frac{\Delta\sigma}{2K'}\right)^{1/n'}\right]$$

[Note that this equation will work only for fully reversed ($R = -1$) constant amplitude loading. For all other cases the procedure described earlier must be used to predict notch stress–strain behavior.] In this equation, E, n', and K' are material properties, and K_t is the theoretical stress concentration factor. It is important to note that K_t and ΔS must both be based on net or gross section area (i.e., $K_{t_{net}}$, S_{net} or $K_{t_{gross}}$, S_{gross}). For most engineering applications, where plastic strains are not large,

$$(K_{t_{net}})(S_{net}) \approx (K_{t_{gross}})(S_{gross})$$

Calculate S_{net} using the applied load and net section area, A_{net}.

$$\text{load } (P_a) = \pm 10 \text{ kips}$$

$$A_{net} = (1 - 0.2)(0.25) = 0.20 \text{ in}^2.$$

$$\frac{\Delta S_{net}}{2} = \frac{P_a}{A_{net}} = \frac{10}{0.20} = 50 \text{ ksi}$$

$$\frac{\Delta S_{net}}{2} = 50 \text{ ksi}$$

Determine notch root stress using the equation developed previously.

$$\frac{K_t^2(\Delta S)}{2}\left[\frac{\Delta S}{2E} + \left(\frac{\Delta S}{2K'}\right)^{1/n'}\right] = \frac{\Delta\sigma}{2}\left[\frac{\Delta\sigma}{2E} + \left(\frac{\Delta\sigma}{2K'}\right)^{1/n'}\right]$$

$$(2.42)^2(50)\left[\frac{50}{30 \times 10^3} + \left(\frac{50}{154}\right)^{1/0.123}\right] = \frac{\Delta\sigma}{2}\left\{\frac{\Delta\sigma}{2(30 \times 10^3)} + \left[\frac{\Delta\sigma}{2(154)}\right]^{1/0.123}\right\}$$

By iterating, we obtain

$$\Delta\sigma = 156 \text{ ksi}$$

Use Eq. (2.36) to determine notch root strain:

$$\frac{\Delta\epsilon}{2} = \frac{156}{2(30 \times 10^3)} + \left[\frac{156}{2(154)}\right]^{1/0.123}$$

$$\frac{\Delta\epsilon}{2} = 0.0065$$

Since the loading is fully reversed, the mean stress, σ_0, is 0.0.

With the notch root strain known, the fatigue life, $2N_f$, can be determined by using the strain–life equation [Eq. (2.41)].

$$\frac{\Delta\epsilon}{2} = \frac{\sigma_f'}{E}(2N_f)^b + \epsilon_f'(2N_f)^c$$

$$0.0065 = \frac{169}{30 \times 10^3}(2N_f)^{-0.081} + 1.14(2N_f)^{-0.670}$$

This equation may be solved with an iteration technique.

$$2N_f = \underline{\underline{5000 \text{ reversals}}}$$

Note that this analysis could be performed using $K_{f_{net}}$. Since in this case $K_{f_{net}} = 2.34 \approx K_{t_{net}}$, the difference in the predicted life would be small.

4.4 FRACTURE MECHANICS APPROACHES

4.4.1 Introduction

In Chapter 3 a method of estimating the fatigue life of a component using fracture mechanics principles was presented. Use of this method results in an estimate of the fatigue life spent in crack propagation. Fracture mechanics approaches may also be used in the fatigue analysis of notched components. In the application of this method to notched components, the near notch stress–strain field must be considered. Important aspects of the analysis include the determination of the size of the region over which this field is effective and the stress intensity factor in this region.

4.4.2 Transition Crack Length

Near the notch, the local notch stress–strain field dominates the stress intensity solution. The notch produces a stress gradient. As the distance from the notch increases, the elevation of the stresses due to the notch decreases. Consequently, at some distance from the notch, the local stress approaches the value of the bulk stress. The crack length corresponding to this distance has been termed the transition crack length, l_t.

Once the crack length is larger than this value, the effective crack length, a, can then be assumed to be the sum of the notch width, D, and the actual crack length, l, growing out of the notch.

Dowling [12] derived an expression for the characteristic length, l_t, by analyzing a circular hole in an infinite plate. He observed that the numerical solution for the stress intensity for this case corresponds to two limiting cases as shown in Fig. 4.21. The short crack limiting case is

$$K_{\text{short}} = 1.12 K_t S \sqrt{\pi l} \tag{4.24}$$

where K_{short} = stress intensity factor for the short crack
S = remote or nominal stress
K_t = elastic stress concentration factor
l = free surface crack length
$f(g)$ = 1.12 = free surface or edge correction factor

For long cracks

$$K_{\text{long}} = S\sqrt{\pi a} \tag{4.25}$$

where K_{long} = stress intensity factor for the long crack
$a = l + D$
D = radius of circular hole as shown in Fig. 4.21

When the crack is small, it behaves like a crack growing from the edge of a plate subjected to the nominal stress of $K_t S$. As the crack becomes longer it

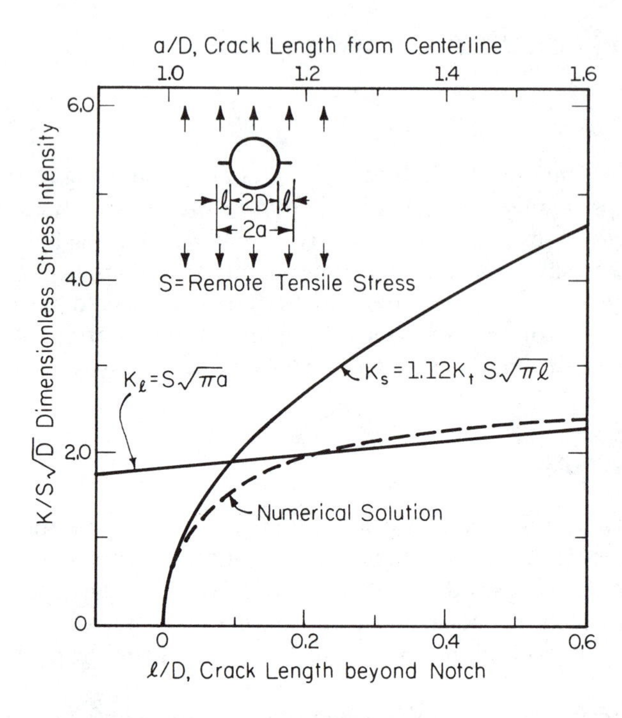

Figure 4.21 Short and long crack limiting cases and numerical solution for cracks growing from a circular hole in an infinite plate. (From Ref. 12.)

grows out of the notch stress–strain field and behaves like a crack whose length includes the notch width ($a = l + D$).

For the general case, all short cracks behave as cracks growing from an edge such that Eq. (4.24) may be assumed to remain valid for any geometry. However, for the long-crack solution the general form of Eq. (4.25) is

$$K_{\text{long}} = f(g)S\sqrt{\pi a} \tag{4.26}$$

where $f(g)$ depends on the specific geometry of the component being analyzed.

Dowling calculated the transition crack length by combining Eqs. (4.24) and (4.26). This is

$$K_{\text{long}} = f(g)S\sqrt{\pi a} = K_{\text{short}} = 1.12K_tS\sqrt{\pi l_t}$$

$$f(g)\sqrt{\pi a} = 1.12K_t\sqrt{\pi l_t}$$

$$f(g)[\pi(D \pm l_t)]^{1/2} = 1.12K_t\sqrt{\pi l_t}$$

$$l_t\left[\frac{1.12K_t}{f(g)}\right]^2 = D \pm l_t$$

$$l_t = \frac{D}{[1.12K_t/f(g)]^2 \mp 1} \tag{4.27}$$

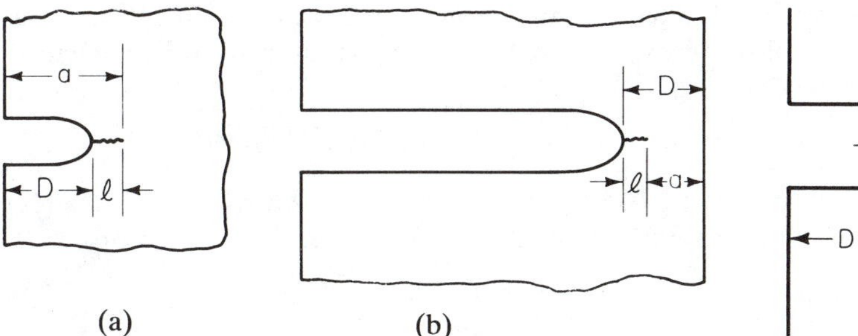

Figure 4.22 (a) Shallow and (b) deep notches. (From Ref. 12.)

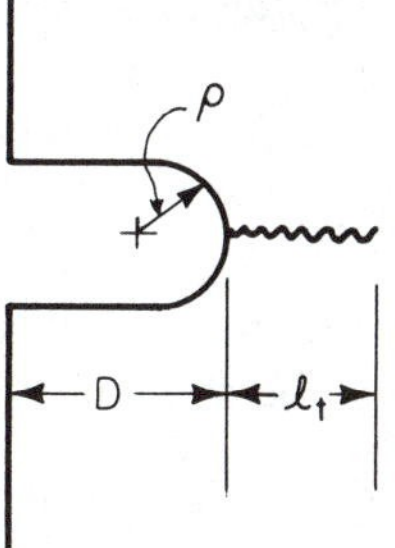

Figure 4.23 Terminology for notch dimensions.

where

$a = D + l$ (Fig. 4.22a)

$a = D - l$ depending on notch configuration (Fig. 4.22b)

He stated that l_t is usually a small fraction of the notch radius, ρ, and generally falls within the range $\rho/20$ to $\rho/4$.

Smith and Miller [13] also developed an expression for the distance, l_t. They fit a curve to a number of numerical solutions to obtain the estimate

$$l_t = 0.13\sqrt{D\rho} \tag{4.28}$$

where l_t = extent of notch stress field
D = depth of an elliptical notch
ρ = actual notch root radius

as shown in Fig. 4.23.

As stated previously, when crack length, l, exceeds this value, the crack is assumed to be outside the notch stress field and the crack growth may then be evaluated using standard fracture mechanics techniques with the crack length equal to $l + D$. This is the sum of the notch depth, D, and the physical crack length, l.

4.4.3 Stress Intensity Factors near a Notch

Stress intensities for a crack growing in the notch stress–strain field, when $l < l_t$, must be obtained in order to use standard LEFM approaches. A limited number of theoretical solutions have been developed for the stress intensity factor near a notch. Examples of several of these are given in Ref. 14.

Usually, the theoretical solution is unavailable, though, and a method of approximating the stress intensity factor near a notch must be used. Using the expression in Eq. (4.28) for the transition crack length, Smith and Miller proposed the following expression for the stress intensity factor when $l < l_t$. For $l < l_t$ or $l < 0.13\sqrt{D\rho}$,

$$\Delta K = \Delta S\sqrt{\pi l}\,[(1 + 7.69\sqrt{D/\rho})^{1/2}] \tag{4.29}$$

When $l > l_t$, standard LEFM techniques may be used with the crack length taken to be the sum of the notch depth, D, and the crack length, l. For $l > 0.13\sqrt{D\rho}$,

$$\Delta K = \Delta S f(g)\sqrt{\pi(l + D)} \tag{4.30}$$

[Recall that $f(g)$ is a correction factor for geometry effects.]

As an example, for a circular notch the range of application of the expressions above is shown in Fig. 4.24. (Note that in this case $D = r$.)

Recall from Eq. (3.10) that life to failure can be determined from

$$N_f = \int_{a_i}^{a_f} \frac{da}{C\,\Delta K^m} \tag{4.31}$$

(*Note*: In the following equations, the final crack length, a_f, is termed l_f, and the initial crack size, a_i, is termed l_i.)

$$N_f = \underbrace{\int_{l_i}^{0.13\sqrt{D\rho}} \frac{da}{C[\Delta S\sqrt{\pi l}\,(1 + 7.69\sqrt{D/\rho})^{1/2}]^m}}_{\text{life of a short crack}} + \underbrace{\int_{0.13\sqrt{D\rho}+D}^{l_f} \frac{da}{C[\Delta S f(g)\sqrt{\pi(l + D)}]^m}}_{\text{life of a long crack}} \tag{4.32}$$

An estimation for final crack size, l_f, is sometimes taken as the crack size at which net section yielding occurs. Recall from Section 3.3.2, though, that the final crack size estimation often does not significantly affect the total life estimate.

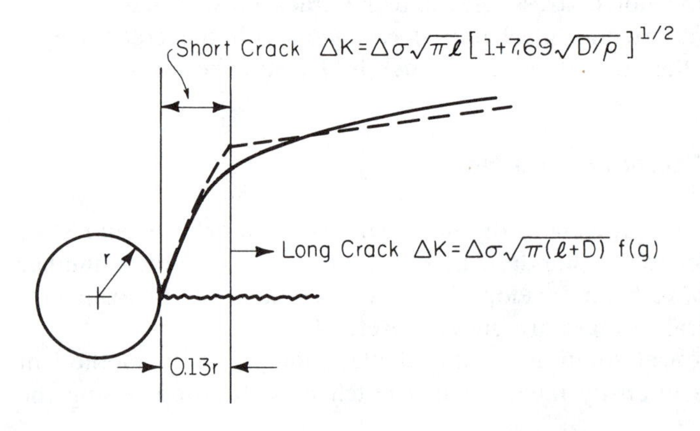

Figure 4.24 Smith and Miller's relations for ΔK for a circular notch. (From Ref. 13.) [The solid line is the theoretical solution from Ref. 14 and the dashed line is the solution predicted using Eqs. (4.29) and (4.30).

However, (4.32) is very sensitive to the value of l_i, which is usually unknown. In fact, this problem plagues all LEFM methods used to determine growth of a crack from the notch when $l < l_t$.

Another expression that has been proposed for approximating the stress intensity factor near a notch is simply to multiply the bulk stress by the theoretical stress concentration factor, K_t, and by 1.12, the free surface (edge) correction factor. This is,

$$\Delta K_{\text{short}} = 1.12 K_t \, \Delta S \sqrt{\pi l} \tag{4.33}$$

Dowling [12], however, showed that Eq. (4.33) is very limited in usefulness for notched fatigue problems because the LEFM assumption of small-scale plasticity (the plastic zone is small compared to the crack length) on which this equation is based is usually violated. In addition, he notes that the "short crack problem" makes Eq. (4.33) useless in predicting short crack growth. This "problem" is described in Section 4.4.4.

Example 4.3

Use Dowling's method to predict the propagation life of the component described in Example 4.1. In this method propagation life is obtained by integrating the Paris relationship from some initial crack length to a final crack length, a_f. The initial crack length is taken to be the sum of the transition crack length, l_t, and the notch depth, D. The steel has a yield strength, σ_y, of 94 ksi. The crack growth constants for the steel are

$$C = 6.157 \times 10^{-10} \qquad m = 2.715$$

Solution Determine the transition crack length, l_t, from Eq. (4.27).

$$l_t = \frac{D}{[1.12 K_t / f(g)]^2 - 1}$$

In this analysis the geometry correction factor, $f(g)$, is approximated as 1.12, the free edge correction factor. The transition crack length, l_t, then becomes

$$l_t = \frac{D}{K_t^2 - 1}$$

where

$$D = 0.1 \text{ in.}$$

$$K_t = 3.03 \quad \text{(based on gross section area)}$$

$$l_t = \frac{0.1 \text{ in.}}{(3.03)^2 - 1} = 0.012 \text{ in.}$$

Determine the final crack length, a_f. Failure is assumed to occur when the net section yields. Net section yielding of the cracked body is expected when the stress in the uncracked ligament reaches the yield stress. The value of load at this time is

$$P_{cr} = \sigma_y (w - 2a_f)(t)$$

where t is the thickness and $w - 2a_f$ is the width of cracked specimen (assuming two equal-length edge cracks).

The value of the peak gross stress (completely reversed) at this load is

$$\frac{P_{cr}}{(w)(t)} = \frac{\Delta S_{\text{gross}}}{2}$$

Therefore, the value of the gross stress, $\Delta S_{\text{gross}}/2$, when net section yielding occurs is

$$\frac{\Delta S_{\text{gross}}}{2} = \frac{\sigma_y(w - 2a_f)}{w}$$

and the value of a_f is

$$a_f = \frac{w}{2}\left(1 - \frac{\Delta S_{\text{gross}}}{2\sigma_y}\right)$$

For this case

$$w = 1.00 \text{ in.}$$
$$\sigma_y = 94 \text{ ksi}$$
$$\text{load } (P_a) = \pm 10 \text{ kips}$$
$$\frac{\Delta S_{\text{gross}}}{2} = \frac{10 \text{ kips}}{(1 \text{ in.})(0.25 \text{ in.})} = 40 \text{ ksi}$$

Therefore,

$$a_f = 0.287 \text{ in.}$$

Determine the stress intensity factor range for the given geometry. For a finite-width edge-cracked plate, an expression for the stress intensity factor range is given in Fig. 3.4c. For a value of b equal to 0.5 in., this is

$$\Delta K = \frac{\Delta S_{\text{gross}}}{2}\sqrt{\pi a}\,(1.12 + 0.4a - 4.8a^2 + 15.44a^3)$$

(This assumes that only tensile stresses cause crack growth.)

Determine the propagation life by integrating the Paris relationship from an initial crack size, $l_t + D$, to a final crack size, a_f.

$$N_p = \int_{l_t+D}^{a_f} \frac{da}{C(\Delta K)^m}$$

where

$$l_t = 0.012 \text{ in.}$$
$$D = 0.10 \text{ in.}$$
$$a_f = 0.287 \text{ in.}$$
$$C = 6.157 \times 10^{-10}$$
$$m = 2.715$$
$$\Delta K = \frac{\Delta S_{\text{gross}}}{2}\sqrt{\pi a}\,(1.12 + 0.4a - 4.8a^2 + 15.44a^3)$$
$$\frac{\Delta S_{\text{gross}}}{2} = 40 \text{ ksi}$$

This equation can be integrated numerically. The propagation life is then obtained as

$$N_p = \underline{\underline{19{,}000 \text{ cycles}}}$$

4.4.4 Short Crack Growth at Notches

A plastic zone often develops in the notch root due to the stress concentrations associated with the notch. As shown in Fig. 4.25, the plastic zone is surrounded by a region of elevated elastic stresses and strains. This region is, in turn, surrounded by the bulk stress–strain field.

In the plastic zone of the notch, LEFM assumptions are violated. In fact, a deviation from linearity often occurs in the log da/dN versus log ΔK plot of crack growth at notches. Even in situations where the strains surrounding the notch remain elastic, the plot is often nonlinear. Three types of deviation from the normal crack growth rate curve are shown in Fig. 4.26. As shown, the three types of anomalous crack growth behavior for cracks at notches are

1. Short crack growth only in a plastic field
2. Short crack growth in both plastic and elastic fields
3. Short crack growth only in an elastic field

In the third case, the totally elastic case, a major reason the deviation from linearity occurs is believed to be due to the "short crack problem" [15]. This "problem" is that short cracks tend to grow faster than long cracks at the same ΔK. In the notch field, a major cause of this "short crack" behavior is the difference in residual plastic deformation in the wake of the crack (see Fig. 4.27).

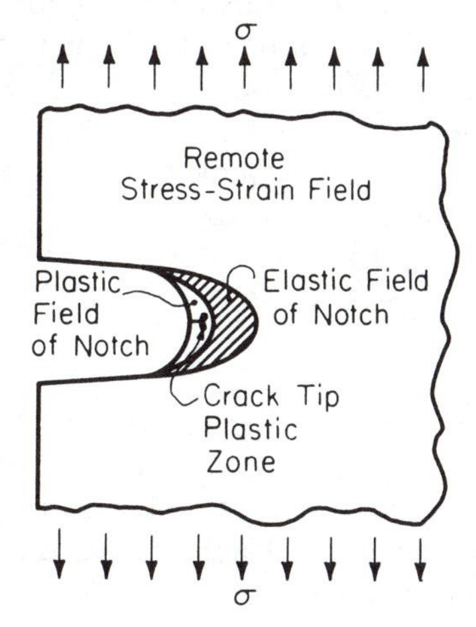

Figure 4.25 Plastic zone at notch root.

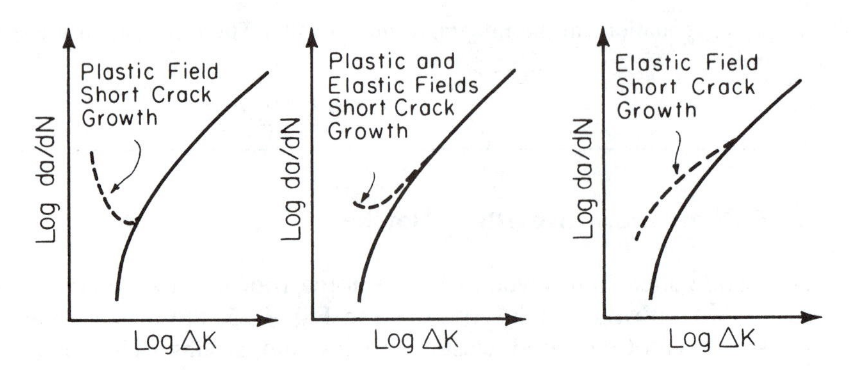

Figure 4.26 Anomalous crack growth at notches.

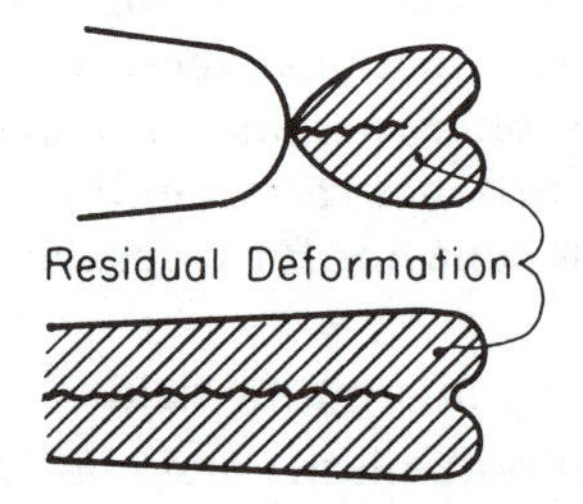

Figure 4.27 Differences in plasticity in wake of long and short cracks.

A higher crack growth rate occurs for the short crack since a smaller amount of residual deformation is developed, resulting in less crack closure and a higher ΔK_{eff} (refer to Section 3.3.6).

For the first two cases, where the crack grows in a plastic field, initially a decrease is seen in the crack growth rate. It is believed that this is due to the rapid decrease in plastic strain, which controls the crack growth, thereby causing the "dip" in the log da/dN versus log ΔK plot.

Attempts to modify LEFM equations have been made so that this plasticity may be accounted for. One of the simplest methods is to replace the ΔS in Eq. (4.33) by $\Delta e\, E$ (see Ref. 16 for further details).

Methods have been developed that circumvent the problems of notch root plasticity, short crack effects, and uncertainties in the estimate of the initial crack size, l_i. These are discussed in the following section.

4.5 COMBINATION METHODS

Dowling [12] proposed a method to estimate the total fatigue life of a notched component. This method combines the local strain approach to predict initiation life and a fracture mechanics approach to predict propagation life. As stated previously, he proposed that within a distance from the notch, l_t, the local notch stress field dominates the stress intensity solution. When the crack is smaller than this length, $(l < l_t)$, crack initiation or early growth from the notch can be estimated using the strain–life approach. Use of this approach to estimate the initiation life avoids the inherent difficulties of using LEFM methods to describe short crack behavior at the notch root. Once the crack is larger than this length $(l > l_t)$, crack growth can be modeled with a standard fracture mechanics approach with the initial crack size taken to be l_t.

$$\text{total life} = \underbrace{\underset{(l < l_t)}{\text{initiation life } (N_I)}}_{\substack{\text{strain–life} \\ \text{approach}}} + \underbrace{\underset{(l > l_t)}{\text{propagation life } (N_p)}}_{\substack{\text{fracture mechanics} \\ \text{approach}}}$$

where

$$N_p = \int_{D+l_t}^{l_f} \frac{da}{C(\Delta K)^m} \tag{4.34}$$

He compared test results with predictions made using this approach for plates with blunt and sharp notches. Good results were obtained between predicted results and test data. [In the strain–life method he used a form of Neuber's rule for nominally elastic behavior, Eq. (4.14). In the fracture mechanics approach for completely reversed loading, he assumed that no damage occurred on the compressive portion of the cycle. Crack propagation was predicted to terminate (l_f) when the remaining uncracked ligament reached a net section stress equal to the yield strength.]

This method requires that strain–life data be available which report life as the number of cycles required to develop a crack of length l_t. However, strain–life data are not often available for a specific crack size. Strain–life data are usually reported as life to failure of a laboratory specimen. Fortunately, Dowling observed that use of smooth specimen total fatigue life data would have predicted reasonable, although somewhat conservative, results for a blunt notched specimen. For sharp notches, Dowling states: "Propagation is expected to dominate at all lives, so that crack initiation data are superfluous." Stated another way, a notch that looks like a crack will behave like a crack and the initiation life is negligible.

A simple combined approach that has been shown to yield very similar results to Dowling's method has been proposed by Socie [17]. This method combines:

1. An initiation life estimate obtained using the standard local strain approach that assumes that the notch is fully effective $(K_f = K_t)$
2. A propagation estimate obtained using a fracture mechanics approach [Eq. (4.31)] that assumes the initial crack size is equal to the depth of the notch (a_i = depth of notch)

These are combined to obtain a total fatigue life estimate.

This method predicts results very close to those obtained from Dowling's method [12]. An advantage of this approach is that it is simple and easy to implement. In addition, the distance, l_t, over which the notch stress field dominates the stress intensity solution does not have to be determined. Predictions using this approach tend to be conservative.

In the two combined approaches discussed above, use of the strain–life approach allows difficulties with strain gradients, plasticity effects, and short crack effects to be avoided when dealing with small cracks. Fracture mechanics methods allow additional life from crack propagation to be included in the total life estimates.

Example 4.4

Use the method proposed by Socie and discussed earlier to predict the total life ($N_p + N_I$) of the component described in Examples 4.1, 4.2, and 4.3.

Solution Determine an initiation life estimate using the strain–life approach assuming that $K_f = K_t$. Using the strain–life method from Example 4.2, the initiation life estimate is

$$N_I = 2500 \text{ cycles}$$

Determine the propagation life using the fracture mechanics approach where the initial crack size, a_i, is taken to be the depth of the notch, D.

$$N_p = \int_{a_i}^{a_f} \frac{da}{C(\Delta K)^m}$$

The values a_f, C, and m and the stress intensity factor relation are the same as those used in Example 4.3. Substituting the necessary values into this equation gives

$$N_p = \int_{0.10}^{0.287} \frac{da}{(6.157 \times 10^{-10})[40\sqrt{\pi a}\,(1.12 + 0.4a - 4.8a^2 + 15.44a^3)]^{2.715}}$$

This equation can be solved using numerical integration techniques. One such technique yields

$$N_p = 21{,}000 \text{ cycles}$$

(Note how this compares to the estimate of propagation life found in Example 4.3 using Dowling's method.)

Determine the total life by summing the initiation and propagation life estimates.

$$N_T = N_I + N_p$$

$$N_T = (2500 + 21{,}000) \text{ cycles}$$

$$= \underline{\underline{23{,}500}} \text{ cycles}$$

4.6 IMPORTANT CONCEPTS

- In general, the effect of a notch on fatigue life is less than that predicted by the theoretical stress concentration factor, K_t. Consequently, the fatigue notch factor, K_f, which is smaller than K_t, is often used in a fatigue analysis.
- Neuber's rule relates the local stress and strain at a notch to the global stress–strain using K_f or K_t.
- In the application of fracture mechanics approaches to the analysis of notches, difficulties arise in determining crack behavior in the vicinity of the notch due to local plasticity at the notch root, "short crack" effects, and difficulties in estimating initial crack length, a_i.
- Fracture mechanics methods are appropriate when the crack grows out of the local notch stress–strain field.
- Combination methods sum life estimates made using strain–life and fracture mechanics approaches. The strain–life approach is used to estimate the life for initiation and early growth of the crack in the notch stress–strain field. Fracture mechanics approaches are used when the crack grows out of this zone.

4.7 IMPORTANT EQUATIONS

Fatigue Notch Factor

$$K_f = \frac{S_e^{(\text{unnotched})}}{S_e^{(\text{notched})}} \tag{4.2}$$

Notch Sensitivity Factors

$$q = \frac{K_f - 1}{K_t - 1} \tag{4.3}$$

$$K_f = 1 + \frac{K_t - 1}{\left(1 + \dfrac{a}{r}\right)} \tag{4.5}$$

$$a = \left[\frac{300}{S_u\ (\text{ksi})}\right]^{1.8} \times 10^{-3}\ \text{in.} \tag{4.6}$$

$$S_u \approx 0.5\text{BHN} \tag{4.7}$$

Neuber Relationship

$$K_t = \sqrt{K_\sigma K_\epsilon} \tag{4.12}$$

$$K_t^2 = \frac{\sigma}{S}\frac{\epsilon}{e} \tag{4.13}$$

Neuber Relationship (Nominally Elastic Behavior)

$$\frac{(K_t S)^2}{E} = \sigma\epsilon \tag{4.14}$$

Fracture Mechanics Approach: Short Crack Solution

$$K_{\text{short}} = 1.12 K_t S\sqrt{\pi l} \tag{4.24}$$

Fracture Mechanics Approach: Long Crack Solution

$$K_{\text{long}} = S\sqrt{\pi a} \tag{4.25}$$

Transition Crack Length Estimates

Smith and Miller's

$$l_t = 0.13\sqrt{D\rho} \tag{4.28}$$

Dowling's

$$l_t = \frac{D}{[1.12K_t/f(g)]^2 \mp 1} \tag{4.27}$$

Combination Approach for Total Life Estimate

$$\text{total life} = \underbrace{\underset{(a < l_t)}{\text{initiation life } (N_I)}}_{\substack{\text{strain–life}\\ \text{approach}}} + \underbrace{\underset{(a > l_t)}{\text{propagation life } (N_p)}}_{\substack{\text{fracture mechanics}\\ \text{approach}}}$$

REFERENCES

1. R. E. Peterson, *Stress Concentration Factors,* Wiley, New York, 1974.
2. J. E. Shigley, *Mechanical Engineering Design,* McGraw-Hill, New York, 1977.
3. R. E. Peterson, "Analytical Approach to Stress Concentration Effects in Aircraft Materials," Technical Report 59–507, *U.S. Air Force—WADC Symp. Fatigue Metals,* Dayton, Ohio, 1959.
4. H. Neuber, *Theory of Notch Stresses: Principle for Exact Stress Calculations,* Edwards, Ann Arbor, Mich., 1946.
5. R. E. Peterson, "Relation between Life Testing and Conventional Tests of Materials," Bulletin ASTM No. 133, Mar. 1945.
6. R. C. Juvinall, *Engineering Considerations of Stress, Strain and Strength,* McGraw-Hill, New York, 1967.
7. Society of Automotive Engineers, *Fatigue Design Handbook,* Vol. 4, SAE, Warrendale, Pa., 1968, p. 29.
8. H. Neuber, "Theory of Stress Concentration for Shear-Strained Prismatical Bodies

with Arbitrary Nonlinear Stress–Strain Laws," *J. Appl. Mech., Trans. ASME,* Vol. E28, 1961, p. 544.

9. T. Seeger and P. Heuler, "Generalized Application of Neuber's Rule," *J. Test. Eval.,* Vol. 8, No. 4, 1980, pp. 199–204.
10. N. E. Dowling, "A Discussion of Methods for Estimating Fatigue Life," *Proc. SAE Fatigue Conf.,* P–109, Society of Automotive Engineers, Warrendale, PA, 1982.
11. T. H. Topper, R. M. Wetzel, and JoDean Morrow, "Neuber's Rule Applied to Fatigue of Notched Specimens," *J. Mater.,* Vol. 4, No. 1, 1969, pp. 200–209.
12. N. E. Dowling, "Fatigue at Notches and the Local Strain and Fracture Mechanics Approaches," in *Fracture Mechanics,* ASTM STP 677, C. W. Smith (ed.), American Society for Testing and Materials, Philadelphia, 1979, pp. 247–273.
13. R. A. Smith and K. J. Miller, "Fatigue Cracks at Notches," *Int. J. Mech. Sci.,* Vol. 19, 1977, pp. 11–22.
14. J. C. Newman, "An Improved Method of Collocation for the Stress Analysis of Cracked Plates with Various Shaped Boundaries," NASA TND-6376, 1971.
15. H. L. Ewalds and R. J. H. Wanhill, *Fracture Mechanics,* Edward Arnold, London, 1985.
16. M. H. El Haddad, K. N. Smith, and T. H. Topper, "A Strain Based Intensity Factor Solution for Short Fatigue Cracks Initiating from Notches," in *Fracture Mechanics,* ASTM STP 677, C. W. Smith (ed.), American Society for Testing and Materials, Philadelphia, 1979, pp. 274–289.
17. D. F. Socie, N. E. Dowling, and P. Kurath, "Fatigue Life Estimation of Notched Members," in ASTM STP 833, *Fracture Mechanics*; Fifteenth Symposium, R. J. Sanford (ed.), American Society for Testing and Materials, Philadelphia, 1984, pp. 284–299.
18. H. F. Moore and J. B. Kommers, "An Investigation of the Fatigue of Metals," *Univ. Ill. Eng. Exp. Stn. Bull.* 124, 1921.
19. H. F. Moore and T. M. Jasper, "An Investigation of the Fatigue of Metals," *Univ. Ill. Eng. Exp. Stn. Bull.* 152, 1925.
20. H. F. Moore and P. E. Henwood, "The Strength of Screw Threads under Repeated Tension," *Univ. Ill. Eng. Exp. Stn. Bull.* 264, 1934.
21. J. W. Fash, "An Evaluation of Damage Development during Multiaxial Fatigue of Smooth and Notched Specimens," Materials Engineering Report No. 123, University of Illinois at Urbana–Champaign, 1985.
22. S. J. Stadnick and JoDean Morrow, "Techniques for Smooth Specimen Simulation of the Fatigue Behavior of Notched Members," in *Testing for Prediction of Material Performance in Structures and Components,* ASTM STP 515, American Society for Testing and Materials, Philadelphia, 1972, pp. 229–252.
23. J. H. Crews, Jr., "Crack Initiation at Stress Concentrations as Influenced by Prior Local Plasticity," in *Achievement of High Fatigue Resistance in Metals and Alloys,* ASTM STP 467, American Society for Testing and Materials, Philadelphia, 1970, pp. 37–52.
24. O. L. Bowie, "Analysis of an Infinite Plate Containing Radial Cracks Originating at the Boundary of an Internal Circular Hole," *Math. Phys.,* Vol. 35, 1956, pp. 60–71.
25. H. Sehitoglu, "Fatigue of Low Carbon Steels as Influenced by Repeated Strain

Aging," Fracture Control Program, Report No. 40, University of Illinois at Urbana–Champaign, June 1981.

26. P. Kurath, "Investigation into a Non-arbitrary Fatigue Crack Size Concept," Theoretical and Applied Mechanics, Report No. 429, University of Illinois at Urbana–Champaign, Oct. 1978.

PROBLEMS

SECTION 4.2

4.1. Determine the fatigue notch factor, K_f, for a wide plate with a center hole ($K_t = 3.0$) under the following conditions:

	Radius of Hole, r (in.)	Ultimate Strength, S_u (ksi)
(a)	0.025	100
(b)	0.025	50
(c)	0.025	200
(d)	0.0025	100
(e)	0.25	100

Discuss the effects of notch root radius, r, and ultimate strength, S_u, on the fatigue notch factor, K_f.

4.2. A notched plate with a hardness of 400 BHN is subjected to a completely reversed nominal stress amplitude of ± 40 ksi in service. Fatigue cracks are found to form in the notch root at approximately 10^6 cycles. Determine the fatigue notch factor. The life of this plate must be extended to meet the service warranty. Suggest several ways to do this.

4.3. Two 0.4-in.-diameter steel shafts contain circumferential notches with the following geometries. The first shaft (shaft A) has a reduced diameter of 0.25 in, a notch root radius, r, of 0.25 in, and a stress concentration factor, K_t, of 1.14. The second shaft (shaft B) also has a reduced diameter of 0.25 in. but it has a much sharper notch with a root radius of 0.005 in. and a stress concentration factor of 4. The properties of the steel are:

$$S_u = 97 \text{ ksi}$$

$$S_e^{(\text{unnotched})} = 48 \text{ ksi}$$

Rotating bending fatigue tests on the notched shafts indicate that the endurance limit value for shaft B ($K_t = 4$) is 19 ksi, while the value for shaft A ($K_t = 1.14$) is 44 ksi. How well do the estimation techniques discussed in Section 4.2 predict the behavior of these shafts? (Data taken from Ref. 18.)

4.4. The effect of notches on the endurance limit values of several steels was investigated by running a series of tests on notched rotating bending specimens. The specimens had a 0.3-in. diameter and contained a radial hole with a diameter of 0.055 in. The stress concentration factor, K_t, for this geometry is 2.05. Given below are the results of these tests. Determine how well the estimation techniques discussed in Section 4.2 predict the behavior of these specimens. (Data taken from Ref. 19.)

Material	Ultimate Strength, S_u (ksi)	Endurance Limit (without Hole), S_e (ksi)	Endurance Limit[a] (with Hole), S_e (ksi)
A	42.5	26	13
B	84	30.5	18
C	97	48	26
D	116	55	33.6
E	140	70	36

[a] Stress based on net area.

4.5. The bar shown below is to be used in a situation where it will be subjected to fully reversed torsional loading of ±800 in.-lb. Determine if this bar will be safe for an infinite life. If it is not, suggest methods to increase its fatigue life.

Material: Steel S_u = 117 ksi

Finish: machined (AA = 16 μ in.)

$D = 1.0$ in.

$d = 0.6$ in.

$r = 0.12$ in.

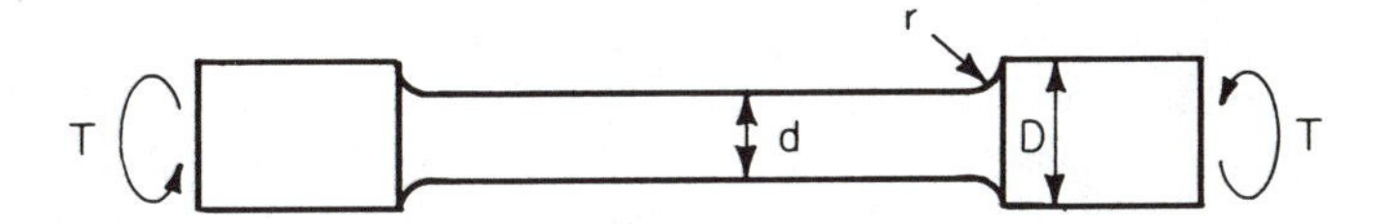

4.6. For the bar in Problem 4.5, plot the allowable fully reversed torsional loading for an infinite life versus notch root radius, r. Plot the values for notch root radius values from 0.01 in. to 0.15 in. Use all other values as given in Problem 4.5.

4.7. The plate shown below is 2 in. wide and has two edge notches. The first is a sharp notch with a notch root radius, r, of 0.005 in. and a stress concentration factor, K_t, of 7. The second notch is larger with a notch root radius of 0.10 in. and a stress concentration factor of 3. The plate is required to withstand long life (>10^6 cycles) fully reversed axial fatigue loading. When the plate is made from a soft steel (S_u = 50 ksi) it is found to fail at the large notch. If a higher strength steel (S_u = 200 ksi) is used, the failures originate at the sharp notch. Explain this behavior. At what ultimate strength value, S_u, would failures be equally likely at the large and small notches?

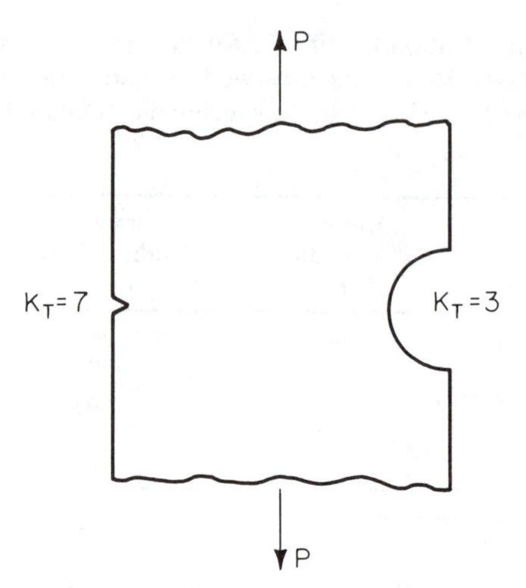

4.8. A valuable concept that can be used to determine the effect of notches on fatigue strength is the idea of a worst-case notch. This can be determined by finding the notch geometry that creates the largest fatigue notch factor, K_f, for a given material. For the case of an edge notch in a wide plate, the stress concentration factor is

$$K_t = 1 + 2\sqrt{t/r}$$

where t is the depth of the notch and r is the notch root radius.

Determine the maximum fatigue notch factor, K_f, for a steel with an ultimate strength, S_u, of 50 ksi and an edge notch of depth, $t = 0.1$ in. At what value of notch root radius, r, does this occur? Repeat this procedure for steels with ultimate strengths of 100, 150, 200, 250, and 300 ksi. Plot the resulting values for maximum fatigue notch factor versus ultimate strength.

4.9. Commonly encountered notched components are threaded fasteners such as bolts. In most cases the cyclic loading on a bolt is zero to maximum ($R = 0$). Given below are the monotonic and fatigue ($R = 0$) data for two steels. Fatigue data are given for both an unnotched specimen and a specimen with Whitworth threads. The Whitworth threads have a stress concentration factor, K_t, of 3.86 and a notch root radius, r, of 0.0086 in. Determine how well the estimation techniques discussed in Section 4.2 predict the behavior of these specimens. For this problem use the Goodman relationship and estimate the fully reversed endurance limit, S_e, as one-half of the ultimate strength, S_u. (Data taken from Ref. 20.)

		Fatigue tests ($R = 0$)	
Material	Ultimate Strength, S_u (ksi)	Unnotched, S_{max} (ksi)	Threaded, S_{max} (ksi)
A	57.4	37	21
B	109	73	22

(Refer to discussion in Problem 4.14.)

4.10. Draw on log–log coordinates the estimated alternating stress versus life to failure (S–N) curves for an unnotched steel with an ultimate strength, S_u, of 150 ksi, an endurance limit, S_e, of 75 ksi and a true fracture strength, σ_f, of 200 ksi. Use the methods shown in Figs. 4.11 and 4.12. Repeat the procedure above for notched members with fatigue notch factors, K_f, of 2, 3, 4, and 5. Compare the curves resulting from the methods shown in Figs. 4.11 and 4.12.

4.11. Draw on log–log coordinates the estimated alternating stress versus life to failure (S–N) curve for a notched member with a stress concentration factor, K_t, of 3 and a notch root radius, r, of 0.075 in. The properties of the steel are

$$S_u = 50 \text{ ksi} \qquad \sigma_f = 100 \text{ ksi} \qquad S_e = 25 \text{ ksi}$$

Use the methods shown in Figs. 4.11 and 4.12. Repeat the procedure above for steels with ultimate strength, S_u, values of 100, 150, 200, 250, and 300 ksi. Estimate the endurance limit, S_e, as one-half of the ultimate strength, and the true fracture strength, σ_f, as the ultimate strength plus 50 ksi. Compare the curves resulting from the methods shown in Figs. 4.11 and 4.12.

4.12. The test results shown below are the fatigue lives of several notched rotating bending specimens. The specimens had a 0.3 in. diameter and contained a radial hole with a diameter of 0.055 in. The stress concentration factor for this geometry, K_t, is 2.05. The properties for this material are

$$S_u = 97 \text{ ksi} \qquad S_e^{(\text{unnotched})} = 48 \text{ ksi} \qquad \sigma_f = 150 \text{ ksi}$$

Determine how well the two estimation techniques shown on Figs. 4.11 and 4.12 predict the behavior of these specimens. (Data taken from Ref. 19.)

Alternating Stress (Based on Net Area), S_a (ksi)	Life to Failure (cycles)
68	1.7×10^4
58	4.3×10^4
44	1.8×10^5
35	7.1×10^5
31.5	2.5×10^6
26	8.6×10^6
26.5	1.0×10^7 [a]
24.6	1.0×10^7 [a]

[a] Specimen did not fail.

4.13. Shown below is a rotating steel shaft that supports a large roller. The configuration of the system causes the shaft to support only the constant radial force, P. This causes the shaft to undergo fully reversed bending. Determine the fatigue life of this shaft at the given load, P, of 2000 lb. Determine the allowable load for an infinite life. Suggest ways that the fatigue behavior of the shaft could be improved. The steel used to make the shaft has an ultimate strength of 100 ksi.

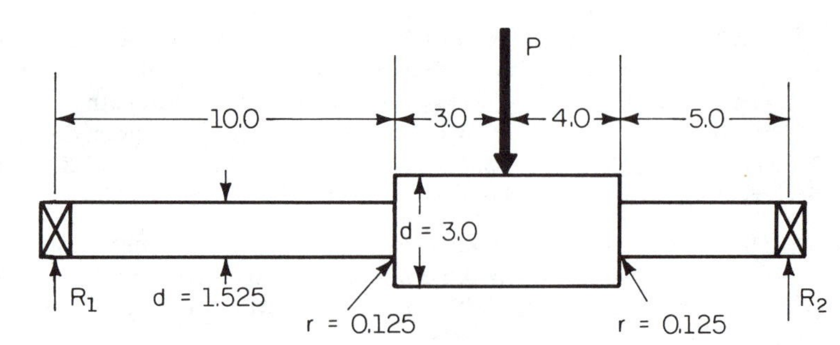

4.14. The effect of mean stress on the fatigue behavior of a notched component can be estimated by constructing a set of constant life lines on a Haigh diagram (Fig. 1.7). For example, the infinite life envelope for steels can be estimated as a curve joining the corrected endurance limit (S_e/K_f) on the alternating stress, σ_a, axis to either the ultimate strength, S_u, or true fracture strength, σ_f, on the mean stress, σ_m, axis. (Typically for ductile materials the stress concentration factor is not applied to the monotonic properties.) Any of the three relationships discussed in Chapter 1 (Goodman, Gerber, or Morrow) can be used to estimate the shape of this curve. Note that in this procedure the fatigue notch factor, K_f, is used as a strength reduction factor.

Construct the infinite-life envelope for a notched specimen with a stress concentration factor, K_t, of 3.3, a notch root radius, r, of 0.01 in. and the following material peroperties:

$$S_u = 158 \text{ ksi} \qquad S_e^{(\text{unnotched})} = 76 \text{ ksi} \qquad \sigma_f = 200 \text{ ksi}$$

Construct on the same diagram the envelopes corresponding to 10^3, 10^4, and 10^5 cycles. For this problem use the Goodman, Gerber, and Morrow relationships. Compare these predicted curves to the master diagram shown in Fig. 1.8.

4.15. A wide sheet of steel with a one-inch-diameter center hole is to be subjected to a zero-to-maximum fatigue loading in tension ($R = 0$). Estimate the cyclic nominal tensile stress that will cause failure at 10^6 cycles and at 10^3 cycles. The material properties are

$$S_u = 135 \text{ ksi} \qquad \sigma_f = 193 \text{ ksi} \qquad S_e^{(\text{unnotched})} = 59 \text{ ksi}$$

Use the Morrow, Goodman, and Gerber relationships for this problem. (Compare these results to the values determined in Problem 4.23.)

4.16. A notched component consists of a bar 0.625 in. wide and 0.25 in. thick with a 0.125 in. diameter center hole. The component is to undergo a zero-to-maximum loading in tension ($R = 0$) which varies between 0 and 7500 lb. Determine the fatigue life of this component. The ultimate strength of the steel used for the component is 82 ksi. Use the method shown in Fig. 4.11 and the Goodman relationship to solve this problem. (Compare the resulting life value to the value determined in Problem 4.24.)

4.17. An axial member is used to suspend a piece of machinery. The machine weighs 5000 lb and produces an alternating force of ±700 lb due to a reciprocating mass.

The axial member is 1 in. wide and has a thickness of 0.375 in. By mistake a 0.25 in. diameter hole is drilled through the center of the bar. Determine if this member is safe for an infinite fatigue life. The material used for the member is a steel with an ultimate strength, S_u, of 56 ksi.

SECTION 4.3

4.18. A notched plate with a net stress concentration factor, K_t, of 3 has the following material properties:

$$E = 200 \text{ GPa} \qquad K' = 1400 \text{ MPa} \qquad n' = 0.14$$
$$S_y = 600 \text{ MPa} \qquad S_u = 930 \text{ MPa}$$

Determine the net stress, S, necessary to:

(a) Reach yield, S_y, at the notch root

(b) Reach a strain at the notch root, ϵ, of 1% (or 0.01)

If a net section stress of 610 MPa is applied to the plate determine, using Eq. (4.13), the resulting stress, σ, and strain, ϵ, at the notch root. How much error is caused by assuming net section elastic behavior and using Eq. (4.14) to determine the notch root stress and strain?

4.19. Referring to Fig. 4.15, determine the zero-to-maximum ($R = 0$) nominal stress, S, which would cause the local stress, σ, to be fully reversed for a notched member with a theoretical stress concentration factor, K_t, of 3. The material stress–strain properties are

$$E = 30 \times 10^3 \text{ ksi} \qquad K' = 154 \text{ ksi} \qquad n' = 0.073$$

4.20. Shown in the figure below are the strain gage measurements of the notch root strain, ϵ, for the notched bar under various bending moments. How well does the Neuber relationship predict the notch root behavior for this case? The material properties are

$$E = 205 \text{ GPa} \qquad S_y = 380 \text{ MPa}$$
$$K' = 1480 \text{ MPa} \qquad n' = 0.256$$

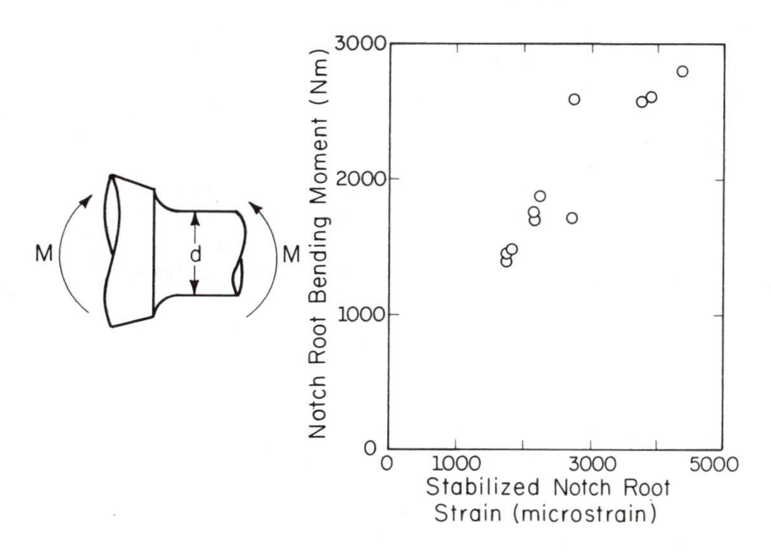

The stress concentration factor, K_t, for the geometry is 1.6. The nominal stress is found using

$$S_{nom} = \frac{32M}{\pi d^3}$$

where M is the bending moment and $d = 40$ mm. Discuss how the error of the Neuber analysis would affect life calculations. (Data taken from Ref. 21.)

4.21. A notched component has a theoretical stress concentration factor, K_t, of 3. The component is loaded to cause a nominal stress, S, of 200 MPa. Determine the resulting notch root stress, σ, and strain, ϵ. The component is then unloaded to a nominal stress of zero. Determine the residual stress at the notch root. What is the fatigue life of the component if it is subjected to a cyclic nominal stress of 0 to 200 MPa ($R = 0$)? Use a Neuber analysis and the Morrow mean stress–strain life relationship [Eq. (2.49)] for this problem. The strain–life properties for this material are

$$E = 100 \text{ GPa}$$

$$\sigma_f' = 1000 \text{ MPa} \qquad b = -0.08$$

$$\epsilon_f' = 1.0 \qquad c = -0.60$$

4.22. A wide sheet of steel with a one inch diameter center hole is to be subjected to a fully reversed fatigue loading in tension ($R = -1$). Estimate the nominal tensile stress levels that will cause failure at 2×10^6 and at 2×10^3 reversals. Use a Neuber analysis for this problem. The material properties are

$$E = 30 \times 10^3 \text{ ksi} \qquad K' = 208 \text{ ksi} \qquad n' = 0.14$$

$$\sigma_f' = 180 \text{ ksi} \qquad b = -0.07$$

$$\epsilon_f' = 0.66 \qquad c = -0.69$$

4.23. Repeat Problem 4.22, but determine the allowable zero-to-maximum ($R = 0$) nominal stress levels that will cause failure at 2×10^6 and at 2×10^3 reversals. Use a Neuber analysis and the Morrow mean stress–strain life relationship [Eq. (2.49)] for this problem. (Compare these values to the life values determined in Problem 4.15.)

4.24. A notched component consists of a bar 0.625 in. wide and 0.25 in. thick with a 0.125 in. diameter center hole. The component is to undergo a zero-to-maximum loading in tension ($R = 0$) which varies between 0 and 7500 lb. Determine the fatigue life of this component. Use a Neuber analysis and the Morrow mean stress–strain life relationship [Eq. (2.49)] for this problem. The material properties are

$$E = 30 \times 10^3 \text{ ksi} \qquad K' = 175 \text{ ksi} \qquad n' = 0.202$$

$$\sigma_f' = 133 \text{ ksi} \qquad b = -0.095$$

$$\epsilon_f' = 0.26 \qquad c = -0.47$$

(Compare the resulting life value to the value determined in Problem 4.16.)

4.25. In many instances the fatigue behavior of a notched component is plotted as applied load, P, versus life to failure, $2N_f$. Develop this type of curve for an axial member with a gross section area of $6.45 \times 10^{-4}\,m^2$ which contains a notch with a gross section stress concentration factor, K_t, of 3. The axial member is made from steel and will undergo fully reversed loading. The plot should go from 1 to 10^6 reversals. The material properties for the member are

$$E = 200\text{ GPa} \qquad K' = 1144\text{ MPa} \qquad n' = 0.121$$

$$\sigma_f' = 1165\text{ MPa} \qquad b = -0.081$$

$$\epsilon_f' = 1.142 \qquad c = -0.670$$

Use a Neuber analysis for this problem.

Develop a load–life (P–N) curve using the stress–life assumptions (Section 4.2). Compare this curve to the one developed using the strain-life approach. The notch root radius, r, of the notch is 6.35 mm and the ultimate strength of the steel, S_u, is 786 MPa.

4.26. As discussed in Section 4.3, a Neuber analysis may be performed using either net section or gross section properties. In some cases (primarily in cases with gross section yield) there will be a difference between life predictions determined using the two different section properties. Determine the fatigue life, N_f, in cycles for a notched component undergoing the fully reversed loading levels listed below. Use a Neuber analysis [Eq. (4.13)] and compare the predictions made using net and gross properties. The geometric and material properties are

$$K_t \text{ (net section)} = 2.16$$

$$K_t \text{ (gross section)} = 4.33$$

$$E = 30 \times 10^3\text{ ksi} \qquad K' = 154\text{ ksi} \qquad n' = 0.123$$

$$\sigma_f' = 169\text{ ksi} \qquad b = -0.081$$

$$\epsilon_f' = 1.142 \qquad c = -0.67$$

Case	Net Section Stress Amplitude (ksi)	Gross Section Stress Amplitude (ksi)
A	60.2	30.1
B	37.8	18.9

The actual fatigue life for case A is 4.2×10^3 cycles and case B is 8.6×10^4 cycles. Compare the predicted and actual life values. (Data taken from Ref. 17.)

4.27. As discussed in Section 4.3, the use of the theoretical stress concentration factor, K_t, in a Neuber analysis tends to predict overly conservative fatigue lives. This is especially true for sharp notches. A recommended modification is to use the fatigue notch factor, K_f, instead of K_t. Determine the fatigue life, N_f, in cycles, for a plate with a sharp edge notch undergoing the fully reversed loading levels listed below. Use a Neuber analysis and compare the predictions made using K_t and K_f. The ultimate strength, S_u, of the steel used for the component is 114 ksi. Use the stress–strain and strain–life properties given in Problem 4.26.

K_t (net section) = 10.7

Notch root radius (r) = 0.0025 in.

Case	Net Section Stress Amplitude (ksi)
A	57
B	15

The actual fatigue life for case A is 4.4×10^3 cycles and case B is 7.7×10^5 cycles. Compare the predicted and actual life values. (Data taken from Ref. 17.)

4.28. An axially loaded steel component containing a notch with a theoretical stress concentration factor, K_t, of 2.5 is found to fail in service after 10^5 reversals. Determine the fully reversed nominal stress, S, that would cause this failure.

A method that is suggested to extend the life of the component is to apply an initial overload to create a beneficial residual stress at the notch root. Should this overload be in tension or compression? It is decided that a notch root stress of 820 MPa needs to be reached at the maximum overload in order to achieve the desired residual stress upon unloading. Determine the maximum nominal stress required during the overload and the residual notch root stress upon unloading. Estimate the increase in life of the component due to the initial overload. For this problem assume that there is no stress relaxation.

Use a Neuber analysis and the Morrow mean stress–strain life relationship [Eq. (2.49)] for this problem. The material properties are

$E = 200$ GPa $K' = 1434$ MPa $n' = 0.14$

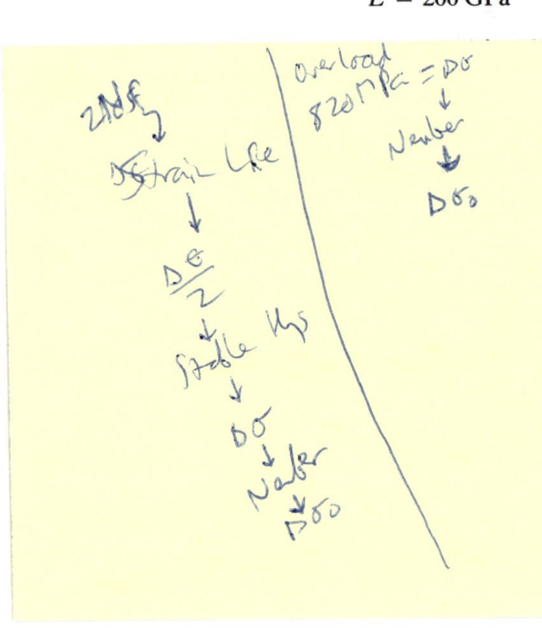

0 MPa $b = -0.07$

6 $c = -0.69$

25 in. wide, 0.10 in. thick, and has a center hole
se the service life, an initial tensile overload of
t before it is released to the customer. Due to an
nts were given a 2360 lb compressive overload.
ife? One way to present the effect of the different
ully reversed axial load amplitude, $\Delta P/2$, versus
nt with the two different overloads. Present these
es ranging 10^3 to 10^8 reversals. The stress–strain
uminum are

$K' = 166$ ksi $n' = 0.08$

47 ksi $b = -0.11$

.21 $c = -0.52$

For this problem use a Neuber analysis and the Morrow mean stress–strain life relationship [Eq. (2.49)].

The following table gives actual test results for the component described above. Compare predicted and actual life values.

Fully Reversed Axial Load Amplitude, $\Delta P/2$ (lb)	Life to Failure, $2N_f$ (reversals)
With tensile 2360-lb overload	
1575	7.9×10^3
1050	1.7×10^5
788	4.2×10^6
656	1.7×10^7
With compressive 2360-lb overload	
1575	8.0×10^3
1050	6.8×10^4
788	2.0×10^5
656	5.5×10^5
525	9.4×10^5

Discuss the effect of preloads on fatigue lives at high and low load amplitudes. What difficulties would there be in analyzing this problem with stress–life concepts? (Data taken from Ref. 22.)

4.30. A notched aluminum component consists of a plate 12.0 in. wide and 1.0 in. thick with a 2.0 in. diameter center hole. Determine the life of the component under the axial load histories shown below. History A is a constant amplitude zero-to-maximum ($R = 0$) axial load. Histories B and C have initial overloads followed by a zero-to-maximum axial load. For this problem use a Neuber analysis and predict

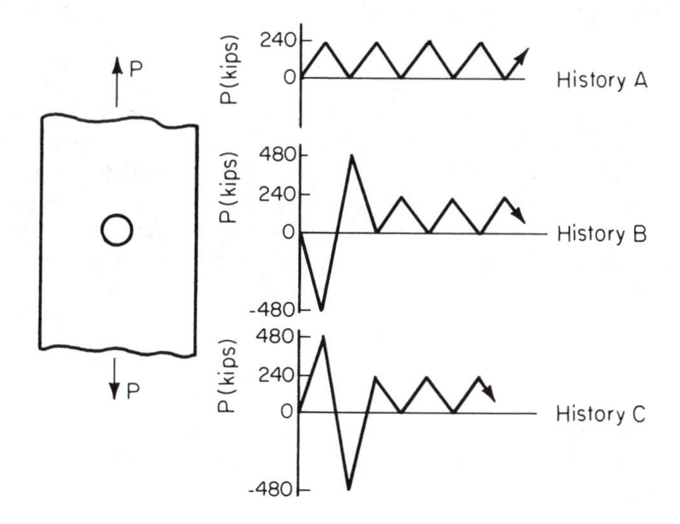

lives using the three mean stress relationships [Eqs. (2.49), (2.50), and (2.52)]. Compare the predictions made using these three relationships. The stress–strain and strain–life properties are

$$E = 10.6 \times 10^3 \text{ ksi} \qquad K' = 95 \text{ ksi} \qquad n' = 0.065$$

$$\sigma_f' = 160 \text{ ksi} \qquad b = -0.124$$

$$\epsilon_f' = 0.22 \qquad c = -0.59$$

Actual test results are given below for the three load histories. Four tests were run for each of the histories. Compare these values to the lives predicted in the first part of this problem. Note that the test results are given for crack initiation life, N_I. This is done since the strain–life approach is considered to give an estimate of crack initiation life. (Data taken from Ref. 23.)

History	Crack Initiation Lives, N_I (cycles)			
A	9.07×10^4	1.13×10^5	1.23×10^5	1.41×10^5
B	2.33×10^5	2.69×10^5	6.96×10^5	1.01×10^6
C	6.18×10^4	6.20×10^4	6.28×10^4	6.54×10^4

SECTION 4.4

4.31. A wide sheet of steel contains a semi-circular edge notch ($K_t = 3.0$) with a radius of 0.5 in. Determine the transition crack length, l_t, for this geometry using both the Dowling [Eq. (4.27)] and Smith–Miller [Eq. (4.28)] approximations. Estimate the crack propagation life, N_p, of the plate at a zero-to-maximum ($R = 0$) nominal stress loading of 0 to 40 ksi. Use Eq. (4.30) for this calculation. Compare the predicted lives found using the two estimates for transition crack length. How do these values compare to a life estimate using the approximation that the initial crack length is equal to the depth of the notch? For all calculations use a value of 1.58 in. for the final crack length, a_f. The crack growth properties of the steel are

$$C = 2.96 \times 10^{-9} \qquad m = 2.385$$

4.32. A notched steel component consists of a bar 1.0 in. wide and 0.25 in. thick with two semi-circular edge notches with radii of 0.1 in. This gives the plate a width at the reduced section of 0.8 in. After a period of cyclic loading, cracks with a length of 0.02 in. are found at the root of each notch. Are these cracks longer than the transition crack length? Determine the remaining life of the component at a zero-to-maximum ($R = 0$) loading of 0 to 6 kips. The necessary material properties are

$$S_y = 94 \text{ ksi} \qquad K_c = 100 \text{ ksi}\sqrt{\text{in.}}$$

$$C = 6.157 \times 10^{-10} \qquad m = 2.715$$

Test results for this situation indicate that the remaining life of the plate is 1.25×10^5 cycles. Compare the estimated life to the actual value. (Data taken from Ref. 17.)

4.33. In Sections 4.4.2 and 4.4.3 two methods are given for predicting the stress intensity for a short crack at a notch [Eqs. (4.24) and (4.29)] and for a long crack at a notch [Eqs. (4.25) and (4.30)]. An alternative method is to use the equation

$$K = f(g)\sigma\sqrt{\pi a}$$

where σ = remote stress
a = crack length
$f(g)$ = geometry correction factor

with the geometry correction factor, $f(g)$, based on the geometry of the cracked body. Shown below is a solution for $f(g)$ for the case of equal-length cracks emanating from a circular hole in an infinite plate ($K_t = 3.0$) under uniaxial loading [24]. The geometry correction, $f(g)$, is plotted as a function of the ratio of crack length, a, to notch radius, r. Compare this solution to the estimates made using Eqs. (4.24), (4.29), (4.25), and (4.30).

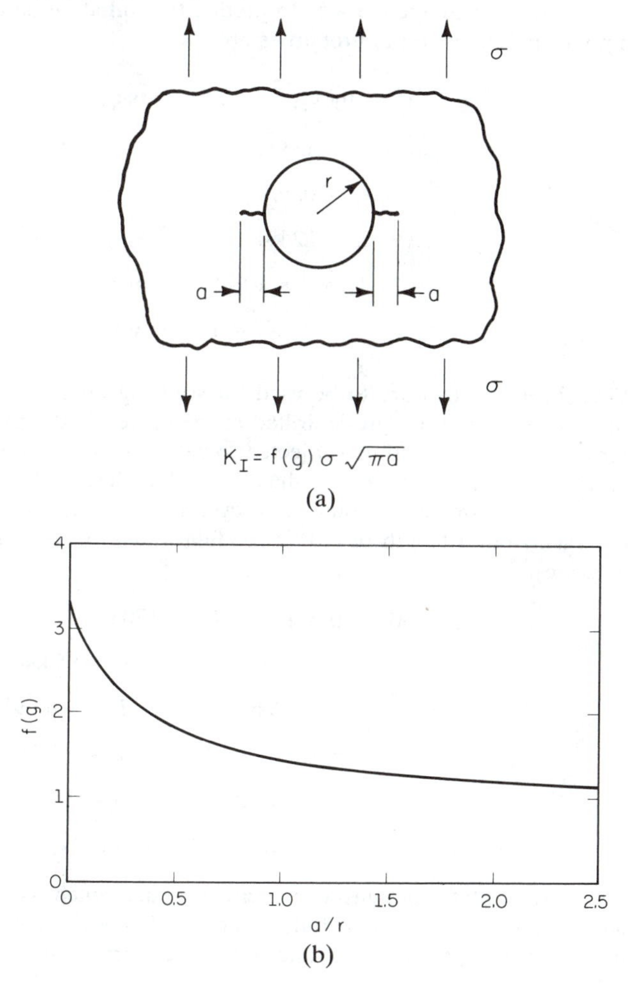

4.34. For the case of an edge notch in a wide plate, the stress concentration factor is

$$K_t = 1 + 2\sqrt{t/r}$$

where t is the depth of the notch and r is the notch root radius.

Determine the transition crack length, l_t, for this geometry as the notch root radius, r, varies from 0.001 to 0.250 and the notch root depth, t, remains constant at 0.25. Use both the Dowling [Eq. (4.27)] and Smith–Miller [Eq. (4.28)] methods and then compare the results.

SECTION 4.5

4.35. A steel component consists of a bar 0.625 in. wide and 0.25 in. thick with a 0.125 in. diameter center hole. The component is to undergo a zero-to-maximum loading in tension ($R = 0$) which varies between 0 and 7500 lb. Use the method proposed by Socie (discussed in Section 4.5) to predict the initiation and propagation life of this component. The material properties are

$$E = 30 \times 10^3 \text{ ksi} \qquad K' = 175 \text{ ksi} \qquad n' = 0.202$$
$$\sigma'_f = 133 \text{ ksi} \qquad b = -0.095$$
$$\epsilon'_f = 0.26 \qquad c = -0.47$$
$$S_u = 82 \text{ ksi} \qquad S_y = 47 \text{ ksi}$$
$$C = 9 \times 10^{-11} \qquad m = 3.4$$
$$K_c = 110 \text{ ksi} \sqrt{\text{in.}}$$

4.36. Wide sheets of steel are to be used for studying fatigue crack propagation. A hole with a diameter of 0.04 in. is drilled in the center of the plates to serve as a "crack starter." The plate is to be subjected to zero-to-maximum loading ($R = 0$) so that a crack will be present in less than 5×10^5 cycles. Determine the nominal stress required. Determine the number of cycles needed at this cyclic load level to grow the crack from a length of 0.01 in. to final fracture. The material properties of the plate are

$$E = 30 \times 10^3 \text{ ksi} \qquad K' = 170 \text{ ksi} \qquad n' = 0.120$$
$$S_u = 115 \text{ ksi} \qquad S_y = 95 \text{ ksi}$$
$$\sigma'_f = 165 \text{ ksi} \qquad b = -0.095$$
$$\epsilon'_f = 1.15 \qquad c = -0.70$$
$$C = 3.6 \times 10^{-10} \qquad m = 3.0$$
$$K_c = 100 \text{ ksi} \sqrt{\text{in.}}$$

4.37. A wide steel plate contains a semi-circular edge notch with a radius of 0.10 in. The plate is to be subjected to a fully reversed ($R = -1$) axial nominal stress. Generate the initiation, propagation, and total life curves on a plot of nominal stress amplitude, $\Delta S/2$, versus life in cycles. Use log–log coordinates. The life axis should go from 1 to 1×10^6 cycles. Compare the relative importance of initiation and propagation life at various stress levels. For these calculations use the method proposed by Socie that was discussed in Section 4.5. The material properties are

$$E = 30 \times 10^3 \text{ ksi} \qquad K' = 170 \text{ ksi} \qquad n' = 0.120$$

$$S_u = 115 \text{ ksi} \qquad S_y = 95 \text{ ksi}$$

$$\sigma_f' = 165 \text{ ksi} \qquad b = -0.095$$

$$\epsilon_f' = 1.15 \qquad c = -0.70$$

$$C = 3.6 \times 10^{-10} \qquad m = 3.0$$

$$K_c = 100 \text{ ksi} \sqrt{\text{in.}}$$

4.38. The next several tables contain the experimental fatigue life data for notched specimens tested at various load levels. Ten geometries and four material types are represented. These experimental data can be compared against predicted life values found using the various analytical techniques discussed in this chapter.

The following information is provided for the fatigue analysis of the notched specimens.

1. Material properties (monotonic, strain–life, crack growth)
2. Specimen geometries
3. Experimental fatigue data (applied load and life to failure)

As discussed in the previous chapters, fatigue analyses can be a multidimensional problem where several techniques, approximations, and so on, can be used to arrive at a life estimate. The data contained in this problem are provided for the user to try out the various analytical techniques, determine how well each method works, and compare the relative accuracy of each approach. There are numerous ways that these data can be utilized. The following are just a few ideas to serve as a starting point:

1. Seven of the specimen types are made from various steels. These data can be used to provide a comparison between the different approaches when they are used for low, medium, and high strength steel components.
2. Five of the data sets contain initiation as well as total life values. Propagation life estimates can be compared to these data.
3. The specimens contain notches ranging from very sharp to blunt. These data can be used to determine the applicability of the Neuber analysis method for different geometries.

The user is encouraged to use these data for other comparisons. (The data for this problem were taken from Refs. 17, 25, and 26.)

Material Properties

Material A: Medium strength steel
Material B: High strength steel
Material C: Low strength steel
Material D: High strength aluminum

Property	Units	Material A	Material B	Material C	Material D
Monotonic properties					
Elastic modulus, E	ksi	30×10^3	30×10^3	30×10^3	10×10^3
Yield strength, S_y	ksi	94	158	51	78
Tensile strength, S_u	ksi	114	168	78	85
Reduction in area, % RA		68	52	67	14
True fracture strength, σ_f	ksi	225	228	170	95
True fracture ductility, ϵ_f		1.139	0.734	1.10	0.145
Cyclic properties					
Fatigue ductility coefficient, ϵ_f'		1.142	0.790	0.338	0.158
Fatigue ductility exponent, c		−0.67	−0.730	−0.480	−0.83
Fatigue strength coefficient, σ_f'	ksi	169	267	162	241
Fatigue strength exponent, b		−0.081	−0.090	−0.110	−0.15
Cyclic strength coefficient, K'	ksi	154	235	194	101
Cyclic strain hardening exponent, n'		0.123	0.112	0.226	0.040
Crack growth properties					
Crack growth coefficient, C		6.157×10^{-10}	2.96×10^{-9}	2.7×10^{-11}	1.18×10^{-8}
Crack growth exponent, m		2.715	2.385	4.12	2.94
Fracture toughness, K_c	ksi $\sqrt{\text{in.}}$	NA	NA	NA	40

NA, not available.

Specimen 1

Material A

Geometry: blunt edge notch

Notch root radius $(r) = 0.10$ in.

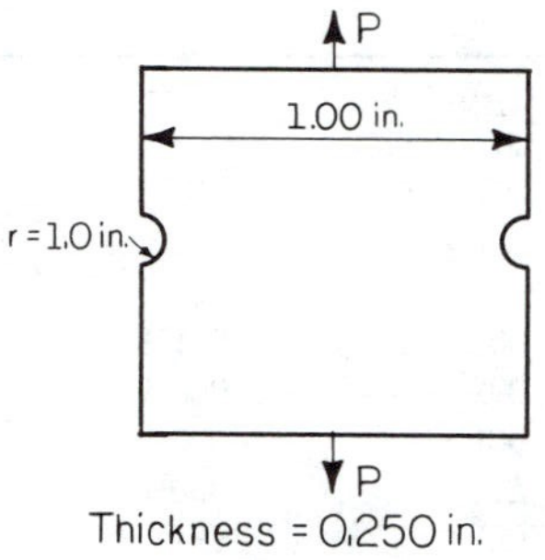

P

1.00 in.

r = 1.0 in.

P

Thickness = 0.250 in.

Spec 1

Test Results[a]

Specimen Number	Load Amplitude (kips)	Life to 0.02-in. Crack, N_I (cycles)	Life to Failure, N_f (*cycles*)
1-1	20	62	83
1-2	16	635	1,023
1-3	14	1,300	3,235
1-4	12	2,400	6,027
1-5	10	6,000	19,500
1-6	10	—	18,700
1-7	8	14,000	46,037
1-8	7	19,000	87,950
1-9	6	100,000	225,000
1-10	6	88,500	213,000
1-11	5.4	1.8×10^6	1.91×10^6
1-12	5.2	500,000	680,000
1-13	5.2	—	5.28×10^6
1-14	5	—	$>1.1 \times 10^7$ [b]

[a] All tests are fully reversed.
[b] Suspended test.

Specimen 2

Material A

Geometry: center hole

Notch root radius $(r) = 0.25$ in.

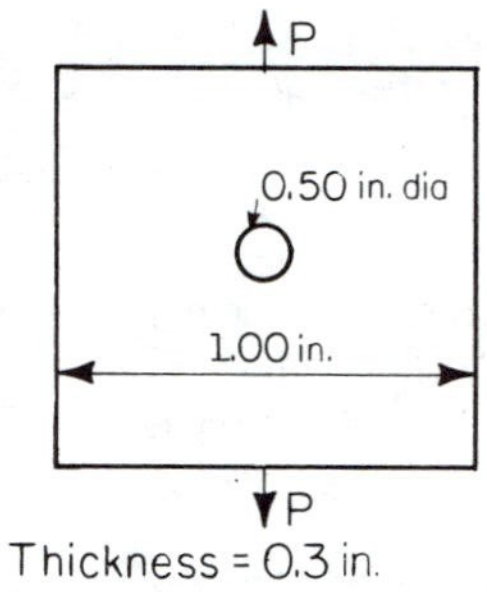

P

0.50 in. dia

1.00 in.

P

Thickness = 0.3 in.

Spec 2

Test Results[a]

Specimen Number	Load Amplitude (kips)	Life to 0.02-in. Crack, N_I (cycles)	Life to Failure, N_f (cycles)
2-1	13.99	68	113
2-2	12.65	190	266
2-3	12.11	265	378
2-4	10.65	1,250	1,936
2-5	9.02	3,600	3,997
2-6	9.03	2,400	4,338
2-7	7.00	11,500	18,715
2-8	5.68	55,400	86,590
2-9	5.39	—	71,981
2-10	4.95	160,780	222,830
2-11	4.70	188,000	254,580
2-12	4.56	—	$>9.7 \times 10^6$ [b]
2-13	4.30	—	$>10^7$ [b]
2-14	3.92	—	$>10^7$ [b]

[a] All tests are fully reversed.
[b] Suspended test.

Specimen 3

Material A

Geometry: sharp edge notch

Notch root radius (r) = 0.0025 in.

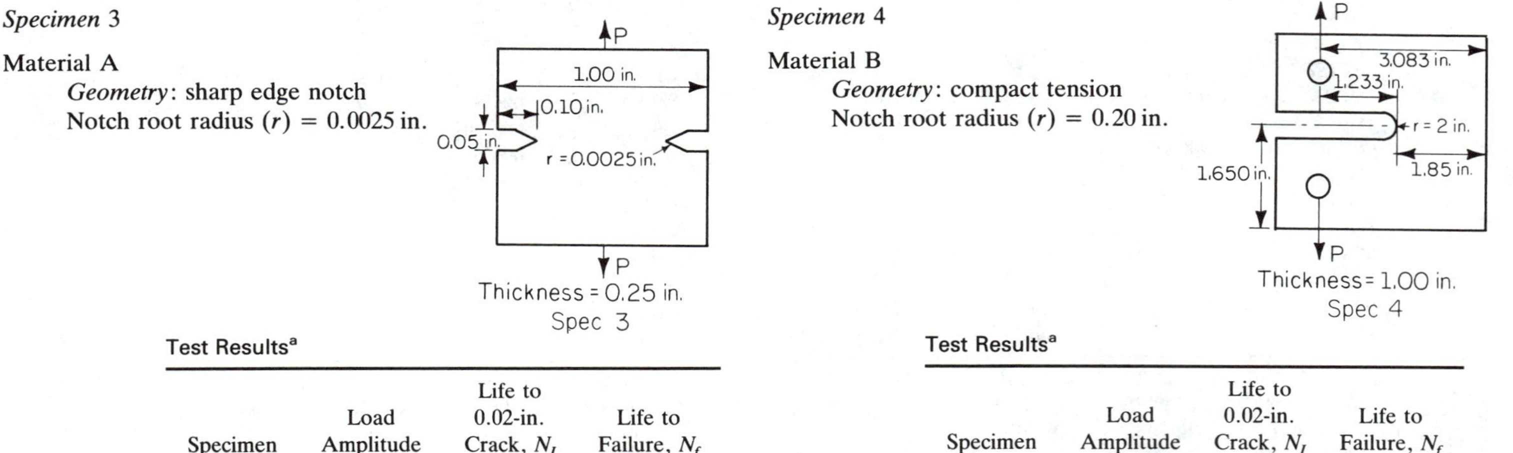

Test Results[a]

Specimen Number	Load Amplitude (kips)	Life to 0.02-in. Crack, N_I (cycles)	Life to Failure, N_f (cycles)
3-1	18	4	125
3-2	14	47	1,881
3-3	11.4	133	4,376
3-4	9	1,000	16,377
3-5	6.6	2,000	43,400
3-6	5	1,000	123,560
3-7	3	11,150	768,700
3-8	2	62,180	1.28×10^6
3-9	1.8	—	1.83×10^6
3-10	1.8	77,000	1.19×10^6
3-11	1.6	—	3.23×10^6
3-12	1.4	—	$>10^7$[b]

[a] All tests are fully reversed.
[b] Suspended test.

Specimen 4

Material B

Geometry: compact tension

Notch root radius (r) = 0.20 in.

P
3.083 in.
1.233 in.
r = 2 in.
1.650 in.
1.85 in.
P
Thickness = 1.00 in.
Spec 4

Test Results[a]

Specimen Number	Load Amplitude (kips)	Life to 0.02-in. Crack, N_I (cycles)	Life to Failure, N_f (cycles)
4-1	40	17	148
4-2	27.5	150	732
4-3	18	742	6,097
4-4	12.5	3,600	24,295
4-5	9.5	32,500	86,446
4-6	8	600,000	691,400
4-7	8	97,000	157,900
4-8	7.5	120,000	240,300
4-9	7	190,000	338,300

[a] All tests are fully reversed.

Specimen 5

Material B
Geometry: center hole
Notch root radius (r) = 0.1025 in.

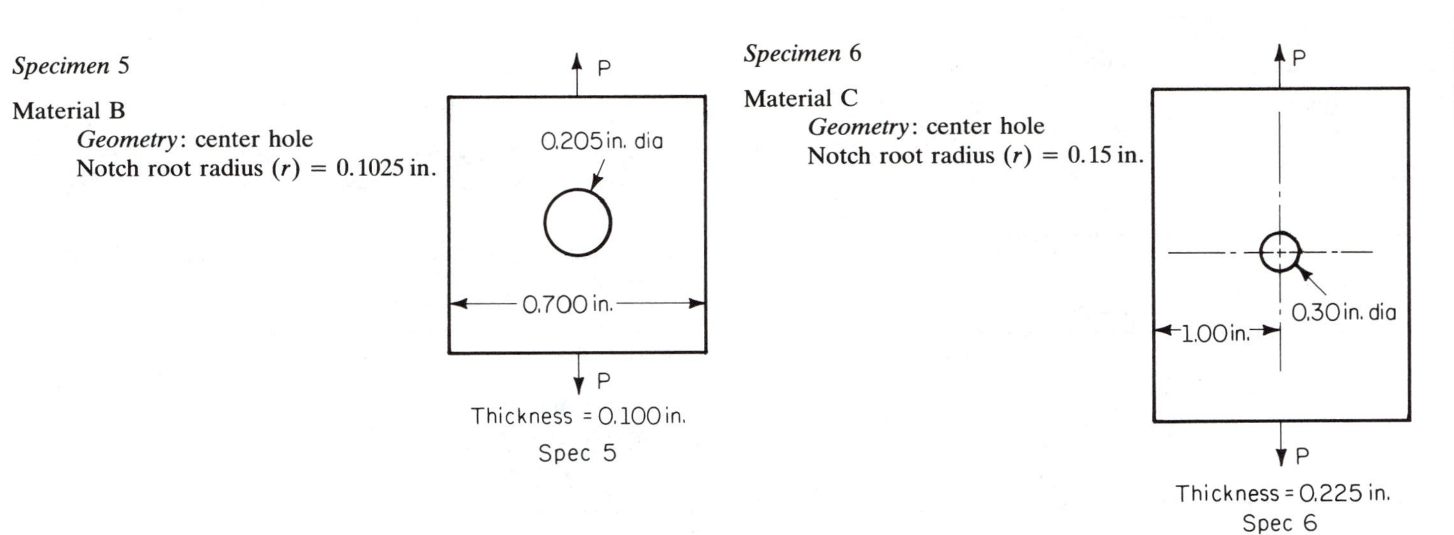

Test Results[a]

Specimen Number	Load Amplitude (kips)	Life to Failure, N_f (cycles)
5-1	4.2	15,000
5-2	3.5	35,000
5-3	2.8	92,000
5-4	2.3	250,000

[a] All tests are fully reversed.

Specimen 6

Material C
Geometry: center hole
Notch root radius (r) = 0.15 in.

Test Results[a]

Specimen Number	Load Amplitude (kips)	Life to Failure, N_f (cycles)
6-1	17.64	3,603
6-2	17.64	2,800
6-3	14.70	13,340
6-4	11.76	51,960
6-5	10.59	129,000
6-6	9.41	230,460
6-7	8.82	424,750
6-8	8.24	589,510
6-9	7.65	1,180,700

[a] All tests are fully reversed.

Specimen 7

Material C

Geometry: center slot

Notch root radius (r) = 0.075 in.

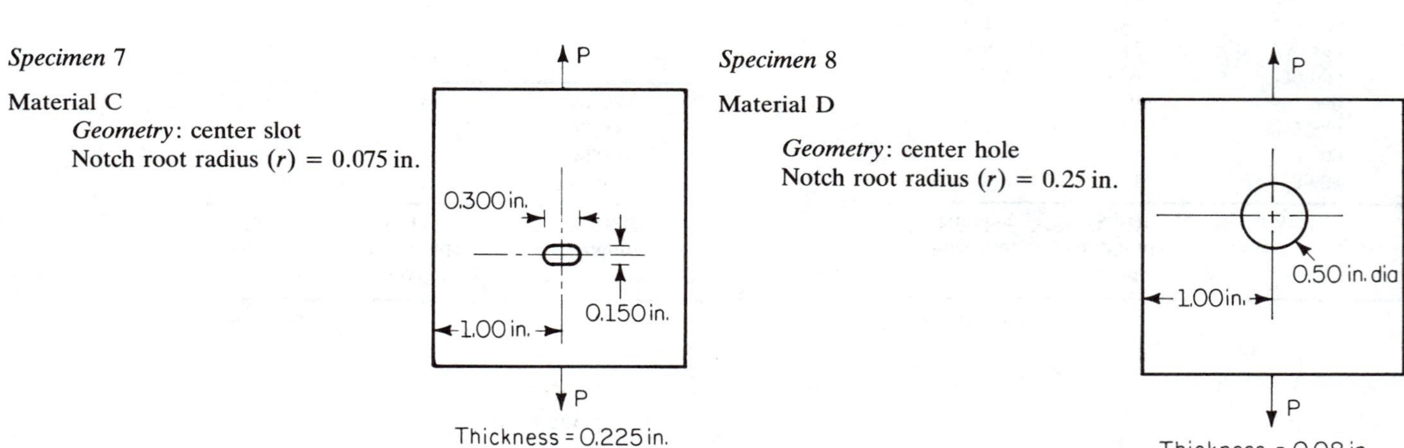

Test Results[a]

Specimen Number	Load Amplitude (kips)	Life to 0.025–in. Crack, N_I (cycles)	Life to Failure, N_f (cycles)
7-1	20.25	924	1,060
7-2	18.00	2,400	3,830
7-3	15.75	5,400	9,900
7-4	12.80	14,000	32,500
7-5	11.25	—	65,400
7-6	11.25	25,000	64,000
7-7	9.00	87,315	215,000
7-8	9.00	—	193,000
7-9	7.43	170,000	443,000
7-10	7.04	—	663,000

[a] All tests are fully reversed.

Specimen 8

Material D

Geometry: center hole

Notch root radius (r) = 0.25 in.

Test Results[a]

Specimen Number	Maximum Load (kips)	Life to Failure, N_f (cycles)
8-1	4.80	4,850
8-2	4.00	8,850
8-3	3.20	22,600
8-4	2.40	238,000
8-5	1.60	2,650,000[b]

[a] All tests performed at $R = 0.1$.

[b] Suspended test.

Specimen 9

Material D

Geometry: center slot

Notch root radius (r) = 0.125 in.

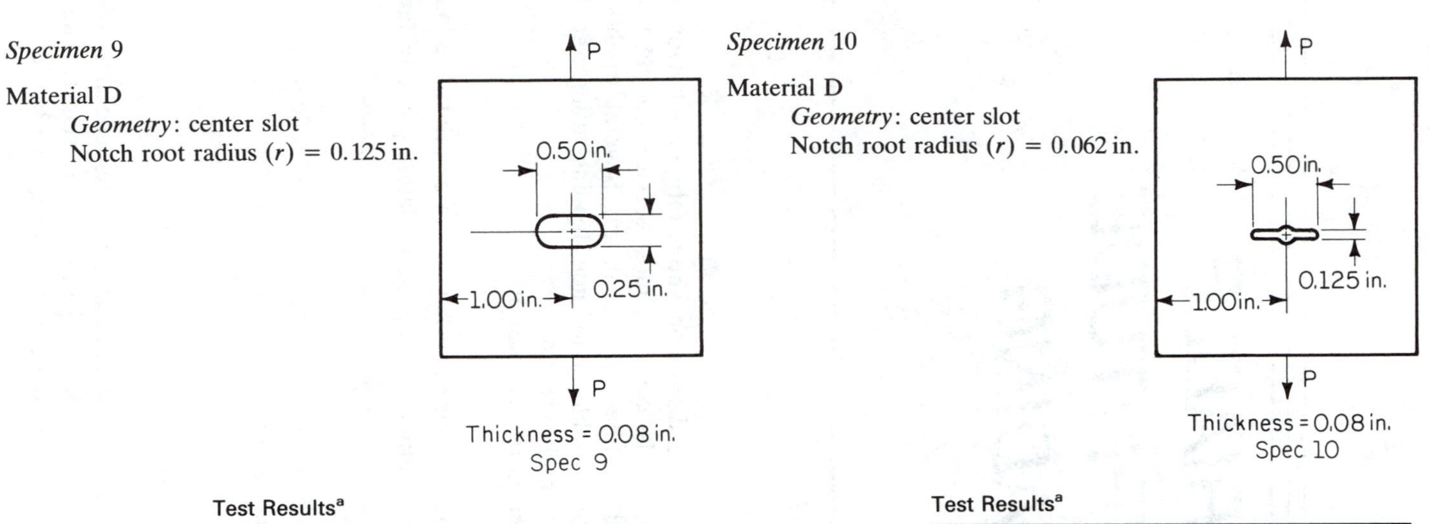

Test Results[a]

Specimen Number	Maximum Load (kips)	Life to Failure, N_f (cycles)
9-1	4.80	2,190
9-2	4.00	5,900
9-3	3.20	10,800
9-4	2.40	27,500
9-5	1.60	2,100,000

[a] All tests performed at $R = 0.1$.

Specimen 10

Material D

Geometry: center slot

Notch root radius (r) = 0.062 in.

Test Results[a]

Specimen Number	Maximum Load (kips)	Life to Failure, N_f (cycles)
10-1	4.80	945
10-2	4.00	1,940
10-3	3.20	4,380
10-4	2.40	10,000
10-5	1.60	70,500

[a] All tests performed at $R = 0.1$.

5 VARIABLE AMPLITUDE LOADING

5.1 INTRODUCTION

Up to this point most of the discussion about fatigue behavior has dealt with constant amplitude loading. In contrast, most service loading histories have a variable amplitude and can be quite complex. Several methods have been developed to deal with variable amplitude loading using the baseline data generated from constant amplitude tests.

The topics discussed in this chapter are:

1. The nature of fatigue damage and how it can be related to load history
2. Damage summation methods during the initiation phase of fatigue
3. Methods of cycle counting which are used to recognize damaging events in a complex history
4. Crack propagation behavior under variable amplitude loading
5. Methods for dealing with service load histories

5.2 DEFINITION OF FATIGUE DAMAGE

There are distinctly different approaches used when dealing with cumulative fatigue damage during the initiation and propagation stages. The differences in these approaches are related to how fatigue damage can be defined during these two stages.

During the propagation portion of fatigue, damage can be directly related to crack length. Methods have been developed which can relate loading sequence to crack extension. These methods are discussed later in this chapter. The important point is that during propagation, damage can be related to an observable and measurable phenomenon. This has been used to great advantage in the aerospace industry, where regular inspection intervals are incorporated into the design of damage-tolerant structures.

The definition of fatigue damage during the initiation phase is much more difficult. During this phase the mechanisms of fatigue damage are on the microscopic level. Although damage during the initiation phase can be related to dislocations, slip bands, microcracks, and so on, these phenomena can only be measured in a highly controlled laboratory environment. Because of this, most damage summing methods for the initiation phase are empirical in nature. They relate damage to the "life used up" for a small laboratory specimen. Life is defined as the separation of this specimen. This definition of damage can be related to a component by using the "equally stressed volume of material" concept discussed in Chapter 2. That is, the separation of a small specimen is equal to the formation of a small crack in a large component or structure.

5.3 DAMAGE SUMMING METHODS FOR INITIATION

In the following discussion the various damage summing methods will be related to the stress–life method. It should be remembered that most of these approaches could also be used in conjunction with the strain–life method.

5.3.1 Linear Damage rule

The linear damage rule was first proposed by Palmgren in 1924 [1] and was further developed by Miner in 1945 [2]. Today this method is commonly known as Miner's rule. The following terminology will be used in the discussion:

$$\frac{n}{N} = \text{cycle ratio}$$

where n is the number of cycles at stress level S and N is the fatigue life in cycles at stress level S.

The damage fraction, D, is defined as the fraction of life used up by an event or a series of events. Failure in any of the cumulative damage theories is assumed to occur when the summation of damage fractions equals 1, or

$$\sum D_i \geq 1 \tag{5.1}$$

The linear damage rule states that the damage fraction, D_i, at stress level S_i is equal to the cycle ratio, n_i/N_i. For example, the damage fraction, D, due to

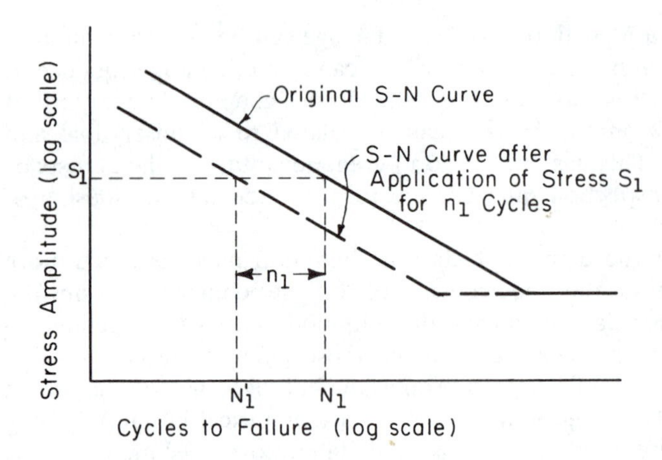

Figure 5.1 Effect of Miner's rule on S–N curve.

one cycle of loading is $1/N$. In other words, the application of one cycle of loading consumes $1/N$ of the fatigue life. The failure criterion for variable amplitude loading can now be stated as

$$\sum \frac{n_i}{N_i} \geq 1 \tag{5.2}$$

The life to failure can be estimated by summing the percentage of life used up at each stress level. The obvious asset of this method is its simplicity.

Miner's rule can also be interpreted graphically by showing its effect on the S–N curve (Fig. 5.1). If n_1 cycles are applied at stress level S_1, the S–N curve is shifted so that it goes through the new life value, N_1'. In this procedure, N_1' is $N_1 - n_1$, and N_1 is the original life to failure at stress level S_1. The S–N curve retains its original slope but is shifted to the left.

Considerable test data have been generated in an attempt to verify Miner's rule. In most cases these tests use a two-step history. This involves testing at an initial stress level S_1 for a certain number of cycles. The stress level is then changed to a second level, S_2, until failure occurs. If $S_1 > S_2$, it is called a high–low test, and if $S_1 < S_2$, a low–high test. The results of Miner's original tests [3] showed that the cycle ratio corresponding to failure ranged from 0.61 to 1.45. Other researchers have shown variations as large as 0.18 to 23.0. Most results tend to fall between 0.5 and 2.0. In most cases the average value is close to Miner's proposed value of 1. There is a general trend that for high–low tests the values are less than 1, and for low–high tests the values are greater than 1. In other words, Miner's rule is nonconservative for high–low tests. One problem with two-level step tests is that they do not relate to many service load histories. Most load histories do not follow any step arrangement and instead are made up of a random distribution of loads of various magnitudes. Tests using random histories with several stress levels show very good correlation with Miner's rule.

An alternative form of Miner's rule has been proposed [4]. This is

$$\sum \frac{n_i}{N_i} \geq X \tag{5.3}$$

where X is selected on a knowledge of the load history or on a desired factor of safety. A value less than 1 is usually used.

The linear damage rule has two main shortcomings when it comes to describing observed material behavior. First, it does not consider sequence effects. The theory predicts that the damage caused by a stress cycle is independent of where it occurs in the load history. An example of this discrepancy was discussed earlier regarding high–low and low–high tests. Second, the linear damage rule is amplitude independent. It predicts that the rate of damage accumulation is independent of stress level. This last trend does not correspond to observed behavior. At high strain amplitudes cracks will initiate in a few cycles, whereas at low strain amplitudes almost all of the life is spent initiating a crack.

5.3.2 Nonlinear Damage Theories

Many nonlinear damage theories have been proposed which attempt to overcome the shortcomings of Miner's rule. Collins describes several of these methods (Henry, Gatts, Corten–Dolan, Marin, Manson–double linear) in Ref. 5. There are some practical problems involved when trying to use these methods:

1. They require material and shaping constants which must be determined from a series of step tests. In some cases this requires a considerable amount of testing.
2. Since some of the methods take into account sequence effects, the number of calculations and the bookkeeping can become a problem in complicated histories.

Another point is that although the nonlinear methods may give better predictions than Miner's rule for two-step histories, it cannot be guaranteed that they will work better for actual service load histories.

The following is a general description of a nonlinear damage approach which is currently of research interest and has some applications in design. This approach was proposed by Richard and Newmark [6] and was developed further by Marco and Starkey [7]. This method predicts the following relationship between damage fraction, D, and cycle ratio, n/N:

$$D = \left(\frac{n}{N}\right)^P \tag{5.4}$$

where the exponent, P, is a function of stress level. The value of P is considered

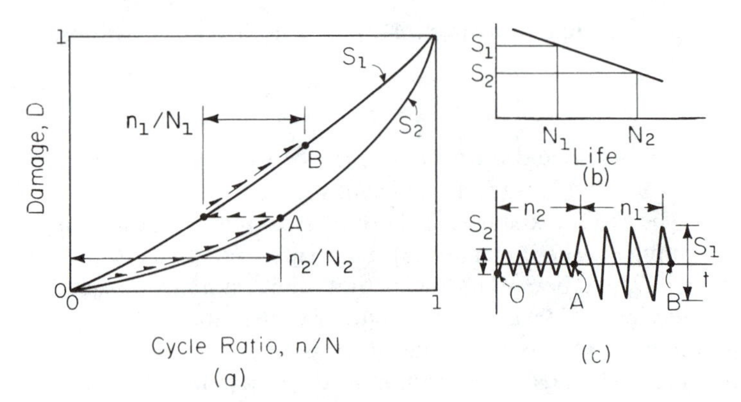

Figure 5.2 Demonstration of nonlinear damage theory.

to fall in the range zero to 1, with the value increasing with stress level. Note that when $P = 1$ the method is equivalent to Miner's rule.

The use of this method is shown in Fig. 5.2a, which is a plot of damage fraction versus cycle ratio for two stress levels, where $S_1 > S_2$. Figure 5.2c shows a low–high stress history, where S_2 is applied for n_2 cycles and then stress S_1 is applied for n_1 cycles. The values N_1 and N_2 are the lives to failure corresponding to S_1 and S_2 on the S–N curve (Fig. 5.2b). Corresponding points on the stress history (Fig. 5.2c) and damage plot (Fig. 5.2a) are labeled 0, A, and B. The damage associated with stress S_2 is found by following the proper damage curve for a horizontal distance n_2/N_2. When the stress level is changed, a transfer is made to the damage curve corresponding to the new stress level, S_1, by following a horizontal line of constant damage. This procedure is continued until failure is predicted ($D \geq 1$).

Note that if the stress blocks are reversed in Fig. 5.2c, giving a high–low test, there will be an increase in the total damage predicted on the damage plot (Fig. 5.2a). Therefore, this method includes both sequence and stress level effects.

The method shown in Fig. 5.2 has good correlation to observed material behavior. It can be also be used to sum damage in high temperature applications where there is interaction between creep and fatigue.

It has the drawback that the family of stress curves on the damage plot must be developed experimentally for a given material. There is also the problem of defining the physical measurement which is used to describe damage. In the case of fatigue, crack length or crack density has been used [8]. In high temperature applications, creep rate or load drop has been used.

5.3.3 Conclusions

For most situations where there is a pseudo-random load history, Miner's rule is adequate for predicting fatigue life. The other nonlinear methods do not give significantly more reliable life predictions. These theories also require material and shaping constants which may not be available.

Miner's rule may also be used in conjunction with the strain–life approach. The only difference is that the life to failure, N, in the cycle ratio, n/N, is taken from the strain–life curve.

Damage summation techniques must account for load sequence effects. One of these is the mean stress effect which is caused by residual stresses. It was shown in Chapter 1 that residual stresses can have a significant effect on the fatigue life of notched components. As mentioned earlier, the stress–life method does not account for residual stresses caused by overloads. The use of the strain–life approach for variable amplitude loading allows the effects of residual stresses due to overloads to be quantified, and subsequently included in fatigue life predictions. Although some nonlinear damage summation techniques have been developed to account for the mean stress effect, they are not the most appropriate approach.

As an example of mean stress effects, consider the situation shown in Fig. 5.3, where the notched plate (2024-T3 aluminum) is subjected to two similar load histories [9]. The only difference is that in history A, the last load at the high level is tensile, while in history B it is compressive. A Miner's analysis using the S–N approach would predict nearly identical lives at the lower stress level for the two histories. In fact, test results show that history A has a life of 460,000 cycles at the low stress level and history B has a life of 63,000 cycles at the same stress level. A test with no preload had a life of 115,000 cycles. The differences in life are due to the beneficial and detrimental residual stresses set up by the final overloads in histories A and B. An analysis using the strain–life method and a Neuber analysis would have determined the residual stress value. The effect of this residual mean stress could then be incorporated into the life prediction. This procedure takes into account sequence effects which the various nonlinear stress–life methods could account for only with empirical constants.

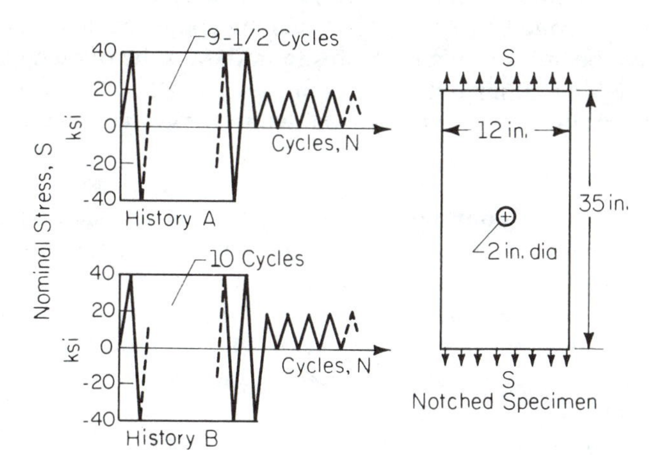

Figure 5.3 Effect of load sequence on fatigue life. (From Ref. 9.)

5.4 CYCLE COUNTING

To predict the life of a component subjected to a variable load history, it is necessary to reduce the complex history into a number of events which can be compared to the available constant amplitude test data. This process of reducing a complex load history into a number of constant amplitude events involves what is termed *cycle counting.*

In the following discussion, cycle counting techniques will be applied to strain histories. In general, these techniques can also be applied to other "loading" parameters, such as stress, torque, moment, load, and so on.

5.4.1 Early Cycle Counting Procedures

Early attempts to reduce complex histories yielded a number of cycle counting procedures, level-crossing counting, peak counting, and simple-range counting being among the most common.

Level-crossing counting. In this procedure the strain axis of the strain–time plot is divided into a number of increments. A count is then recorded each time a positively sloped portion of the strain history crosses an increment located above the reference strain. Similarly, each time a negatively sloped portion of the strain history crosses an increment located below the reference strain, a count is made. In addition, crossings at the reference strain by a positively sloped portion of the strain history are also counted. Figure 5.4a shows a sample strain history and the resulting level-crossing counts. In this particular example, zero strain has been used as the reference strain value.

Once the counts are determined, they must be combined to form completed cycles. A variety of methods are available for the combining of counts to obtain completed cycles. The most damaging combination of counts, from a fatigue point of view, is obtained by first forming the largest possible cycle. The next largest cycle possible is then formed by using the remaining counts available, and so on, until all counts have been used. Figure 5.4b shows the results of using this procedure on the level-crossing count obtained from the strain history given in Fig. 5.4a. In this example, strain reversals have been defined as occurring midway between increment levels.

Peak counting. The peak counting method is based on the identification of local maximum and minimum strain values. To begin, the strain axis is divided into a number of increments. The positions of all local maximum (peak) strain values above the reference strain are tabulated, as are the positions of all local minimum (valley) strain values below the reference strain. An example of this procedure is shown in Fig. 5.5a. In this example zero strain has once again been used as the reference strain value.

As with the level-crossing method, once counts have been obtained, they

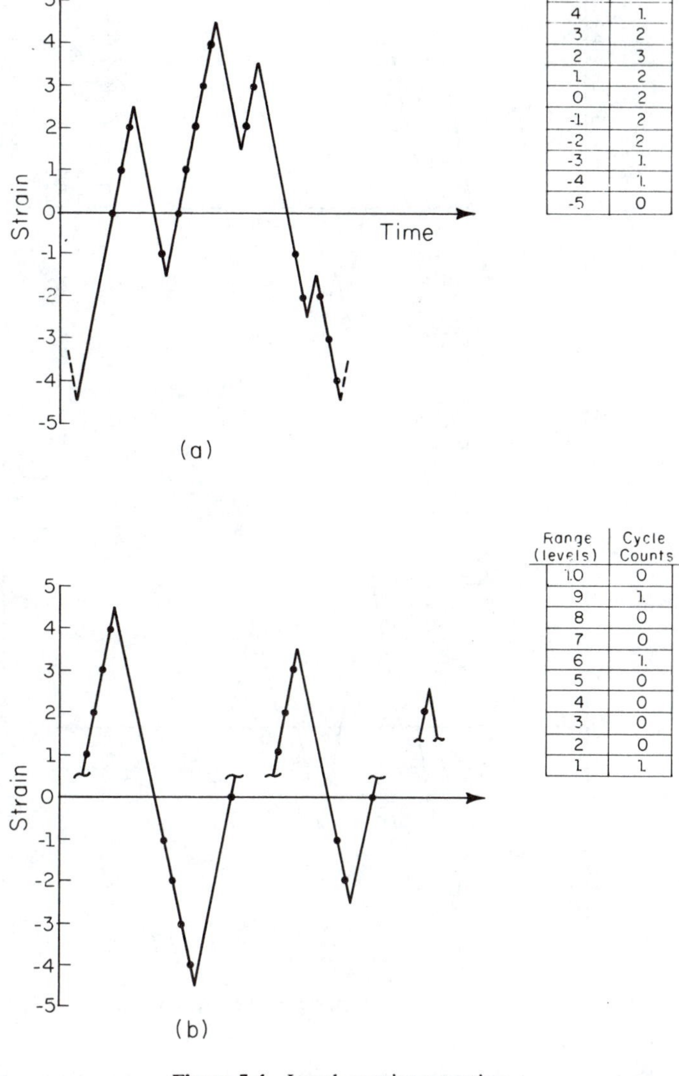

Level	Counts
5	0
4	1
3	2
2	3
1	2
0	2
-1	2
-2	2
-3	1
-4	1
-5	0

Range (levels)	Cycle Counts
10	0
9	1
8	0
7	0
6	1
5	0
4	0
3	0
2	0
1	1

Figure 5.4 Level-crossing counting.

must be combined to form complete cycles in order to perform a fatigue analysis. The most damaging history, in terms of fatigue, is obtained by first combining the largest peak with the largest valley. The second largest cycle is then formed by combining the largest peak and valley of the remaining counts. This process is continued until all counts have been used. Figure 5.5b shows the resulting completed cycles obtained from using this procedure for the peak count obtained for the strain history in Fig. 5.5a.

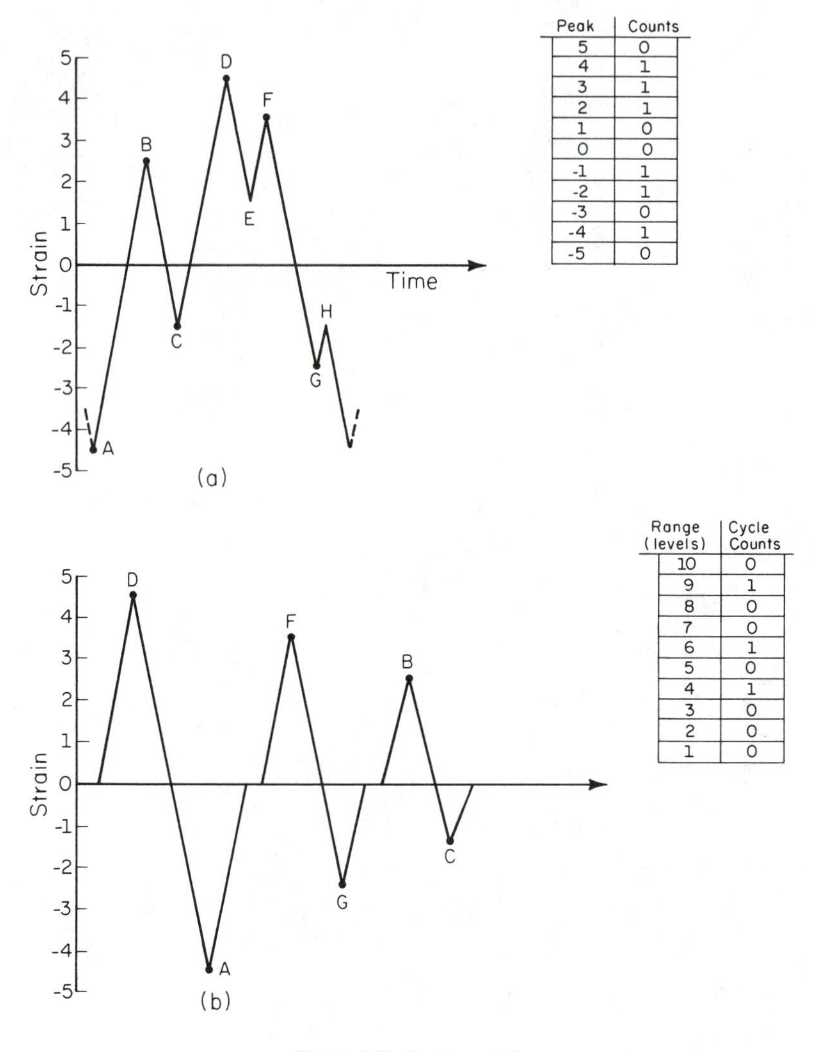

Peak	Counts
5	0
4	1
3	1
2	1
1	0
0	0
-1	1
-2	1
-3	0
-4	1
-5	0

Range (levels)	Cycle Counts
10	0
9	1
8	0
7	0
6	1
5	0
4	1
3	0
2	0
1	0

Figure 5.5 Peak counting.

Simple-range counting. With this method the strain range between successive reversals is recorded. In determining counts, if both positive ranges (valleys followed by peaks) and negative ranges (peaks followed by valleys) are included, each range is considered to form one-half cycle. If just positive or negative ranges are recorded, each is considered to form one full cycle. A cycle count completed using this method for a sample strain history is shown in Fig. 5.6. In this example both positive and negative ranges were counted. As noted above and shown in the cycle count results in Fig. 5.6b, each range is considered to form one-half cycle.

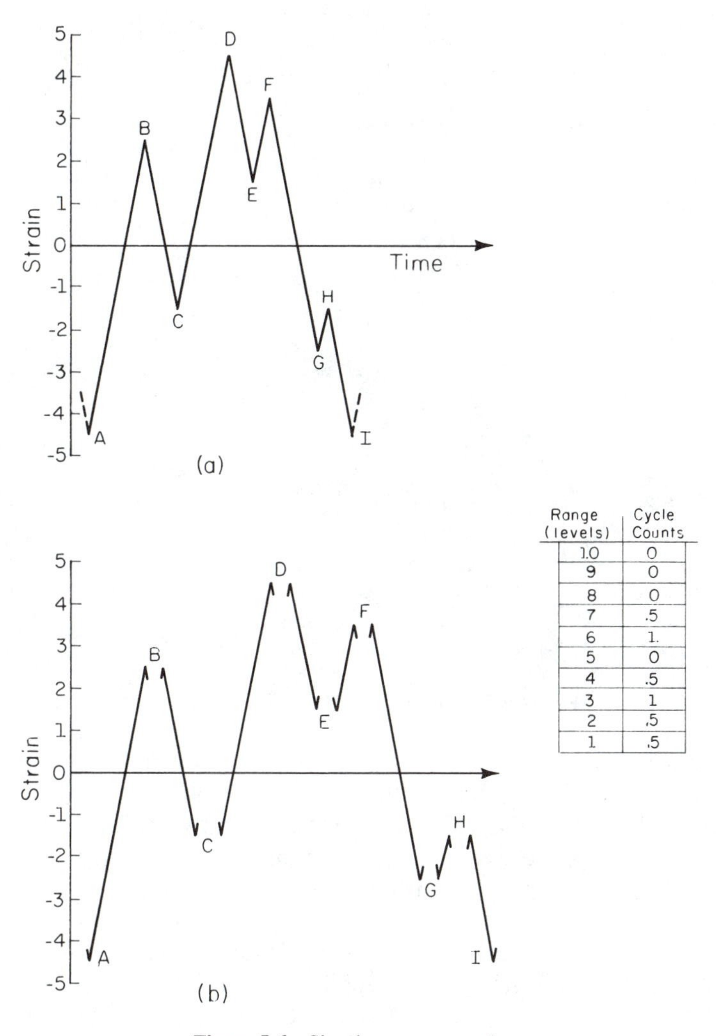

Range (levels)	Cycle Counts
10	0
9	0
8	0
7	.5
6	1.
5	0
4	.5
3	1
2	.5
1	.5

Figure 5.6 Simple-range counting.

In using the simple-range counting method, the mean value of each range is often recorded. This information is then combined with the determined range values in the form of a two-dimensional matrix. In performing a fatigue analysis mean stress effects are then included. When mean values are recorded, this cycle counting procedure is then called the simple-range-mean counting method.

5.4.2 Sequence Effects

All three of the cycle counting methods described above make no consideration of the order in which cycles are applied. Due to the nonlinear relationship

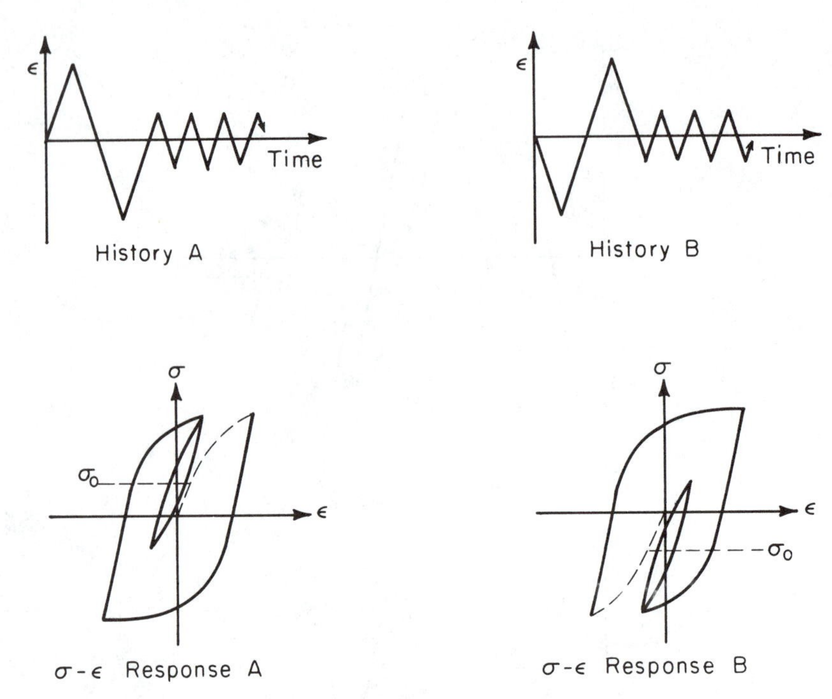

Figure 5.7 Loading sequence effects on material stress–strain response.

between stress and strain (plastic material behavior), the order in which cycles are applied has a significant effect on the stress–strain response of a material to applied loading. For example, consider the two strain histories shown in Fig. 5.7. Due to sequence effects these strain–time histories will yield very different stress–strain responses. In the case of history A, the occurrence of a compressive overload immediately preceding the application of the smaller strain cycles results in the development of tensile mean stresses. For history B, the occurrence of a tensile overload prior to the application of the smaller strain cycles results in the development of compressive mean stresses. Since mean stresses have a significant effect on the fatigue life of a material these two strain histories would experience very different lives. Thus sequence effects should be considered in fatigue life predictions. (Note that this effect is similar to the one shown in Fig. 5.3.)

As a further example, consider the two strain histories shown in Fig. 5.8. As shown, these two histories yield identical cycle counts when using the level-crossing and peak cycle counting methods. The cyclic stress–strain responses of the two histories are recognized as being very different. These histories would therefore be expected to produce very different fatigue lives. Therefore, any fatigue analysis technique should account for these differences. Failure of early cycle counting methods to predict the fatigue life of components subjected to variable amplitude histories has in part been attributed to the fact that these sequence effects have not been included.

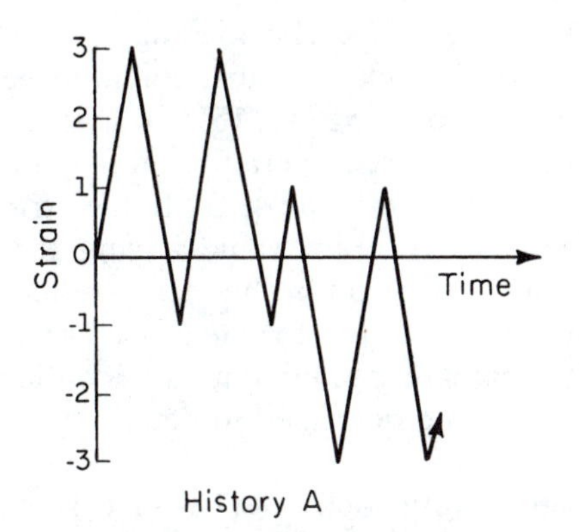

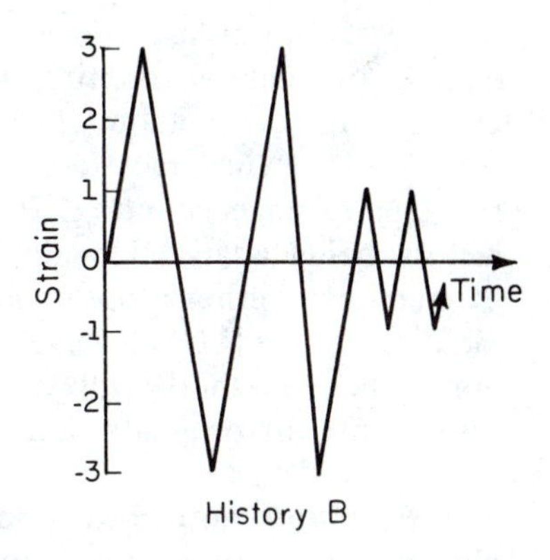

Level-Crossing Counting

Level	Counts A	Counts B
3	2	2
2	2	2
1	4	4
0	4	4
-1	4	4
-2	2	2
-3	2	2

Range (Levels)	Cycle Counts
6	2
5	0
4	0
3	0
2	2
1	0

Peak Counting

Peak	Counts A	Counts B
3	2	2
2	0	0
1	2	2
0	0	0
-1	2	2
-2	0	0
-3	2	2

Range (Levels)	Cycle Counts
6	2
5	0
4	0
3	0
2	2
1	0

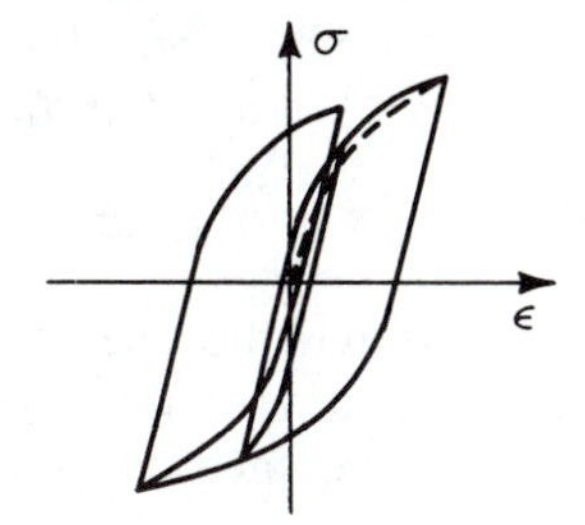

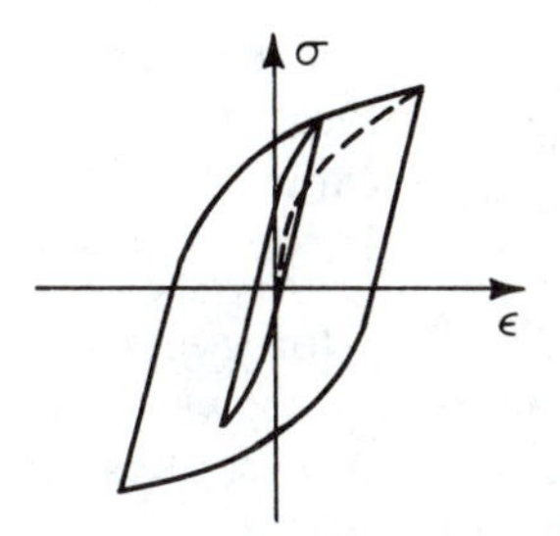

Figure 5.8 Insensitivity of the level-crossing and peak cycle counting methods to loading sequence effects.

5.4.3 Rainflow Counting

The original rainflow method of cycle counting derived its name from an analogy used by Matsuishi and Endo in their early work on this subject [10]. Since that time "rainflow counting" has become a generic term that describes any cycle counting method which attempts to identify closed hysteresis loops in the stress–strain response of a material subjected to cyclic loading. At present a number of rainflow counting techniques are in use, the most common being the original rainflow method [10–14], range-pair counting [11, 15, 16], hysteresis loop counting [17], the "racetrack" method [18], ordered overall range counting [19], range-pair–range counting [20], and the Hayes method [21]. If the strain–time history being analyzed begins and ends at the strain value having the largest magnitude, whether it occurs at a peak or a valley, all of the methods above yield identical results [11]. However, if an intermediate strain value is used as a starting point, these methods will yield results that are similar but not identical in all cases. Only the original rainflow method will be described in detail.

Rainflow counting ("falling rain" approach). The first step in implementing this procedure is to draw the strain–time history so that the time axis is oriented vertically, with increasing time downward. One could now imagine that the strain history forms a number of "pagoda roofs." Cycles are then defined by the manner in which rain is allowed to "drip" or "fall" down the roofs. (This is the previously mentioned analogy used by Matsuishi and Endo, from which the rainflow method of cycle counting received its name.) A number of rules are imposed on the dripping rain so as to identify closed hysteresis loops. The rules specifying the manner in which rain falls are as follows:

1. To eliminate the counting of half cycles, the strain–time history is drawn so as to begin and end at the strain value of greatest magnitude.
2. A flow of rain is begun at each strain reversal in the history and is allowed to continue to flow unless;
   a. The rain began at a local maximum point (peak) and falls opposite a local maximum point greater than that from which it came.
   b. The rain began at a local minimum point (valley) and falls opposite a local minimum point greater (in magnitude) than that from which it came.
   c. It encounters a previous rainflow.

The foregoing procedure can be clarified through the use of an example. Figure 5.9 shows a strain history and the resulting flow of rain. The following discussion describes in detail the manner in which each rainflow path was determined.

As shown in Fig. 5.9, the given strain–time history begins and ends at the strain value of greatest magnitude (point A). Rainflow is now initiated at each reversal in the strain history.

A. Rain flows from point A over points B and D and continues to the end of the history since none of the conditions for stopping rainflow are satisfied.

B. Rain flows from point B over point C and stops opposite point D, since both B and D are local maximums and the magnitude of D is greater than B (rule 2a above).

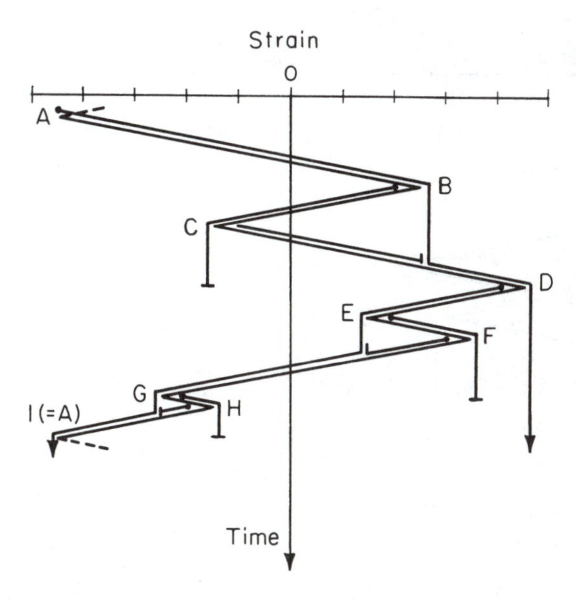

Figure 5.9 Rainflow counting ("falling rain" approach).

C. Rain flows from point C and must stop upon meeting the rain flow from point A (rule 2c).

D. Rain flows from point D over points E and G and continues to the end of the history since none of the conditions for stopping rainflow are satisfied.

E. Rain flows from point E over point F and stops opposite point G, since both E and G are local minimums and the magnitude of G is greater than E (rule 2b).

F. Rain flows from point F and must stop upon meeting the flow from point D (rule 2c).

G. Rain flows from point G over point H and stops opposite point A, since both G and A are local minimums and the magnitude of A is greater than G (rule 2b).

H. Rain flows from point H and must stop upon meeting the rainflow from point D (rule 2c).

Having completed the above, we are now able to combine events to form completed cyles. In this example events A–D and D–A are combined to form a full cycle. Event B–C combines with event C–B (of strain range C–D) to form an additional cycle. Similarly, cycles are formed at E–F and G–H.

The use of the rainflow method of cycle counting in recognizing closed hysteresis loops is clearly seen upon examination of the stress–strain response of the material from the given strain history. Figure 5.10 shows the stress–strain response for the current example. Since point A represents the largest strain magnitude in the given history, in a stress–strain plot it will be located at the tip of a hysteresis loop. Furthermore, all loading from this point on will follow the

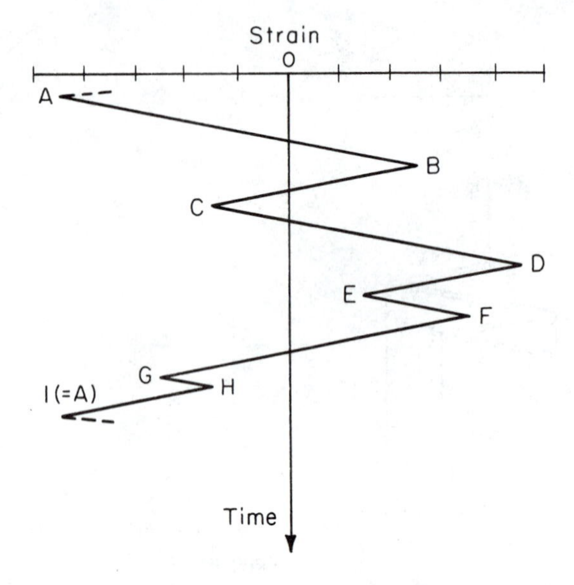

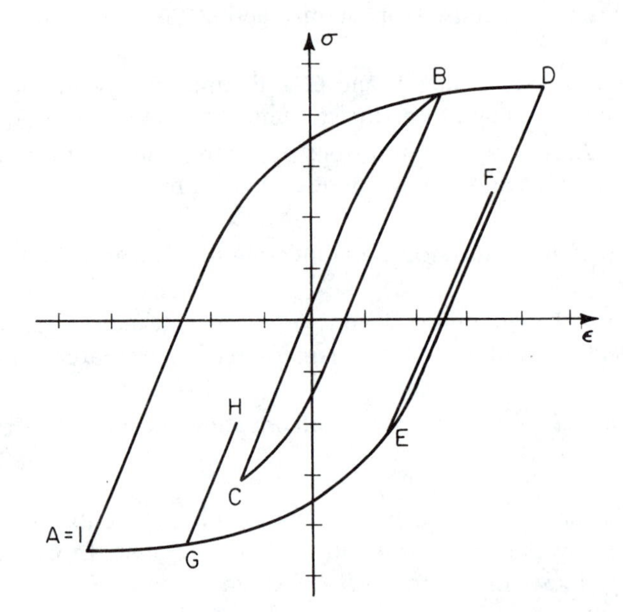

Figure 5.10 Material stress–strain response to given strain history.

hysteresis curve. This is shown to occur as one moves from point A to point B. After reaching point B the strain is then decreased to point C, following a path defined by the hysteresis loop shape. Upon reloading after reaching point B, the material continues to point D along the hysteresis path starting from point A, as though event cycle B–C had never occurred. This behavior of the material of "remembering" its prior state of deformation is known as *material memory*. In the current example, material memory is also recognized as occurring at points E and G.

In the current example four events resembling constant amplitude behavior are recognized. These events, A–D, B–C, E–F, and G–H, occur as closed hysteresis loops, each having its own strain range and mean stress values. Note that these hysteresis loops correspond to the cycles obtained from the rainflow cycle count.

When used in conjunction with the predicted stress–strain response of a material, rainflow counting provides valuable insight into the effect of a given strain history on material response. When used alone, rainflow counting only gives the strain ranges of closed hysteresis loops. When the stress–strain response of the material is considered the mean stress of these loops can also be determined.

Once the closed hysteresis loops have been determined, a fatigue life analysis can be performed on a variable amplitude history by using a strain-life equation that incorporates mean stress effects, such as suggested by Morrow [Eq. (2.49)]. This equation is

$$\frac{\Delta\epsilon}{2} = \frac{\sigma_f' - \sigma_0}{E}(2N_f)^b + \epsilon_f'(2N_f)^c \tag{5.5}$$

If the value of strain range, $\Delta\epsilon$, and mean stress, σ_0, for the hysteresis loop are input into the equation, it can be solved for life to failure, N_f. The reciprocal of this, $1/N_f$, is the cycle ratio corresponding to one cycle. If Miner's linear damage rule is used, this value, $1/N_f$, corresponds to the damage fraction for the hysteresis loop. Life to failure will be predicted when the sum of the damage fractions of the individual hysteresis loops is greater than or equal to 1.

$$\sum \frac{1}{N_f} \geq 1 \tag{5.6}$$

Rainflow counting (ASTM standard). While rainflow counting can be completed manually for relatively simple load histories, for more complex loading histories the method is better implemented through the use of computers. At present a number of computer algorithms for rainflow cycle counting have been developed [22]. Several such rainflow counting algorithms that may easily be adapted into computer programs are given by the American Society for Testing and Materials (ASTM) in its *Annual Book of ASTM Standards* [11]. Given below is an algorithm taken from Ref. 22 which was developed by Downing. This algorithm determines load cycles for closed hysteresis loops in a loading history.

To begin, arrange the history to start with either the maximum peak or the minimum valley. Let X denote range under consideration; and Y, previous range adjacent to X.

1. Read the next peak or valley. If out of data, *stop*.

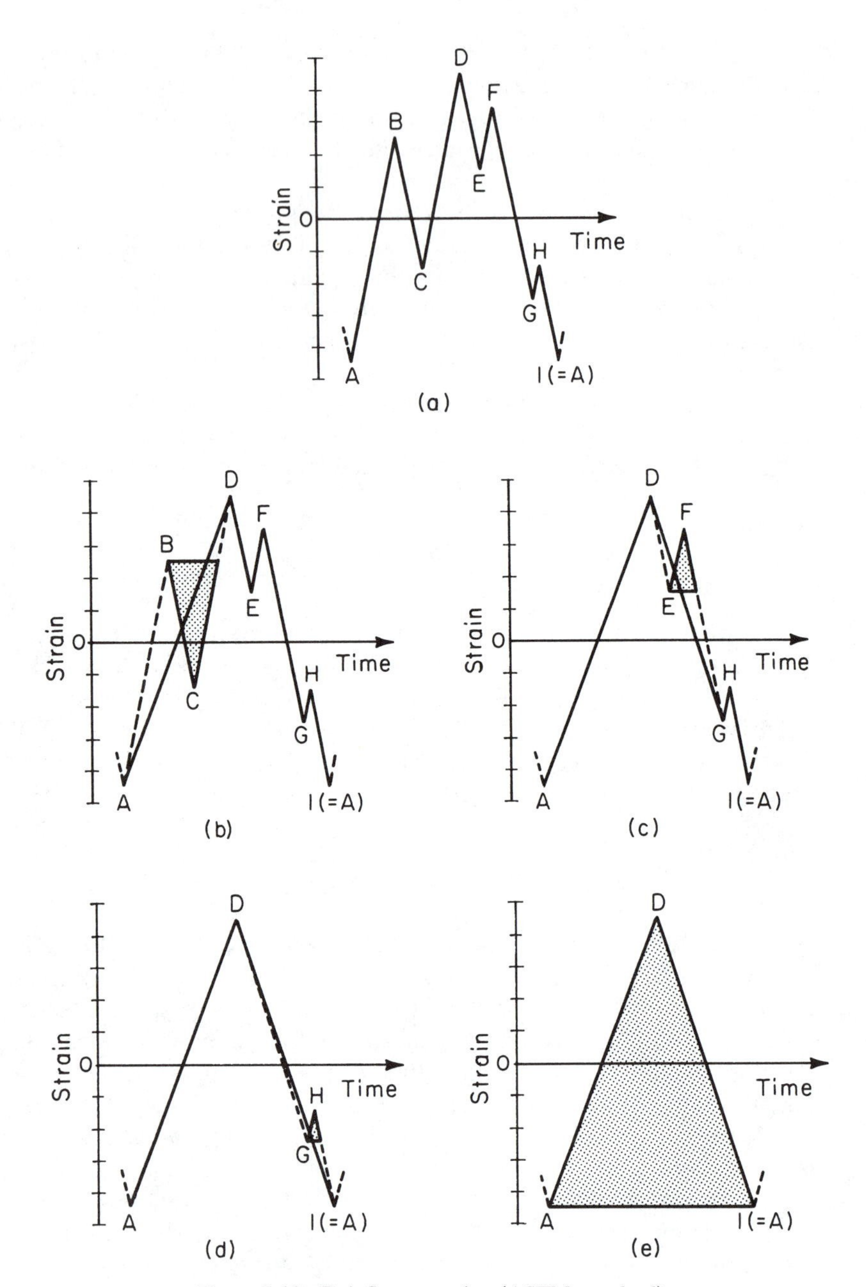

Figure 5.11 Rainflow counting (ASTM standard).

2. If there are less than three points, go to step 1. Form ranges X and Y using the three most recent peaks and valleys that have not been discarded.
3. Compare the absolute values of ranges X and Y.
   a. If $X < Y$, go to step 1.
   b. If $X \geq Y$, go to step 4.
4. Count range Y as one cycle; discard the peak and valley of Y; and go to step 2.

This algorithm is now used in determining closed hysteresis loops for the strain history given in Fig. 5.11a.

1. $Y = |A-B|$; $X = |B-C|$; $X < Y$.
2. $Y = |B-C|$; $X = |C-D|$; $X > Y$. Count $|B-C|$ as one cycle and discard points B and C (see Fig. 5.11b). Note that a cycle is formed by pairing range $B-C$ and a portion of range $C-D$.
3. $Y = |A-D|$; $X = |D-E|$; $X < Y$.
4. $Y = |D-E|$; $X = |E-F|$; $X < Y$.
5. $Y = |E-F|$; $X = |F-G|$; $X > Y$. Count $|E-F|$ as one cycle and discard points E and F (see Fig. 5.11c).
6. $Y = |A-D|$; $X = |D-G|$; $X < Y$.
7. $\mathrm{Y} = |D-G|$; $X = |G-H|$; $X < Y$.
8. $Y = |G-H|$; $X = |H-A|$; $X > Y$. Count $|G-H|$ as one cycle and discard points G and H (see Fig. 5.11d.)
9. $Y = |A-D|$; $X = |D-A|$; $X = Y$. Count $|A-D|$ as one cycle and discard points A and D (see Fig. 5.11e).
10. End of counting.

In this example closed hysteresis loops were identified as occurring at events $A-D$, $B-C$, $E-F$, and $G-H$ in the strain history. These results are identical to those obtained using the method shown in Fig. 5.9.

Given below is a Fortran listing of a program that will perform a rainflow count on a load or strain history [22].

```
C     RAINFLOW ALGORITHM I
C
C     THIS PROGRAM RAINFLOW COUNTS A HISTORY OF PEAKS
C     AND VALLEYS IN SEQUENCE WHICH HAS BEEN REARRANGED
C     TO BEGIN AND END WITH THE MAXIMUM PEAK (OR MINIMUM
C     VALLEY). STATEMENT LABELS CORRESPOND TO THE STEPS IN
C     THE RAINFLOW COUNTING RULES.
C
```

```
      DIMENSION E(50)
      N = 0
    1 N = N + 1
      CALL DATA(E(N),K)
      IF (K.EQ.1) STOP
    2 IF (N.LT.3) GO TO 1
      X = ABS(E(N) - E(N - 1))
      Y = ABS(E(N - 1) - E(N - 2))
    3 IF (X.LT.Y) GO TO 1
    4 RANGE = Y
      XMEAN = (E(N - 1) + E(N - 2))/2.
      N = N - 2
      E(N) = E(N + 2)
      GO TO 2
      END
```

This program, when used in conjunction with the stress–strain response of a material, provides strain range and mean stress values of closed hysteresis loops of a strain history. This information can then be evaluated through the use of a strain–life relation corrected for mean stress [such as that given by Eq. (5.5)] to obtain the amount of fatigue damage caused by each closed hysteresis loop. The damage caused by each hysteresis loop can then be summed to determine the total fatigue damage caused by the strain history.

Typically, the following steps are used to determine the total fatigue life of a component under variable amplitude loading:

1. Assume that the actual service history can be modeled by a repeating block of strain history (i.e., loading block).
2. Determine the damage caused per loading block by using the procedure outlined in the preceding paragraph.
3. Calculate the total fatigue life, stated in terms of blocks, by taking the reciprocal of the damage per block.

5.5 CRACK PROPAGATION UNDER VARIABLE AMPLITUDE LOADING

5.5.1 Introduction

Fatigue crack growth tests and predictions performed under constant amplitude loading often differ considerably from variable amplitude loading conditions. In contrast to constant amplitude loading where the increment of crack growth, Δa, is dependent only on the present crack size and the applied load, under variable amplitude loading the increment of fatigue crack growth is also dependent on the preceding cyclic loading history. This is known as load interaction. Load interaction or sequence effects significantly affect the fatigue crack growth rate and consequently, fatigue lives. In the following section we first discuss observed load interaction behavior and then review several types of models developed to predict fatigue crack growth under variable amplitude loading.

5.5.2 Load Interaction Effects

Observed behavior. In the early 1960s, interaction effects were first recognized [23–26]. The application of a single overload was observed to cause a decrease in the crack growth rate, as shown in Fig. 5.12. This phenomenon is termed *crack retardation*. If the overload is large enough, crack arrest can occur and the growth of the fatigue crack stops completely. (Jokingly, it has been suggested that an airline passenger spotting a crack in the wing of the plane while in flight should order the pilot to perform a 360° loop. This would overload the wing and thereby cause crack arrest to occur.)

Crack retardation remains in effect for a period of loading after the overload. The number of cycles in this period has been shown to correspond to the plastic zone size developed due to the overload. The larger the overload plastic zone, the longer the crack growth retardation remains in effect [27, 28]. Recalling Eq. (3.5), the relation for the plastic zone size is

$$r_y = \frac{1}{\beta\pi}\left(\frac{K}{\sigma_y}\right)^2 \tag{5.7}$$

where $\beta = 2$ for plane stress and $\beta = 6$ for plane strain.

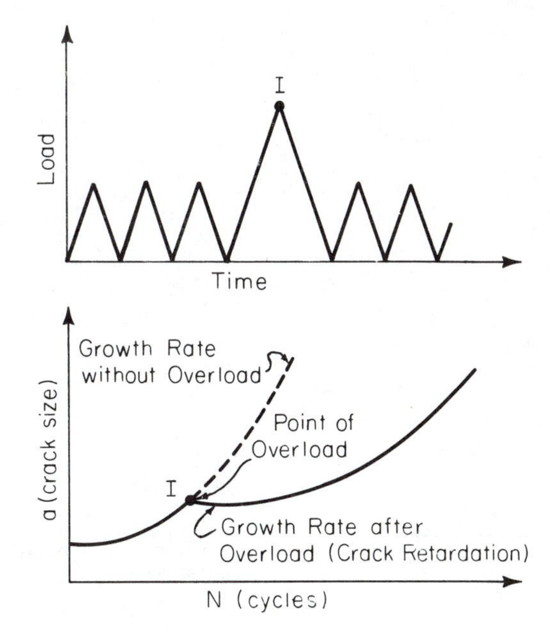

Figure 5.12 Crack growth retardation after an overload.

The period of crack growth retardation is longer for materials that develop larger plastic zone sizes, such as thin specimens ($\beta = 2$) or low yield strength materials. It is a general belief that a return to normal growth rate occurs when the crack grows out of the overload zone.

The crack growth rate does not reach a minimum immediately after the overload is applied (see Fig. 5.13). Rather, the minimum is reached after the crack has grown a distance approximately one-eighth to one-fourth of the distance into the overload plastic zone. This behavior is known as *delayed retardation*.

A single compressive overload (termed *underload*), as shown in Fig. 5.14, generally causes an acceleration in the crack growth rate. In addition, when an underload follows an overload, as shown in Fig. 5.15, the amount of crack growth retardation is significantly diminished.

In high-to-low loading sequences (Fig. 5.16), crack retardation also occurs. Retardation increases for an increasing number of tensile overloads, P, until a limiting "saturation" of overloads is reached. Spacing (M in Fig. 5.16) between overloads is critical for this trend. Overloads too closely spaced will eliminate the benefits of crack retardation as the material response tends toward constant amplitude crack growth corresponding to the overloads.

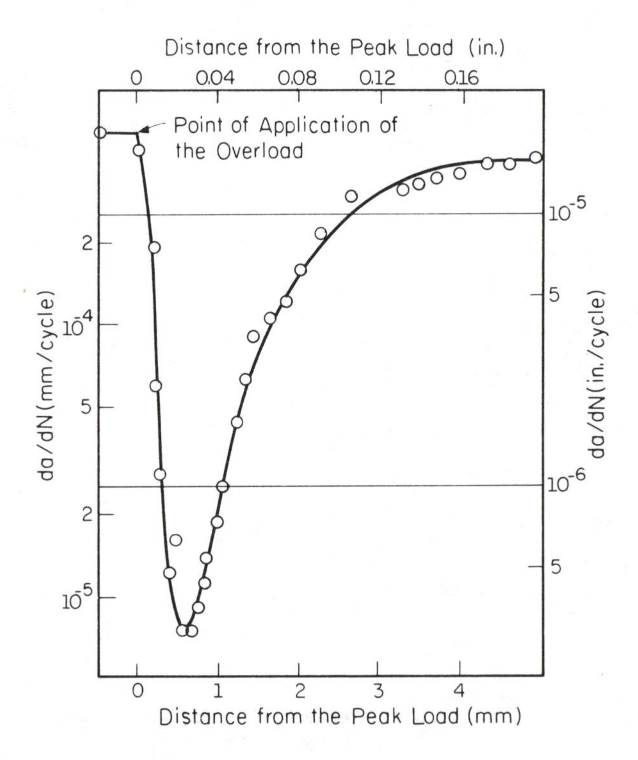

Figure 5.13 Delayed retardation. (From Ref. 27.)

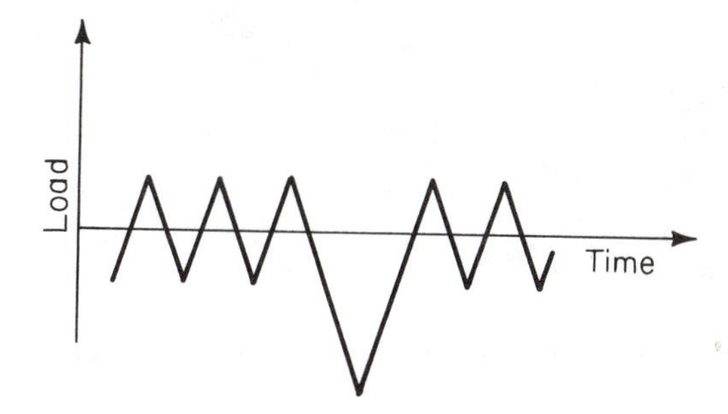

Figure 5.14 Single compressive overload (termed underload).

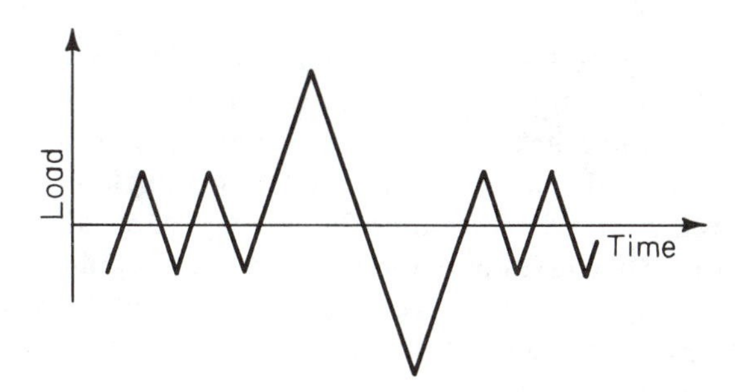

Figure 5.15 Overload followed by an underload.

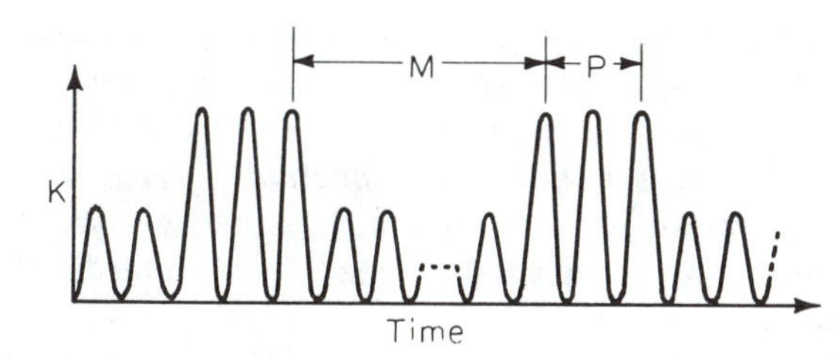

Figure 5.16 Spacing, *M*, between multiple overloads, *P*, is critical in producing maximum retardation. (From Ref. 30.)

Finally, low-to-high sequences (Fig. 5.17) can cause crack growth acceleration. Fortunately, this acceleration stabilizes quickly when compared to retardation effects.

It should be noted that periodic overloads are not always beneficial. In some low cycle fatigue (LCF) tests, periodic overloads have also been found to cause crack growth acceleration [31].

Load interaction models. Several theories have been developed to explain crack retardation. These include:

1. Crack-tip blunting
2. Compressive residual stresses at crack tip
3. Crack closure effects

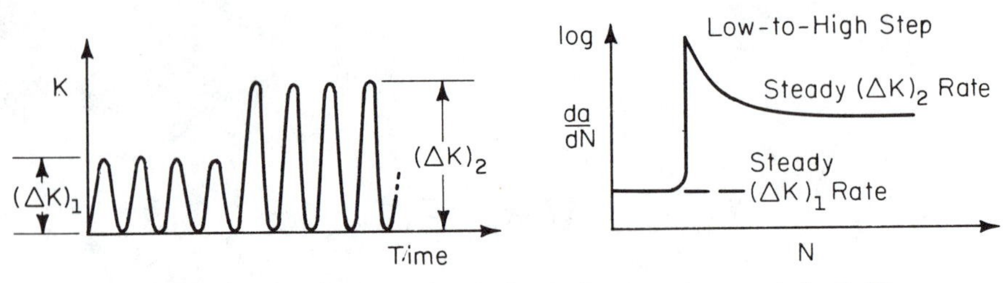

Figure 5.17 Acceleration in crack growth rate due to an increase in load. (From Ref. 30.)

Crack-tip blunting is proposed to occur as a result of the overload. This theory states that as the crack-tip blunts during the overload, the stress concentration associated with the crack becomes less severe, resulting in a slower crack growth rate. Unfortunately, this theory does not explain and is not consistent with the observed delayed retardation behavior. It would be expected that the retardation would be a maximum immediately after application of the overload since crack-tip blunting would be greatest at this time. Instead, maximum retardation occurs after the crack has propagated a portion of the way through the overload zone (delayed retardation).

A second model proposed to account for load interaction effects states that after an overload, compressive residual stresses are developed at the crack tip due to the large plastic zone. Compressive stresses are developed as the elastic body surrounding the crack tip "squeezes" the overload plastic zone once the overload is removed. These compressive stresses reduce the effective stress at the crack tip, causing a reduction in the crack growth rate. Again, this theory does not predict delayed retardation. Instead, it predicts that maximum retardation occurs immediately after the overload.

Crack closure models assume that crack retardation and acceleration are caused by crack closure effects (see Section 3.3.6.) Crack closure effects cause variations in the opening stress, σ_{op}, and the effective stress intensity, ΔK_{eff}, with changes in loads. As discussed in Chapter 3, residual displacements in the wake of a crack cause the crack faces to contact or close before the tensile load is removed. This is termed *crack closure.* Crack closure arguments are successful in predicting load interaction behavior.

Crack closure arguments predict crack acceleration in low-to-high loading sequences and crack retardation in high-to-low loading sequences. In the low-to-high loading, at the beginning of a high block cycle crack growth rate acceleration is a transient increase in the crack growth rate which eventually stabilizes. Similarly, in high-to-low loading, crack retardation is a transient decrease in crack growth rate. As shown in Fig. 5.18, the stabilized closure stress intensity associated with the lower load level is K_A and for the higher load level,

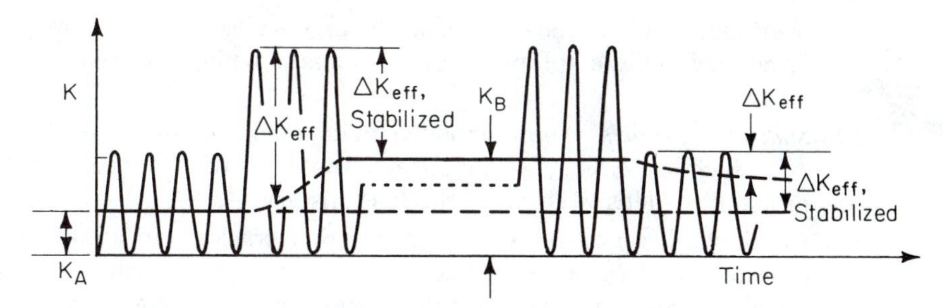

Figure 5.18 Variation in crack closure stress intensity factor (and variation in ΔK_{eff}) with variation in load level. (From Ref. 30.)

K_B. As the load is increased to the high load level, the closure level adjusts from that corresponding to the lower level to that of the higher level. During this transition period, crack growth acceleration will occur because the transient ΔK_{eff} is larger than the stabilized value at the higher load. Upon continued cycling at the higher load level, ΔK_{eff} stabilizes and the crack growth rate reaches the value associated with the higher load. In a similar manner, in a high-to-low loading sequence, a transient crack growth retardation occurs at the beginning of the low block cycle (see Fig. 5.18). In this transient period, ΔK_{eff} is less than its stabilized value at the lower load.

Crack closure arguments successfully predict the observed delayed retardation behavior after a single overload. The elastic body surrounding the overload plastic zone develops compressive residual stresses. While this plastic zone is in front of the crack tip, these compressive stresses do not affect the crack opening stress. Once the crack has propagated into the overload zone, though, the compressive stresses act on the crack surfaces. Consequently, the load applied to the crack tip is reduced due to these compressive stresses and crack growth retardation occurs. As the crack grows into the overloaded plastic zone, the crack propagates at a decreasing rate until it reaches a minimum as shown in Fig. 5.13. As it begins to grow out of this zone the crack growth rate increases until it resumes its normal rate.

5.5.3 Prediction Methods

Methods to predict fatigue crack growth under variable amplitude loading have been developed that attempt to account for load interaction effects. In general, they are based upon linear elastic fracture mechanics concepts and may be divided into the following three categories:

1. *Crack-tip plasticity models.* These assume that load interaction effects (crack growth retardation) occur due to the large plastic zone developed during the

overload. The effects remain active as long as the crack-tip plastic zones developed on the following cycles remain within the plastic zone of the overload.

2. *Statistical models.* These models relate crack growth rate to an effective ΔK, such as ΔK_{rms}, a statistical parameter that is characteristic of the probability–density curve of the load history.
3. *Crack closure models.* These assume that crack retardation/acceleration is caused by crack closure effects (see Section 3.3.6), which cause variations in opening stress, σ_{op}, and effective stress intensity, ΔK_{eff}, with variation in loads.

Crack-tip plasticity models. Crack-tip plasticity models are based on the assumption that crack growth rates under variable amplitude loading can be related to the interaction of the crack-tip plastic zones. Well-known models of this type were developed by Wheeler [32] and Willenborg [33]. These are reviewed individually below.

Wheeler Model. The Wheeler model predicts that retardation in the crack growth rate following an overload may be predicted by modifying the constant amplitude growth rate. Recall from Chapter 3 that the constant amplitude growth rate is usually modeled as

$$\frac{da}{dN} = f(\Delta K)$$

Using the Paris relation [Eq. (3.8)], this is

$$\frac{da}{dN} = C(\Delta K)^m$$

Wheeler's model modifies the constant amplitude growth rate by an empirical retardation parameter, C_p:

$$\frac{da}{dN_i} = (C_p)_i\left(\frac{da}{dN}\right)_{CA_i} \tag{5.8}$$

where $(da/dN)_{CA_i}$ = constant amplitude growth rate appropriate to the stress intensity factor range ΔK_i, and stress ratio R_i, of load cycle i or

$$\frac{da}{dN_i} = (C_p)_i[C(\Delta K_i)^m] \tag{5.9}$$

This retardation parameter, $(C_p)_i$, is a function of the ratio of the current plastic zone size to the plastic zone size created by the overload.

$$(C_p)_i = \left(\frac{r_{yi}}{a_p - a_i}\right)^p \tag{5.10}$$

where r_{yi} = cyclic plastic zone size due to the ith loading cycle
a_p = sum of the crack length at which the overload occurred and the overload plastic zone size
a_i = crack length at ith loading cycle
p = empirically determined shaping parameter

These terms are defined graphically in Fig. 5.19.

This model predicts that retardation decreases proportionally to the penetration of the crack into the overload zone with maximum retardation occurring immediately after the overload. The values of $(C_p)_i$ range from 0.0 to 1.0. Crack retardation is assumed to occur as long as the current plastic zone is within the plastic zone created by the overload. As soon as the boundary of the current plastic zone touches the boundary of the overload zone, retardation is assumed to cease and $(C_p)_i = 1.0$ (see Fig. 5.20). In terms of the parameters above, retardation ceases and $(C_p)_i = 1.0$ if

$$a_i + r_{yi} \geq a_p \tag{5.11}$$

or, rearranging,

$$\frac{r_{yi}}{a_p - a_i} \geq 1 \tag{5.12}$$

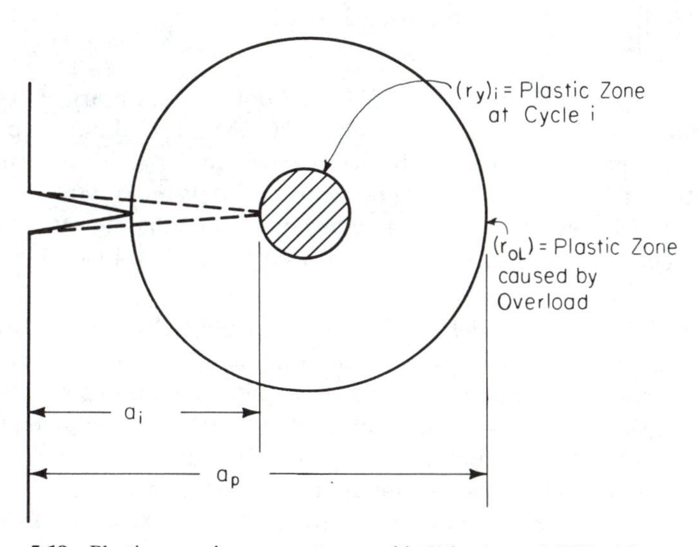

Figure 5.19 Plastic zone size parameters used in Wheeler and Willenborg models.

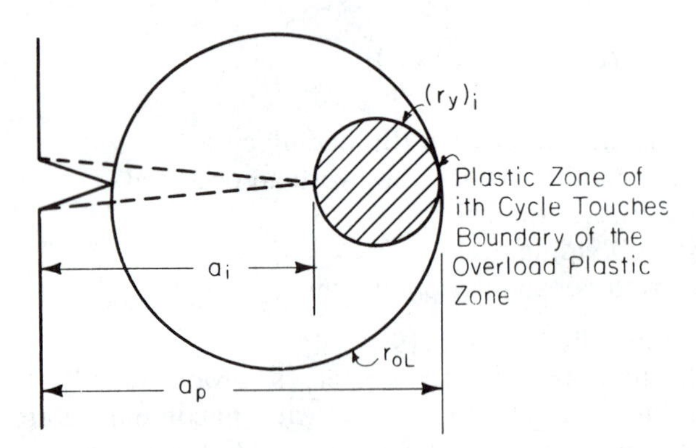

Figure 5.20 Retardation ceases when plastic zone of ith cycle touches the boundary of the overload plastic zone.

Maximum retardation occurs as $(C_p)_i$ approaches 0.0.

Using this model, crack growth is summed as follows:

$$a_r = a_0 + \sum_{i=1}^{r} \left(\frac{da}{dN}\right)_i \tag{5.13}$$

where a_0 = initial crack length

a_r = crack length after r cycles

$\left(\frac{da}{dN}\right)_i$ = crack growth during ith cycle

A major disadvantage of this model is the empirical constant, p, which is required to "shape" the parameter $(C_p)_i$ to test data. It must be determined experimentally and cannot be determined in advance of testing. In addition, in contrast to the observed phenomenon of delayed retardation discussed earlier, this model predicts maximum retardation immediately after the overload. It also neglects the counteracting effect of a negative peak load in crack retardation.

Willenborg Model. The crack-tip plasticity model developed by Willenborg is based on the assumption that crack growth retardation is caused by compressive residual stresses acting on the crack tip. These stresses are developed due to the elastic body surrounding the overload plastic zone, which causes this zone to be put into compression after the overload is removed. This model uses an effective stress, which is the applied stress reduced by the compressive residual stress, to determine the crack-tip stress intensity factor for the subsequent loading cycles. This model is outlined below.

1. *Determine a_p.* Similar to Wheeler's model, a_p is the sum of the initial crack length (the crack length when the overload was applied) and the plastic zone due to the overload.

$$a_p = a_0 + r_{0L} \tag{5.14}$$

$$r_{0L} = \frac{1}{\beta\pi}\left(\frac{K_{0L}}{\sigma_y}\right)^2 \tag{5.15}$$

where $\beta = 2$ for plane stress and $\beta = 6$ for plane strain. Again, retardation gradually decreases until the sum of the current crack length, a_i, and its associated plastic zone, r_{yi}, is equal to or larger than a_p. In other words, when the boundary of the current plastic zone touches the boundary of the overload zone, retardation ceases (see Fig. 5.20).

2. *Calculate the required stress, $(\sigma_{req})_i$.* This is the stress required to produce a yield zone, $(r_{req})_i$, whose boundary just touches the overload plastic zone boundary.

$$a_i + (r_{req})_i = a_p \tag{5.16}$$

$$a_i + \frac{1}{\beta\pi}\left[\frac{(K_{req})_i}{\sigma_y}\right]^2 = a_p \tag{5.17}$$

$$a_i + \frac{1}{\beta\pi}\left[\frac{f(g)(\sigma_{req})_i\sqrt{\pi a_i}}{\sigma_y}\right]^2 = a_p \tag{5.18}$$

$$(\sigma_{req})_i = \sqrt{\frac{(a_p - a_i)\beta}{a_i}}\,\frac{\sigma_y}{f(g)} \tag{5.19}$$

3. *Determine the compressive stress, $(\sigma_{comp})_i$.* The model states that the compressive stress due to the elastic body surrounding the overload is the difference between the maximum stress occurring at the ith cycle, $(\sigma_{max})_i$, and the corresponding value of the "required" stress, $(\sigma_{req})_i$.

$$(\sigma_{comp})_i = (\sigma_{req})_i - (\sigma_{max})_i \tag{5.20}$$

4. *Determine an "effective" stress, σ_i^e.* The actual maximum and minimum stress at the ith loading cycle are reduced by $(\sigma_{comp})_i$ to obtain the "effective" stress, σ_i^e.

$$(\sigma_{max}^e)_i = (\sigma_{max})_i - (\sigma_{comp})_i \tag{5.21a}$$

$$(\sigma_{max}^e)_i = 2(\sigma_{max})_i - (\sigma_{req})_i \tag{5.21b}$$

$$(\sigma_{min}^e)_i = (\sigma_{max})_i + (\sigma_{min})_i - (\sigma_{req})_i \tag{5.21c}$$

If either of these effective stresses is less than zero, it is set equal to zero and $\Delta\sigma_i^e$ is then calculated.

$$\Delta\sigma_i^e = (\sigma_{max}^e)_i - (\sigma_{min}^e)_i \tag{5.21d}$$

5. *Compute* $(\Delta K_{eff})_i$ *[and* $(R_{eff})_i$ *if needed].*

$$(\Delta K_{eff})_i = \Delta\sigma_i^e\sqrt{\pi a_i}\, f(g) \tag{5.22}$$

$$(R_{eff})_i = \frac{(K_{min})_i^{eff}}{(K_{max})_i^{eff}} \tag{5.23}$$

Substitute these into chosen crack growth law (i.e., Paris or Forman's relations). For the Paris relation [refer to Eq. (3.8)],

$$\frac{da}{dN_i} = C(\Delta K_{eff})_i^m \tag{5.24}$$

For the Forman relation [refer to Eq. (3.20)],

$$\frac{da}{dN_i} = \frac{C(\Delta K_{eff})_i^m}{[1 - (R_{eff})_i]K_c - (\Delta K_{eff})_i} \tag{5.25}$$

Willenborg's model predicts that crack arrest will occur if the ratio of the overload stress intensity, K_{0L}, to the stress intensity of the subsequent lower load levels, K_{max}, is larger or equal to 2.0. Nelson [30] tabulated values of K_{0L}/K_{max} for various materials. He reported that values of this ratio were between 2.0 and 2.7 when crack arrest occurred. Thus he stated that the Willenborg value of 2.0 is a "fairly reasonable proposition in view of the test data."

Like the Wheeler model, the Willenborg model predicts that maximum retardation occurs immediately after the overload. It fails to predict the observed delayed retardation effect. It also fails to predict a decrease in retardation due to underloads. The most significant difference between the Wheeler and Willenborg models is that the Willenborg model uses only constant-amplitude crack growth data and does not require a "shaping" constant.

Statistical Models. Statistical approaches have also been used to predict crack growth under variable amplitude loading. Barsom [34] developed a model based on the root-mean-square stress intensity factor, ΔK_{rms}:

$$\Delta K_{rms} = \sqrt{\sum_{i=1}^{n} \frac{\Delta K_i^2}{n}} \tag{5.26}$$

where n is the number of cycles and ΔK_i is the stress intensity range associated with ith cycle. An average fatigue crack growth rate can be predicted from constant amplitude data using the ΔK_{rms} value in the following equation:

$$\frac{da}{dN} = C(\Delta K_{rms})^m \tag{5.27}$$

where C and m are material constants determined from constant amplitude data.

Statistical approaches have been shown to be applicable only to short spectra, in which load effects are minimized, since they do not account for load sequence effects such as retardation and acceleration. Chang, Szamossi, and Liu [35] found that fatigue life predictions were highly conservative for random spectra consisting of a majority of tension–tension cycles ($R > 0$) when interaction effects were not considered. Alternatively, when interaction effects were neglected for spectra consisting of predominantly tension–compression cycles ($R < 0$), predictions were nonconservative. It is suggested that use of an rms-type statistical approach be limited to load sequences that can be described by a continuous, unimodal distribution [30].

Crack closure models. As discussed previously, the theory of crack closure accounts for delayed retardation and the effect of underloads by the variation in σ_{op} and hence ΔK_{eff}. In recent years, several crack closure models have been developed [36–39]. The difficulty with crack closure models is in determining the opening stress, σ_{op}, for variable amplitude loading (see references above for details). Once this is determined, the procedure for determining crack growth rate predictions is as follows:

1. Determine $(\Delta\sigma_{eff})_i$ and consequently ΔK_{eff}:

$$(\Delta\sigma_{eff})_i = (\sigma_{max})_i - (\sigma_{op})_i$$

2. Calculate Δa_i:

$$\Delta a_i = \left(\frac{da}{dN}\right)_i = f(\Delta K_{eff})_i$$

That is, for Paris law,

$$\Delta a_i = \left(\frac{da}{dN}\right)_i = C_0(\Delta K_{eff})_i^m$$

where C_0 and ΔK_{eff} correspond to the same closure level.

3. Determine a_{i+1}:

$$a_{i+1} = a_i + \Delta a_i$$

Repeat these steps until final crack length is reached.

Since this must usually be done cycle by cycle and due to the difficulty of determining σ_{op}, large computer programs with long run times are often required. However, good correlations have been obtained between predicted and experimental results [36, 38].

Note that when the Paris equation is used with a crack closure model (as in step 2 above), the crack growth coefficient, C_0, must correspond to the same closure level as the effective stress intensity factor term, ΔK_{eff}. Most tabulated values of C do not consider crack closure effects. In other words, $\sigma_{op} = \sigma_{min}$. The constant C can be corrected for a new closure level with the relationship

$$C_0 = \frac{C}{(U)^m}$$

where

$$U = \frac{\Delta K_{\text{eff}}}{\Delta K}$$

$$\Delta K_{\text{eff}} = K_{\max} - K_{\text{op}}$$

$$\Delta K = K_{\max} - K_{\min}$$

For example, if the crack growth coefficient, C, is determined using $R = 0$ data (i.e., $K_{\min} = 0$) and the opening level, K_{op}, is assumed to be $0.3\,K_{\max}$, the corrected C_0 value is

$$C_0 = \frac{C}{(0.7)^m}$$

The crack growth exponent, m, does not need to be modified to account for crack closure effects.

Block loading. An approximate method of summing crack growth has been suggested [40] that results in a considerable savings of time. Instead of calculating crack growth by summing growth for each cycle, the crack growth per loading block, $\Delta a/\Delta B$, is determined. The block loading method is limited to short spectra loading histories. In these histories, the crack growth per block is not large enough to grow out of the plastic zone formed by the largest load cycle in the block. This method is not intended for long spectra loading. Long spectra loading is where a large overload cycle is followed by thousands of smaller load cycles. When using the block loading method, the following assumptions are made:

- Assume that the actual loading history can be modeled by a repeating block of loading.
- Assume that crack length is fixed during a loading block, ($\Delta a/\Delta B \ll a$).

In this method, damage is assumed to occur only when the crack is open. The level at which crack opening occurs, σ_{op}, must be determined. Once determined, this value is considered to remain constant during the entire loading block. The opening stress level can be used to determine the effective stress intensity factor, which in turn is used in crack growth rate calculations. Two assumptions that allow σ_{op} to be determined are given in step 1 below.

The change in crack length per block is calculated by the following steps:

1. Determine $(\Delta\sigma_{eff})_i$ by one of the following two methods.

a. Assuming that only tensile loads cause crack extension

$$\Delta\sigma_{eff} = \begin{cases} \sigma_{max} - \sigma_{min} & \text{if } \sigma_{min} > 0 \\ \sigma_{max} & \text{if } \sigma_{min} < 0 \\ 0 & \text{if } \sigma_{max} < 0 \end{cases}$$

(In this assumption, if $\sigma_{min} \leq 0$, $\sigma_{op} = 0$.) This assumption is applicable when there are no large overloads in the loading history and the loading tends to be completely reversed.

b. Assuming that crack opening stress, σ_{op}, is greater than zero

$$\Delta\sigma_{eff} = \sigma_{max} - \sigma_{op} \tag{5.28}$$

In many cases, assumption **a** may be too conservative. (Examples are loading histories with periodic overloads or situations where the component is in a state of plane stress.) For these cases, a crack opening stress, σ_{op}, may be estimated. Newman [36] estimated that values of the ratio of the opening stress to the maximum stress, σ_{op}/σ_{max}, range from 0.0 to 0.7. Thus, for $\sigma_{op}/\sigma_{max} = 0.0$,

$$\Delta\sigma_{eff} = \sigma_{max}$$

and for $\sigma_{op}/\sigma_{max} = 0.7$,

$$\begin{aligned} \Delta\sigma_{eff} &= \sigma_{max} - 0.7\sigma_{max} \\ &= 0.3\sigma_{max} \end{aligned}$$

(In assumption **a**, $\sigma_{op}/\sigma_{max} = 0.0$.)

Bounds on the value of $\Delta\sigma_{eff}$ or ΔK_{eff} may therefore be obtained using values of opening stress ranging from 0.0 to $0.7\sigma_{max}$. This results in bounds for the value of change in effective stress, $\Delta\sigma_{eff}$, of $1.0\sigma_{max}$ to $0.3\sigma_{max}$. Values of σ_{op} tending toward $0.0\sigma_{max}$ ($\Delta\sigma_{eff}$ tending towards $1.0\sigma_{max}$) are the most conservative.

For block loading histories that contain a periodic overload, the opening stress, σ_{op}, corresponding to the overload may dominate the block history due to crack closure effects. As stated previously, the opening stress remains constant for the block and

$$(\Delta\sigma_{eff})_i = (\sigma_{max})_i - \sigma_{op} \tag{5.29}$$

where σ_{op} is the opening stress associated with the overload. Only loads above this value, σ_{op}, are assumed to cause crack extension on subsequent loading cycles. This is illustrated in Fig. 5.21.

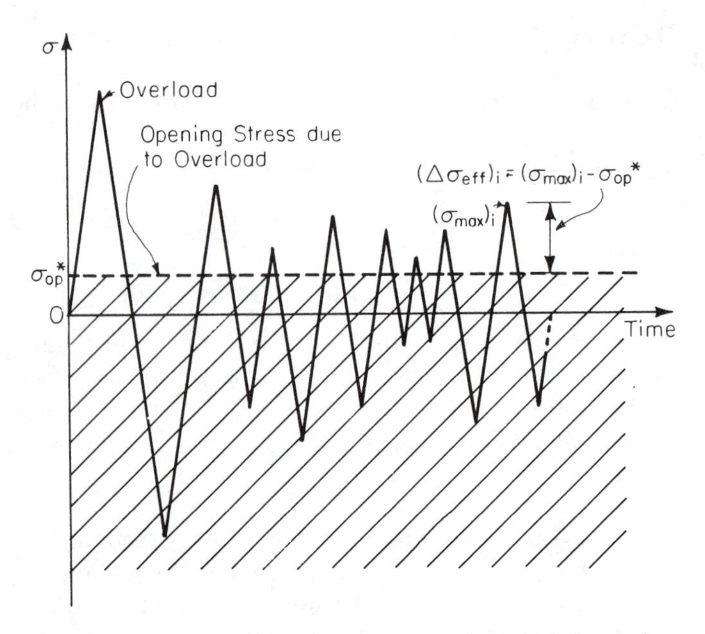

Figure 5.21 For block history only loads above σ_{op} (non-shaded region) assumed to cause crack extension.

2. Determine $(\Delta K_{eff})_i$. Once $(\Delta\sigma_{eff})_i$ is determined, $(\Delta K_{eff})_i$ is evaluated by

$$(\Delta K_{eff})_i = (\Delta\sigma_{eff})_i\sqrt{\pi a}\,f(g) \tag{5.30}$$

3. Calculate the change in crack length per block, $\Delta a/\Delta B$. The change in crack length per block, $\Delta a/\Delta B$, is obtained by summing the change in crack length due to each loading cycle, i.

$$\frac{\Delta a}{\Delta B} = \sum_{i=1}^{n} \Delta a_i \tag{5.31}$$

where n is the number of cycles/block. Using Paris' law,

$$\Delta a_i = \left(\frac{da}{dN}\right)_i = C_0(\Delta K_{eff})_i^m \tag{5.32}$$

(where C_0 and ΔK_{eff} correspond to the same closure level), then

$$\frac{\Delta a}{\Delta B} = \sum_{i=1}^{n} C_0(\Delta K_{eff})_i^m \tag{5.33}$$

$$\frac{\Delta a}{\Delta B} = C_0[f(g)\sqrt{\pi a}]^m \sum_{i=1}^{n} (\Delta\sigma_{eff})_i^m \tag{5.34}$$

Since it was assumed that the loading history can be modeled by a repeating block

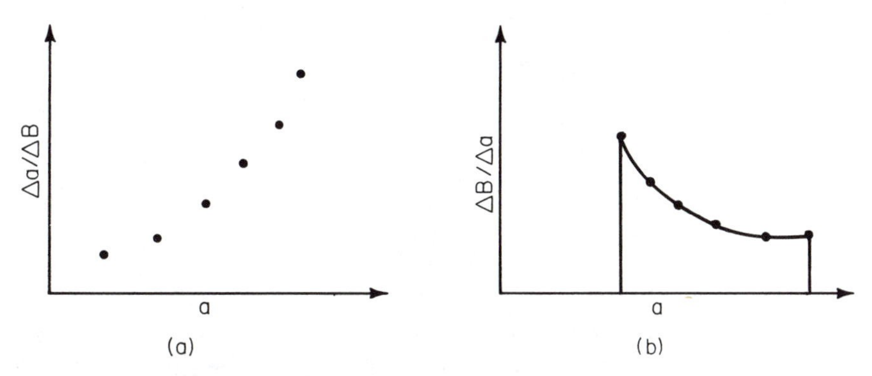

Figure 5.22 (a) Crack growth per block, $\Delta a/\Delta B$, plotted for values of crack length, a; (b) blocks to failure obtained by integrating $\Delta B/\Delta a$.

history, the term

$$\sum_{i=1}^{n} (\Delta\sigma_{\text{eff}})_i^m \tag{5.35}$$

remains constant for every block during the entire life analysis. It does not change with crack length. Thus, to calculate the crack growth per block for the next loading block, only the $f(g)$ terms need to be reevaluated.

4. Determine blocks to failure. Once the crack growth per block, $\Delta a/\Delta B$, for several crack lengths has been determined, fatigue lives may be calculated by integrating the reciprocal, $\Delta B/\Delta a$, to obtain blocks to failure (see Fig. 5.22).

$$B_f = \int_{a_i}^{a_f} \frac{\Delta B}{\Delta a}\, da \tag{5.36}$$

A simple numerical scheme, such as Simpson's rule, may be used to integrate Eq. (5.36).

Example 5.1

An "infinite" double-edge-cracked panel is subjected to the gross tensile stresses shown in the given block of load history (Fig. E5.1). The panel contains two initial cracks of length $a = 0.25$ in. each. The panel is made of a material having the following properties:

$$K_c = 70 \text{ ksi}\sqrt{\text{in.}}$$

$$m = 3.5$$

$$C = 5.0 \times 10^{-10}$$

$$p = 2.0$$

Estimate the life of the panel in loading blocks using Wheeler's model to account for crack growth retardation effects.

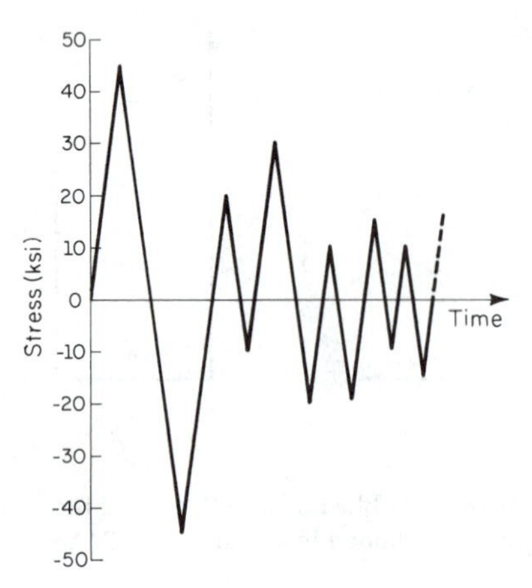

Figure E5.1 Variable-amplitude load history.

Solution In the following application of Wheeler's model to a block loading situation, the assumption is made that the overload is repeatedly applied due to the repetitive nature of the block loading history. Consequently, the crack length, a_i, remains relatively close to the crack length at which the overload was applied, a_0 ($a_i \approx a_0$). In other words, an overload is reapplied before the crack length, a_i, becomes much greater than the initial length, a_0.

The assumption above implies two points. First, the crack length, a, remains constant during a block of loading. Second, the change in crack length per block of loading is small compared to the overload plastic zone size. These points simplify the calculation procedures and allows a diagram such as Fig. 5.22 to be developed. From this the fatigue life (in blocks to failure) can then be determined.

Wheeler's model predicts crack growth retardation through a modification of constant amplitude crack growth data. From Eq. (5.8),

$$\frac{da}{dN_i} = (C_p)_i \left(\frac{da}{dN}\right)_{\text{CA}_i}$$

where $(C_p)_i$ is the Wheeler crack growth retardation parameter. From Eq. (5.10),

$$(C_p)_i = \left(\frac{r_{yi}}{a_p - a_i}\right)^p$$

Defining a_0 as the crack length at which an overload occurs, $a_p = a_0 + r_{0L}$, where r_{0L} is the overload plastic zone size. The retardation parameter, $(C_p)_i$, can then be defined as

$$(C_p)_i = \left(\frac{r_{yi}}{a_0 + r_{0L} - a_i}\right)^p$$

From the assumption above, $a_0 + r_{0L} - a_i \approx r_{0L}$, $(C_p)_i$ can be given as

$$(C_p)_i = \left(\frac{r_{yi}}{r_{0L}}\right)^p$$

Since the equation for the plastic zone size [Eq. (3.5)] is

$$r_y = \frac{1}{\pi\beta}\left(\frac{\Delta K}{\sigma_y}\right)^2$$

and $\Delta K = 1.12\Delta\sigma\sqrt{\pi a}$ for a double-edge-cracked infinite plate, $(C_p)_i$ becomes

$$(C_p)_i = \left(\frac{\Delta K_i}{K_{0L}}\right)^{2p}$$

$$= \left(\frac{\Delta\sigma_i}{\sigma_{0L}}\right)^{2p}\left(\frac{a_i}{a_{0L}}\right)^{p}$$

Since, as assumed above, the change in crack length per block of loading is small in comparison to the overall crack length, $a_i \approx a_0$,

$$(C_p)_i = \left(\frac{\Delta\sigma_i}{\sigma_{0L}}\right)^{2p}$$

Since the constant amplitude crack growth rate is given by

$$\left(\frac{da}{dN}\right)_{\mathrm{CA}_i} = C(\Delta K_i)^m$$

and $\Delta K_i = 1.12\Delta\sigma_i\sqrt{\pi a_i}$ for the given geometry, the retarded crack growth rate is given as

$$\frac{da}{dN_i} = \left(\frac{\Delta\sigma_i}{\sigma_{0L}}\right)^{2p} C(\Delta K_i)^m$$

$$= \left(\frac{\Delta\sigma_i}{\sigma_{0L}}\right)^{2p} C(1.12\Delta\sigma_i\sqrt{\pi a_i})^m$$

The change in crack length per block of loading is equal to the sum of crack growth caused by each cycle in the block. Thus

$$\frac{da}{dB} = \sum_{i=1}^{n}\frac{da}{dN_i} \qquad \text{where } n = \text{number of cycles in the block}$$

$$= \sum_{i=1}^{n}\left(\frac{\Delta\sigma_i}{\sigma_{0L}}\right)^{2p} C(1.12\Delta\sigma_i\sqrt{\pi a})^m$$

(*Note:* $a \approx$ constant over a given loading block.)

$$= \frac{C(1.12\sqrt{\pi})^m}{\sigma_{0L}^{2p}} a^{m/2}\sum_{i=1}^{n}(\Delta\sigma_i)^{2p+m}$$

$$= 5.51 \times 10^{-9}\sigma_{0L}^{-4}a^{1.75}\sum_{i=1}^{n}(\Delta\sigma_i)^{7.5}$$

As discussed in Section 5.5.3 a crack-opening level, σ_{op}, can be assumed to also account for crack retardation effects. A conservative estimate is that $\sigma_{\mathrm{op}} = 0.0$, which implies that only tensile stresses cause crack extension. Therefore,

$$\Delta\sigma = 45, 20, 30, 10, 15, 10 \text{ ksi}$$

and

$$\sigma_{0L} = \sigma_{max} = 45 \text{ ksi}$$

Substituting these values into the relation above gives

$$\frac{da}{dB} = 3.54 \times 10^{-3} a^{1.75}$$

Integrating yields

$$B_f = 282 \int_{a_0}^{a_f} a^{-1.75}\, da$$

$$= 282\left(-\frac{4}{3}a^{-0.75}\right)_{a_0}^{a_f}$$

The initial crack length, a_0, was given as 0.25 in. The final crack length, a_f, is determined from the fracture toughness of the material.

$$K_c = 1.12\sigma_{max}\sqrt{\pi a_f}$$

So

$$a_f = \frac{1}{\pi}\left(\frac{K_c}{1.12\sigma_{max}}\right)^2 = \frac{1}{\pi}\left(\frac{70}{1.12 \times 45}\right)^2 = 0.614 \text{ in.}$$

So total blocks to failure is

$$B_f = \underline{\underline{523 \text{ blocks}}}$$

In cases where the subcycles are much smaller than the overload, the assumptions on which this problem are based are most valid ($a_i \approx a_0$). However, in this case the subcycles do very little damage. A fatigue life estimate based on a constant amplitude history is very close to the estimate obtained in the approach described above. (*Note:* The constant amplitude is equal to the overload amplitude.) In cases where the subcycles significantly contribute to the damage developed, the assumptions are less valid. Consequently, unless detailed calculations are to be done, it may be most appropriate to predict the fatigue life using a constant amplitude crack growth analysis.

5.6 DEALING WITH SERVICE HISTORIES

5.6.1 Introduction

One of the major problems encountered when analyzing a component for fatigue is that the actual service load history may be unknown. In most situations a representative service history, or loading block, is obtained from field tests. Life is then predicted in terms of a number of these blocks, where a block may represent a particular loading event, hours of operation, and so on.

5.6.2 SAE Cumulative Damage Test Program

The Society of Automotive Engineers (SAE) coordinated an extensive study of the fatigue of a component under variable amplitude loading. The results of this study are covered in Ref. 41. Contained in the reference are the results of the test program and several reports that compare these results to life predictions using various analysis techniques.

The "component" used in the study was the notched specimen shown in Fig. 5.23. Two common steels were used in the study, ManTen (S_u = 80 ksi) and RQC-100 (S_u = 120 ksi).

The variable amplitude histories used in the study are shown in Fig. 5.24. The suspension history has a compressive mean value with maneuvering forces superimposed over a random load. The transmission history contains large changes in the mean load value. The bracket history has a pseudorandom distribution.

A series of tests were conducted to provide the following information:

1. Baseline material strain–life and crack growth data.

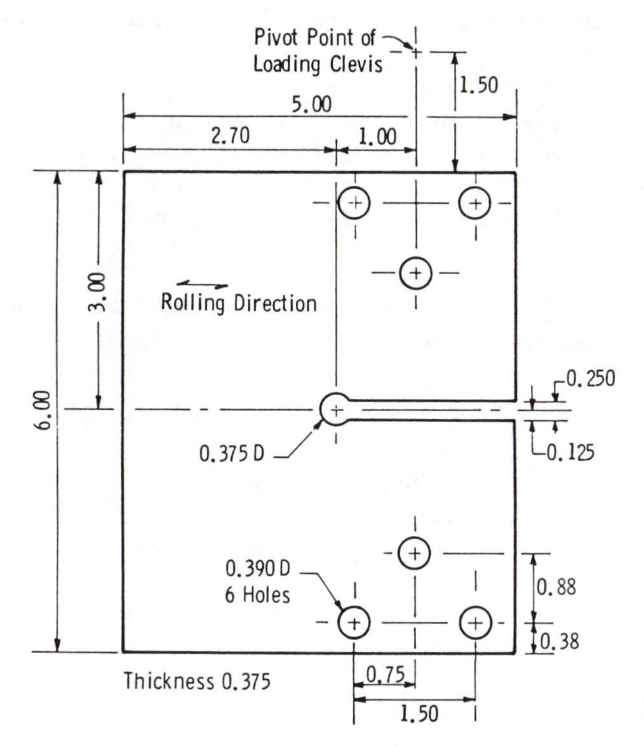

Figure 5.23 SAE fatigue specimen. (From Ref. 42.)

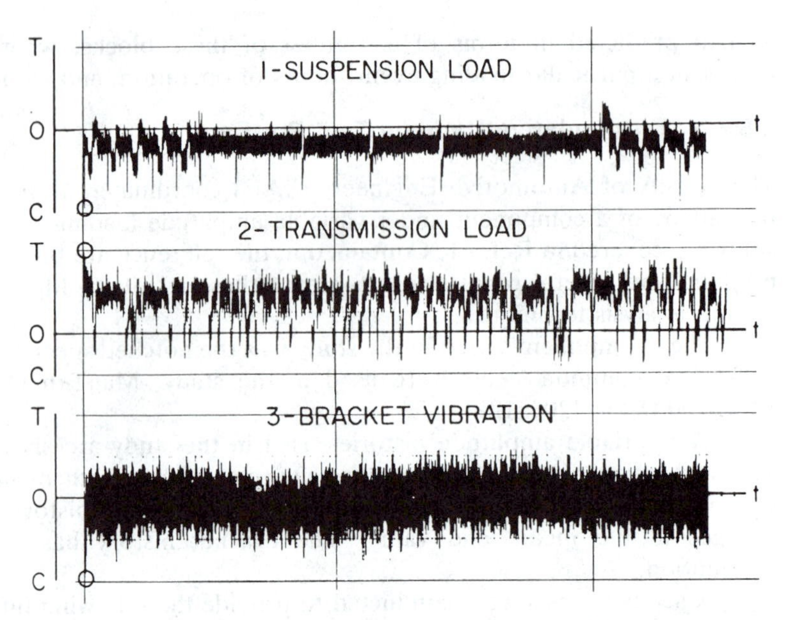

Figure 5.24 SAE load histories. (From Ref. 42.)

2. Constant amplitude component load–life data. These data cover four decades, 100 to 1,000,000 reversals.
3. Variable amplitude component data. These tests used the three load histories, which were scaled to various levels to give a wide range in resulting lives. These lives ranged from 3 blocks to tests which were suspended after 1×10^5 blocks. In each case lives were reported for crack initiation (formation of 0.1 in. crack) and failure. A total of 57 tests were performed.

Numerous analysis techniques were used to predict lives. The various elements of these techniques are summarized below:

1. Rainflow or range pair counting was used to find load ranges in each of the histories.
2. Damage summation was performed using Miner's rule.
3. The life analysis of the specimen was carried out using:
   - **a.** Stress–life method and the fatigue strength reduction factor, K_f.
   - **b.** Load–life curves.
   - **c.** Local strain approach. The local strain was related to load using:
     - **(1)** Neuber analysis using K_f
     - **(2)** Finite element results
     - **(3)** Assumption of elastic strain behavior

(4) Load–strain calibration curves using strain gage measurements

d. Analyses were made in which both mean stress effects were considered (Smith–Watson–Topper and Morrow methods) and ignored.

4. Techniques were also used to condense load histories. These methods allow a shortcut to be taken, where only the critical cycles in a history are included in an analysis.

5. No analysis was made of crack propagation lives.

Results of the various predictions were presented on a graph similar to Fig. 5.25. This graph displays predicted life versus actual life on log–log coordinates. The solid line represents perfect correlation between actual and predicted values. The dashed lines represent a factor of three scatter. Data points that fall below the solid line represent conservative estimates, and points above represent nonconservative predictions. The data plotted on this curve represent two materials and three load histories. A plot such as Fig. 5.25 is valuable for judging the general applicability of a particular analysis procedure.

A large amount of information is presented in Ref. 43. Although it cannot all be reported here, several trends are of interest.

1. There was not a significant difference in the predictions made by any method that used a reasonable estimate of notch root stress–strain behavior.

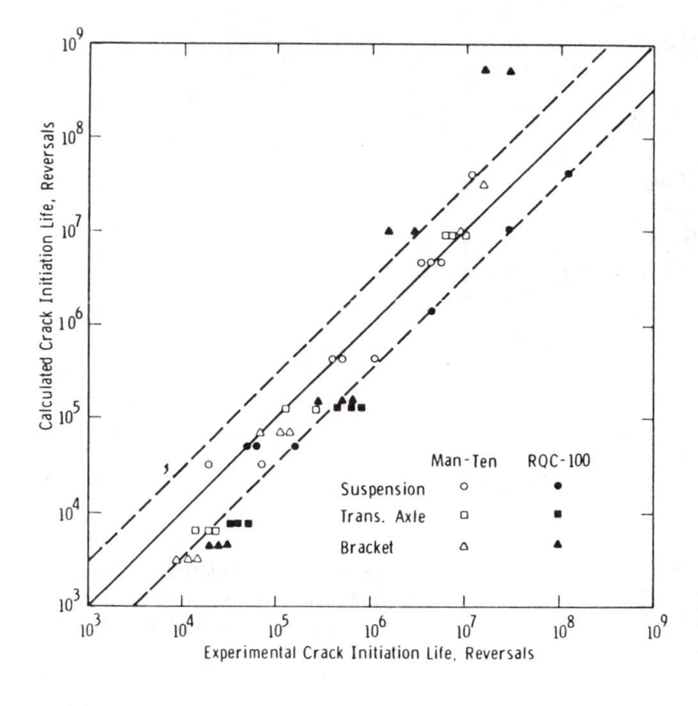

Figure 5.25 Comparison of predicted values using Neuber analysis to actual values. (From Ref. 43.)

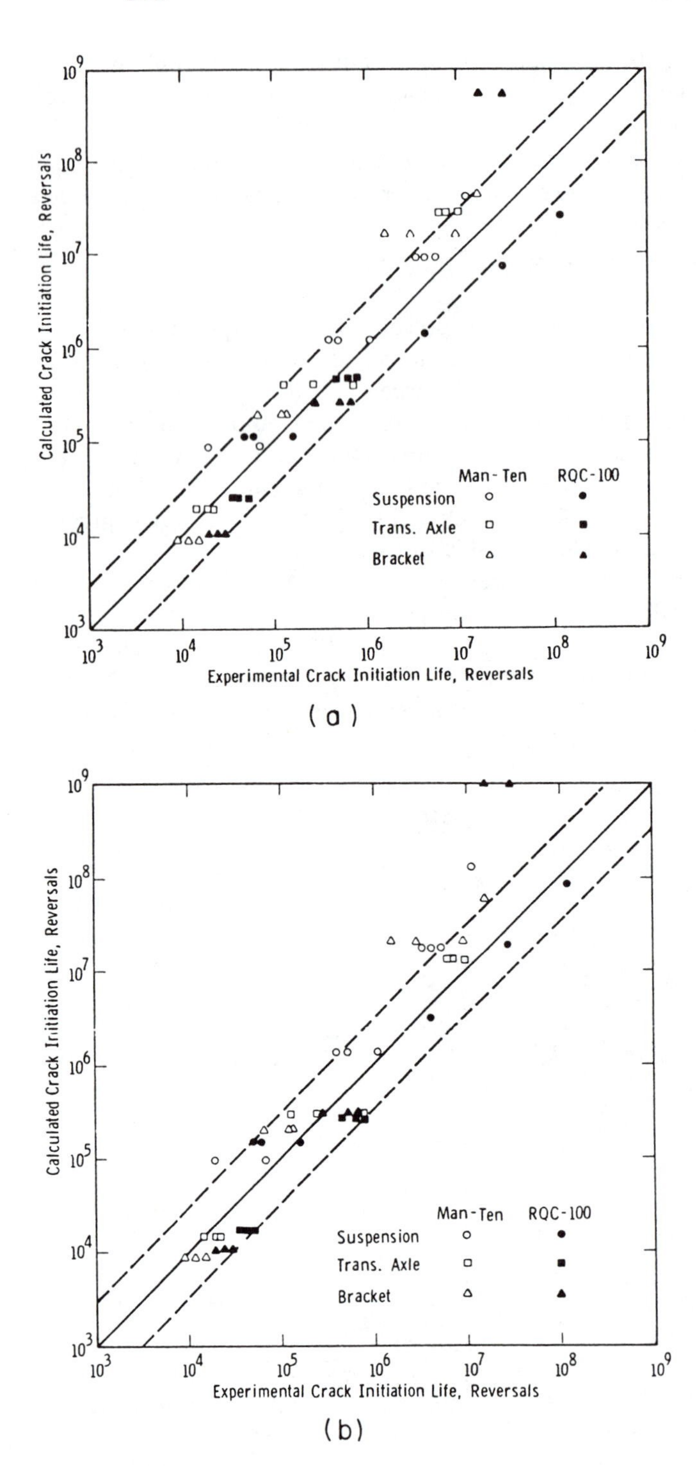

Figure 5.26 Effect of mean stress on life predictions: (a) mean stress ignored; (b) mean stress considered. (From Ref. 43.)

2. Good predictions were made using the Neuber approach (see Fig. 5.25). This method tended to be slightly conservative.
3. There was not a large difference between predictions which included and excluded mean stresses (see Fig. 5.26).

An interesting point is that predictions made using the simple stress–life method showed correlation which was as good as those predicted by more complicated techniques (see Fig. 5.27). The bars on this graph represent the scatter of experimental results and the numbers represent specimen ID numbers.

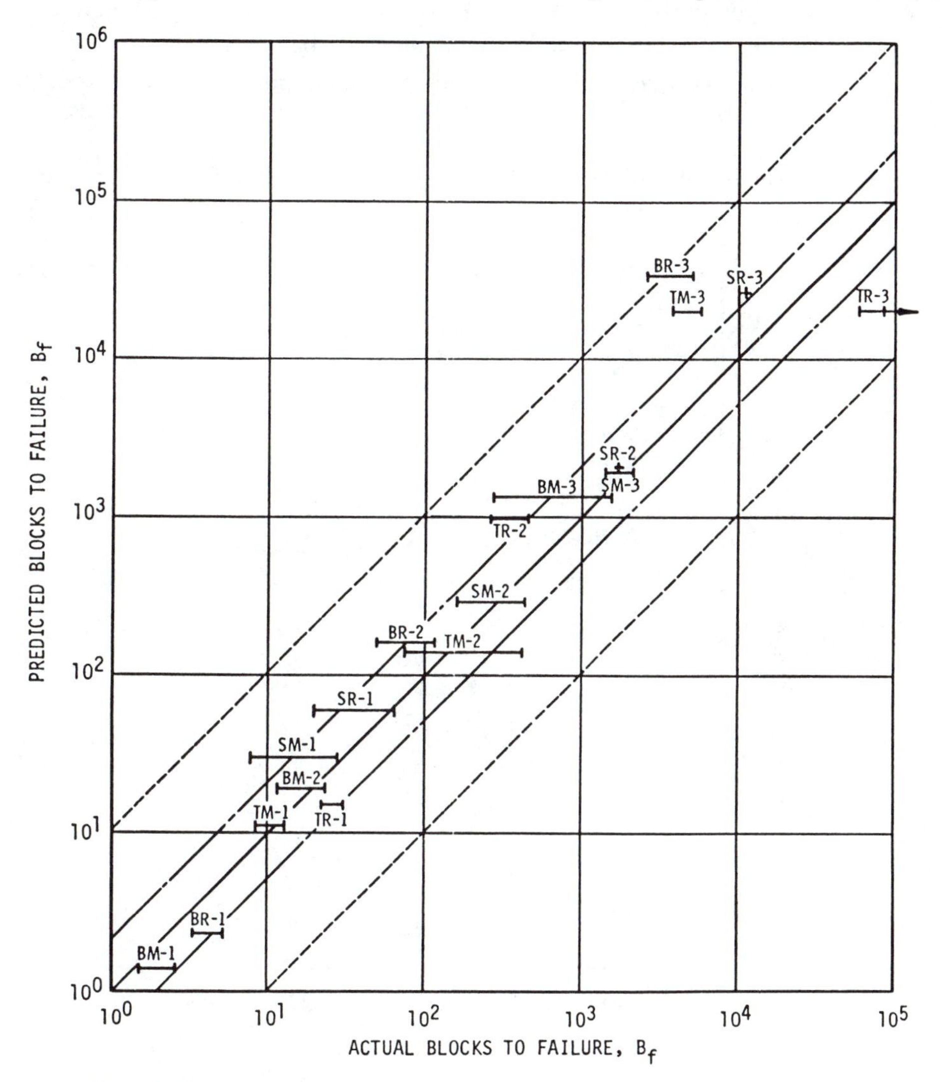

Figure 5.27 Comparison of predicted values to actual values using stress-life approach. (From Ref. 44.)

A separate study [45] compares predicted propagation lives to the test results of the SAE cumulative damage program. The prediction technique used was the block loading method outlined in Section 5.5.3. The assumptions made in the analysis were:

1. Crack opening levels, in terms of load, were determined from finite element models.
2. Damaging events were determined by using a rainflow counting procedure on the portion of the load history above the crack opening load.
3. The initial crack length was assumed to be 0.1 in.
4. The final crack length was determined using the fracture toughness of the material.

The results of the analysis are shown in Fig. 5.28. As can be seen, this

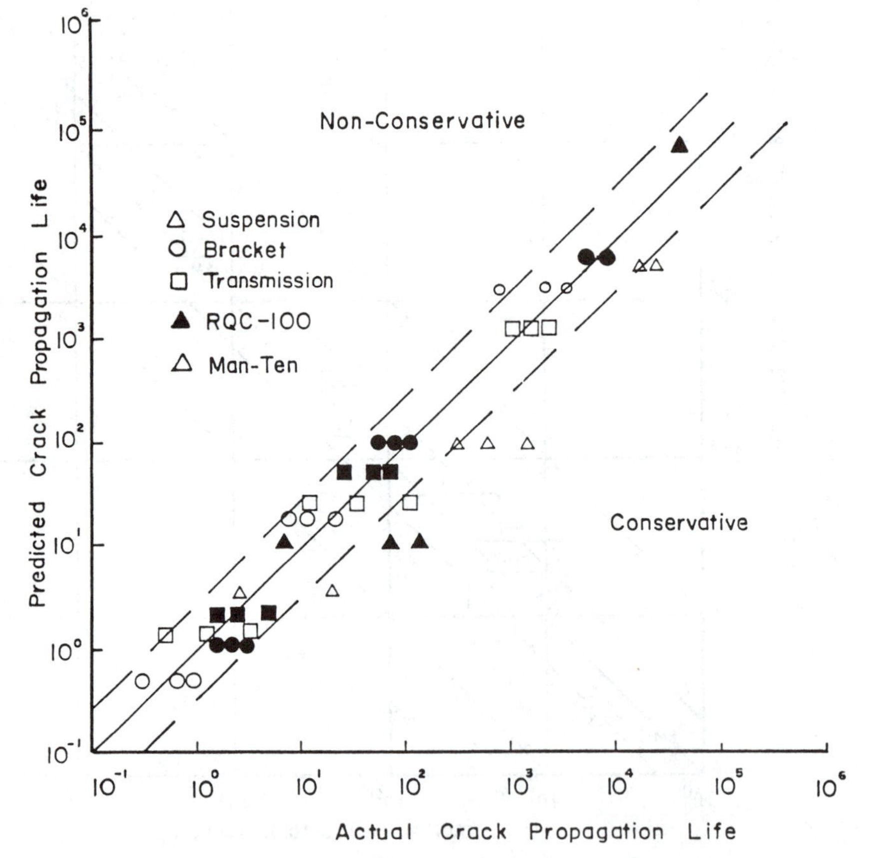

Figure 5.28 Comparison of predicted propagation lives and actual values. (From Ref. 45.)

method gives good estimates of propagation life. These predictions are good even at high load levels where LEFM concepts (small-scale yielding) are violated.

5.7 IMPORTANT CONCEPTS

- The linear damage rule (Miner's rule) provides reasonable life estimates for a wide variety of loading histories. This is accomplished even though the method does not consider sequence effects and is amplitude independent.
- The most effective cycle counting procedures relate damaging events to the stress–strain response of the material. These counting procedures define a damaging event as a hysteresis loop and determine the strain range and mean stress corresponding to the loop.
- Variable amplitude loading histories can frequently be represented as a repeating block of load cycles. Life may then be determined by calculating the damage per block and then summing the number of blocks to failure.
- Application of a large tensile overload may cause crack growth retardation.

5.8 IMPORTANT EQUATIONS

Miner's rule

$$\sum D_i \geq 1 \tag{5.1}$$

$$\sum \frac{n_i}{N_i} \geq 1 \tag{5.2}$$

Change in Crack Length per Block

$$\frac{\Delta a}{\Delta B} = \sum_{i=1}^{n} \Delta a_i \tag{5.31}$$

Blocks to Failure

$$B_f = \int_{a_i}^{a_f} \frac{\Delta B}{\Delta a}\, da \tag{5.36}$$

REFERENCES

1. A. Palmgren, "Durability of Ball Bearings," *ZVDI,* Vol. 68, No. 14, 1924, pp. 339–341 (in German).
2. M. A. Miner, "Cumulative Damage in Fatigue," *J. Appl. Mech.*, Vol. 12, *Trans. ASME,* Vol. 67, 1945, pp. A159–A164.

3. G. Sines and J. L. Waisman, (eds.), *Metal Fatigue,* McGraw-Hill, New York, 1959.
4. A. F. Madayag (ed.), *Metal Fatigue: Theory and Design,* Wiley, New York, 1969.
5. J. A. Collins, *Failure of Materials in Mechanical Design,* Wiley-Interscience, New York, 1981.
6. F. E. Richart and N. M. Newmark, "An Hypothesis for the Determination of Cumulative Damage in Fatigue," *Am. Soc. Test. Mater. Proc.,* Vol. 48, 1948, pp. 767–800.
7. S. M. Marco and W. L. Starkey, "A Concept of Fatigue Damage," *Trans. ASME,* Vol. 76, No. 4, 1954, pp. 627–632.
8. C. T. Hua, *Fatigue Damage and Small Crack Growth during Biaxial Loading,* Materials Engineering-Material Behavior Report No. 109, University of Illinois at Urbana–Champaign, July 1984.
9. J. H. Crews, Jr., "Crack Initiation at Stress Concentrations as Influenced by Prior Local Plasticity," in *Achievement of High Fatigue Resistance in Metals and Alloys,* ASTM STP 467, American Society for Testing and Materials, Philadelphia, 1970, p. 37.
10. M. Matsuishi and T. Endo, "Fatigue of Metals Subjected to Varying Stress," paper presented to Japan Society of Mechanical Engineers, Fukuoka, Japan, March 1968.
11. American Society for Testing and Materials, *Annual Book of ASTM Standards,* Section 3: Metals Test Methods and Analytical Procedures, Vol. 03.01-Metals-Mechanical Testing; Elevated and Low-Temperature Tests, ASTM, Philadelphia, 1986, pp. 836–848.
12. T. Endo et al., "Damage Evaluation of Metals for Random or Varying Loading," *Proceedings of the 1974 Symposium on Mechanical Behavior of Materials,* Vol. 1, The Society of Materials Science, Kyoto, Japan, 1974, pp. 371–380.
13. H. Anzai and T. Endo, "On-Site Indication of Fatigue Damage under Complex Loading," *Int. J. Fatigue,* England, Vol. 1, No. 1, 1979, pp. 49–57.
14. T. Endo and H. Anzai, "Redefined Rainflow Algorithm: *P/V* Difference Method," *J. Jpn. Soc. Mater. Sci.,* Vol. 30, No. 328, 1981, pp. 89–93.
15. A. Burns, "Fatigue Loadings in Flight: Loads in the Tailpipe and Fin of a Varsity," Tech. Report C.P. 256, Aeronautical Research Council, London, 1956.
16. Vickers-Armstrongs Ltd., "The Strain Range Counter," VTO/M/416, Vickers-Armstrongs Ltd. (now British Aircraft Corporation Ltd.), Technical Office, Weybridge, Surrey, England, Apr. 1955.
17. F. D. Richards, N. R. LaPointe, and R. M. Wetzel, "A Cycle Counting Algorithm for Fatigue Damage Analysis," Paper No. 740278, *Automot. Eng. Congr.* Society of Automotive Engineers, Detroit, Mich., Feb. 1974.
18. D. V. Nelson and H. O. Fuchs, "Predictions of Cumulative Fatigue Damage Using Condensed Load Histories," in *Fatigue under Complex Loading: Analyses and Experiments,* Vol. AE-6, R. M. Wetzel (ed.), The Society of Automotive Engineers, Warrendale, Pa., 1977, pp. 163–187.
19. H. O. Fuchs et al., "Shortcuts in Cumulative Damage Analysis," in *Fatigue under Complex Loading: Analyses and Experiments,* Vol. AE-6, R. M. Wetzel (ed.), The Society of Automotive Engineers, Warrendale, Pa., 1977, pp. 145–162.
20. G. M. van Dijk, "Statistical Load Data Processing," *6th ICAF Symp.,* Miami, Fla.,

May 1971; see also NLR MP 71007U, National Aerospace Laboratory, Amsterdam, 1971.

21. J. E. Hayes, "Fatigue Analysis and Fail-Safe Design," in *Analysis and Design of Flight Vehicle Structures,* E. F. Bruhn (ed.), Tristate Offset Co., Cincinnati, Ohio, 1965, pp. C13-1–C13-42.
22. S. D. Downing and D. F. Socie, "Simplified Rainflow Counting Algorithms," *Int. J. Fatigue,* Vol. 4, No. 1, 1982, pp. 31–40.
23. J. Schijve, "Fatigue Crack Propagation in Light Alloy Sheet Material and Structures," Report MP-195, National Aerospace Laboratory, Amsterdam, August 1960.
24. C. M. Hudson and H. F. Hardrath, "Investigation of the Effects of Variable Amplitude Loadings on Fatigue Crack Propagation Patterns," NASA TN-D-1803, 1963.
25. C. M. Hudson and H. F. Hardrath, "Effects of Changing Stress Amplitude on the Rate of Fatigue Crack Propagation in Two Aluminum Alloys," NASA TN-D-960, Sept. 1961.
26. R. H. Christensen, *Proc. Crack Propagation Symp.,* Cranfield, England, 1961.
27. E. J. F. von Euw, R. W. Hertzberg, and R. Roberts, "Delay Effects in Fatigue Crack Propagation," ASTM STP 513, American Society for Testing and Materials, Philadelphia, 1972.
28. W. J. Mills and R. W. Hertzberg, *Eng. Fract. Mech.,* Vol. 8, 1976.
29. R. W. Hertzberg, *Deformation and Fracture Mechanics of Engineering Materials,* 2nd ed., Wiley, New York, 1983.
30. D. V. Nelson, "Review of Fatigue Crack-Growth Prediction under Irregular Loading," *Spring Meet.,* Society for Experimental Stress Analysis, 1975.
31. R. C. McClung and H. Sehitoglu, "Closure Behavior of Small Cracks under Strain Fatigue Histories," *ASTM Symp. Fatigue Crack Closure,* Charleston, S.C., May 1986 (to be published in ASTM STP 982 "Mechanics of Fatigue Crack Closure," 1988, pp. 279–299.)
32. O. E. Wheeler, "Spectrum Loading and Crack Growth," *J. Basic Eng., Trans. ASME,* Vol. D94, No. 1, 1972 pp. 181–186.
33. J. Willenborg, R. M. Engle, and H. A. Wood, "A Crack Growth Retardation Model Using an Effective Stress Concept," AFFDL TM-71-1-FBR, Jan. 1971.
34. J. M. Barsom, "Fatigue Crack Growth under Variable Amplitude Loading in ASTM 514-B Steel," in *Progress in Flaw Growth and Fracture Toughness Testing,* ASTM STP 536, American Society for Testing and Materials, Philadelphia, 1973, pp. 147–167.
35. J. B. Chang, M. Szamossi, and K.-W. Liu, "Random Spectrum Fatigue Crack Life Predictions with or without Considering Load Interactions," in *Methods and Models for Predicting Fatigue Crack Growth under Random Loading,* ASTM STP 748, American Society for Testing and Materials, Philadelphia, 1981, pp. 115–132.
36. J. C. Newman, Jr., "A Crack Closure Model for Predicting Fatigue Crack Growth under Aircraft Spectrum Loading," in *Methods and Models for Predicting Fatigue Crack Growth under Random Loading,* ASTM STP 748, American Society for Testing and Materials, Philadelphia, 1981, pp. 53–84.
37. J. C. Newman, Jr., "Prediction of Fatigue Crack Growth under Variable-Amplitude

and Spectrum Loading Using a Closure Model," in *Design of Fatigue and Fracture Resistant Structures,* ASTM STP 761, American Society for Testing and Materials, Philadelphia, 1982, pp. 255–277.

38. H. D. Dill and C. R. Saff, "Spectrum Crack Growth Prediction Methods Based on Crack Surface Displacement and Contact Analyses," in *Fatigue Crack Growth under Spectrum Loads,* ASTM STP 595, American Society for Testing and Materials, Philadelphia, 1976, pp. 306–319.
39. H. D. Dill, C. R. Saff, and J. M. Potter, "Effects of Fighter Attack Spectrum on Crack Growth," ASTM STP 714, American Society for Testing and Materials, Philadelphia, 1980, pp. 205–217.
40. T. R. Brussat, "An Approach to Predicting the Growth to Failure of Fatigue Cracks Subjected to Arbitrary Uniaxial Cyclic Loading," in *Damage Tolerance in Aircraft Structures,* ASTM STP 486, American Society for Testing and Materials, Philadelphia, 1971, pp. 122–143.
41. R. M. Wetzel (ed.), *Fatigue under Complex Loading: Analyses and Experiments,* Society of Automotive Engineers, Warrendale, Pa., 1977.
42. L. Tucker and S. Bussa, "The SAE Cumulative Fatigue Damage Test Program," in *Fatigue under Complex Loading: Analysis and Experiments,* R. M. Wetzel (ed.), Society of Automotive Engineers, Warrendale, Pa., 1977, p. 1.
43. N. E. Dowling, W. R. Brose, and W. K. Wilson, "Notched Member Fatigue Life Predictions by the Local Strain Approach," in *Fatigue under Complex Loading: Analysis and Experiments,* R. M. Wetzel (ed.), Society of Automotive Engineers, Warrendale, Pa., 1977, p. 55.
44. L. Tucker, S. Downing, and L. Camillo, "Accuracy of Simplified Fatigue Prediction Methods," in *Fatigue under Complex Loading: Analysis and Experiments,* R. M. Wetzel (ed.), Society of Automotive Engineers, Warrendale, Pa., 1977, p. 137.
45. D. F. Socie, "Estimating Fatigue Crack Initiation and Propagation Lives in Notched Plates under Variable Loading Histories," T.&A.M. Report No. 417, University of Illinois, June, 1977.
46. S. J. Stadnick and JoDean Morrow, "Techniques for Smooth Specimen Simulation of the Fatigue Behavior of Notched Members," in *Testing for Prediction of Material Performance in Structures and Components,* ASTM STP 515, American Society for Testing and Materials, Philadelphia, 1972, pp. 229–252.

PROBLEMS

SECTION 5.3

5.1. The safe operation of a military aircraft, in terms of fatigue life, depends on the type of service for which it is used. One particular type of aircraft has a rated life of 1000, 600, or 2000 hours when used for combat, training, or surveillance missions, respectively. One of these aircraft has seen 50 hours of combat service, 300 hours of training service, and 400 hours of surveillance service. A second of these aircraft has seen 500 hours, 150 hours, and 300 hours of combat, training, and surveillance service, respectively. Using Miner's rule, answer the following questions.

(a) How many hours of training service life may each aircraft be used for before it must be overhauled?

(b) Combat and surveillance missions, each of 200 hours, must be completed. Which aircraft should be sent on each mission, and how much useful combat life will each aircraft have remaining when it returns?

5.2. Using the strain–life relationship, determine the number of blocks to failure for components made from SAE 1020 (100 BHN) and SAE 1040 (400 BHN) carbon steel subjected to the strain history listed below. Neglect sequence (mean stress) effects. (*Hint:* The BHN of each material can be used to obtain approximate values of the constants in the strain–life relation.)

Block loading

Strain Range	Number of Cycles
0.02	10
0.01	20
0.006	200
0.003	1000

(a) What percentage of damage does each type of cycle contribute to a block of loading?

(b) What is the constant amplitude strain range at which these two materials have equal lives? What is this life value? What would this suggest regarding the general nature of the block loading?

(c) What is the maximum constant amplitude strain range at which each material could survive 100 cycles? Which material is better suited to operation in this loading regime?

5.3. A steel component has a 1000 cycle fatigue strength of 90 ksi and an endurance limit of 50 ksi. Using Miner's rule, answer the following questions.

(a) What is the fatigue life of this component at 60 ksi and at 80 ksi?

(b) If the component is first subjected to 1000 cycles at a stress amplitude of 80 ksi and then to 50,000 cycles at a stress amplitude of 60 ksi, what is the cumulative damage of this load history?

(c) Plot the shift in the S–N curve due to the applied load history using the method discussed in Section 5.3. Using this method does the final result depend on the order in which the cycles are applied?

(d) What is the remaining life of the component if it is subjected to a stress amplitude of 70 ksi?

5.4. A component is subjected to two blocks of cycles, n_1 and n_2, at two different stress levels, S_1 and S_2, respectively. Show algebraically that the graphical interpretation of Miner's rule (a shift in the S–N curve with no change in slope) provides identical results irrespective of the order in which the cycles are applied.

SECTION 5.4

5.5. For the given block of strain history shown below:

(a) Obtain a cycle count using the level-crossing counting method.

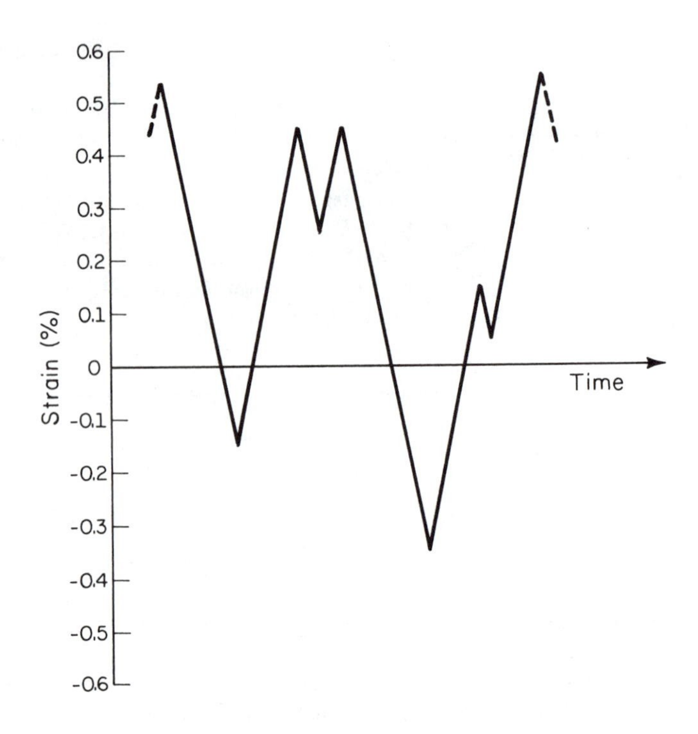

(b) Perform a life analysis for a soft steel ($\sigma_f' = 100$ ksi, $b = -0.10$, $\epsilon_f' = 1.2$, $c = -0.5$) using Miner's linear damage rule. (Ignore mean stress effects.)

(c) What percentage of damage does each cycle contribute to the block loading? What does this suggest regarding the effects of very small subcycles on the overall life of a variable amplitude loading history?

5.6. For the block of strain history given in Problem 5.5:

(a) Obtain a cycle count using the peak counting method.

(b) Perform a life analysis for a medium strength steel ($\sigma_f' = 175$ ksi, $b = -0.085$, $\epsilon_f' = 0.8$, $c = -0.6$) using Miner's linear damage rule. (Ignore mean stress effects.)

(c) What percentage of damage does each cycle contribute to the block loading? What does this suggest regarding the effects of very small subcycles on the overall life of a variable amplitude loading history?

5.7. For the block of strain history given in Problem 5.5:

(a) Obtain a cycle count using the simple-range counting method.

(b) Perform a life analysis for a hard steel ($\sigma_f' = 250$ ksi, $b = -0.07$, $\epsilon_f' = 0.5$, $c = -0.7$) using Miner's linear damage rule. (Ignore mean stress effects.)

(c) What percentage of damage does each cycle contribute to the block loading? What does this suggest regarding the effects of very small subcycles on the overall life of a variable amplitude loading history?

5.8. For the given block (see figure) of load history, obtain a cycle count using the rainflow counting method. For each cycle, determine the range and mean values. How would one edit this history, while still maintaining its damage content, to allow

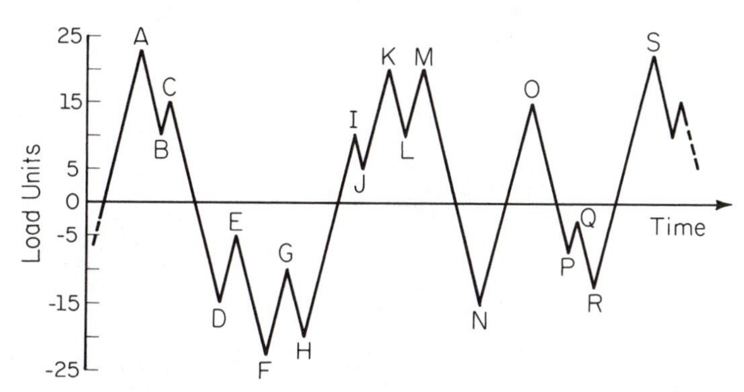

a life prediction to be made from a relatively few number of constant amplitude tests? Sketch the σ–ϵ response of the block loading, indicating closed hysteresis loops. (Assume that the load units given are strain increments.)

5.9. A critical location on a steel component is repeatedly subjected to the following high cycle fatigue block loading of tensile stresses (ksi):

$$0, 40, 20, 30, 5, 55, 20, 30, 10, 45, \text{ repeat}$$

The material has the following fatigue properties:

$$E = 30{,}000 \text{ ksi} \qquad \sigma_f' = 150 \text{ ksi} \qquad \epsilon_f' = 0.9$$

$$b = -0.095 \qquad c = -0.55$$

(a) Obtain a cycle count of the load history using the rainflow counting method.
(b) Estimate the life of the component in blocks to failure, neglecting mean stress effects.
(c) Estimate the life of the component in blocks to failure, including mean stress effects.
(d) How do the two previously determined life predictions compare?
(e) When is use of the method of life analysis in step (b) not appropriate?

5.10. Two cyclically stabilized smooth steel specimens are subjected to the given strain history (see the figure below). The steel has the following fatigue properties:

$$E = 30{,}000 \text{ ksi}$$

$$\sigma_f' = 150 \text{ ksi} \qquad \epsilon_f' = 0.8 \qquad n' = 0.20$$

$$b = -0.10 \qquad c = -0.5 \qquad K' = 157 \text{ ksi}$$

(a) Plot the σ–ϵ response of each load history.
(b) Calculate the number of reversals to failure for each strain history. (In each history neglect the damage caused by the initial setup cycle.) Use Morrow's modification to the strain–life equation [Eq. (2.49)] to account for mean stress effects.
(c) Calculate the number of blocks to failure for each history if a large strain cycle occurs after every 1000 reversals of the smaller reversals. Use Morrow's modification to the strain–life equation to account for mean stress effects.
(d) How do these two life predictions compare?

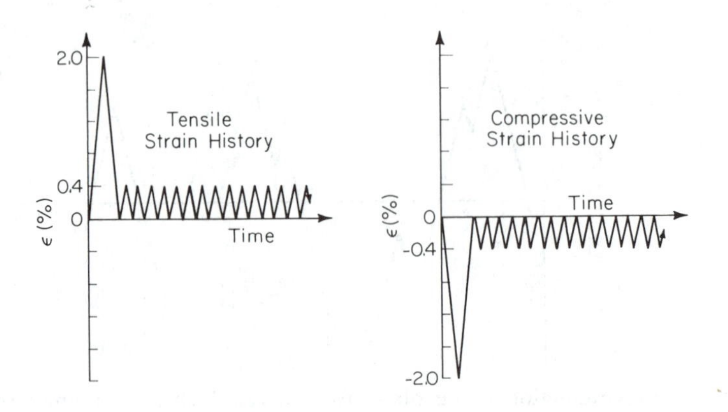

5.11. For the block of strain history given in Problem 5.5:

(a) Obtain a cycle count of the strain history using the rainflow counting method.

(b) Plot the σ–ϵ response of each strain history.

(c) Perform a life analysis for a medium strength steel ($\sigma_f' = 175$ ksi, $b = -0.085$, $\epsilon_f' = 0.8$, $c = -0.6$, $K' = 181$ ksi, $n' = 0.14$) using Miner's linear damage rule. Use Manson and Halford's modification to the strain–life equation [Eq. (2.50)] to account for mean stress effects.

(d) What percentage of damage does each cycle contribute to the block loading? What does this suggest regarding the effects of very small subcycles on the overall life of a variable amplitude loading history?

5.12. Notched ($K_f = 1.95$) aluminum specimens were subjected to the variable load history shown below. Tests were run with the history amplified to three different levels. The peak values for these three cases are given in the table in terms of nominal stress. Determine the life, in terms of blocks, of the notched specimens for the three cases. Use a Neuber analysis, Morrow mean stress–strain life relationship [Eq. (2.49)], rainflow count, and Miner's rule for these calculations. The stress–strain and strain–life properties of the aluminum are

$$E = 10 \times 10^3 \text{ ksi} \qquad K' = 243 \text{ ksi} \qquad n' = 0.146$$

$$\sigma_f' = 191 \text{ ksi} \qquad b = -0.126$$

$$\epsilon_f' = 0.19 \qquad c = -0.52$$

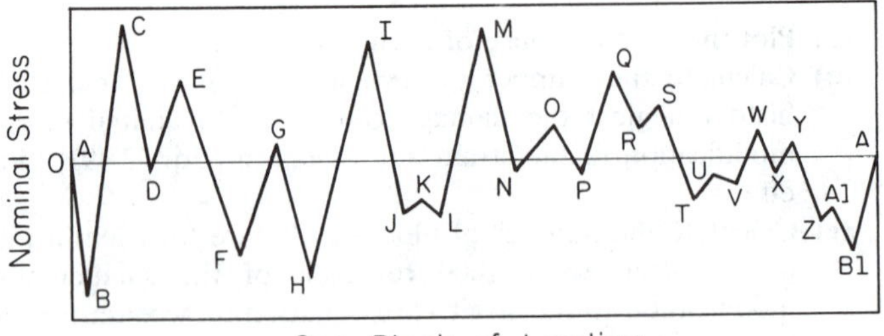

Peak	Nominal Stress (ksi)		
	Level A	Level B	Level C
A	0	0	0
B	−50.5	−48.0	−45.5
C	51.6	49.0	46.4
D	−3.2	−3.0	−2.9
E	30.5	29.0	27.4
F	−35.8	−34.0	−32.2
G	5.3	5.0	4.8
H	−46.3	−44.0	−41.7
I	45.3	43.0	40.8
J	−20.0	−19.0	−18.0
K	−15.8	−15.0	−14.2
L	−22.1	−21.0	−19.9
M	49.5	47.0	44.6
N	−4.9	−4.7	−4.4
O	11.6	11.0	10.4
P	−4.2	−4.0	−3.8
Q	32.6	31.0	29.3
R	11.6	11.0	10.4
S	18.9	17.9	17.0
T	−15.8	−15.0	−14.2
U	−8.4	−8.0	−7.6
V	−10.5	−10.0	−9.5
W	8.4	8.0	7.6
X	−6.3	−6.0	−5.7
Y	4.2	4.0	3.8
Z	−25.3	−24.0	−22.8
A1	−20.0	−19.0	−18.0
B1	−35.8	−34.0	−32.2

The actual test results for the notched specimens are given below. Compare the predicted and actual lives. (Data for the problem taken from Ref. 46.)

	Life to Failure in Blocks		
	Level A	Level B	Level C
Notched specimen	261	420	597

SECTION 5.5

5.13. A very thin component with a crack is subjected to a 0 to 20 ksi alternating stress. When the crack in the component reaches a given length (a_0), a 40 ksi tensile overload is applied. The structure of the component is such that the stress intensity is given by the relation $\Delta K = 1.2\Delta\sigma\sqrt{\pi a}$. The component material properties are

$$\sigma_y = 90 \text{ ksi}$$

$$p = 2.0 \text{ (shaping parameter)}$$

$$m = 4.0 \text{ (Paris exponent)}$$

Plot the retarded crack growth rate (normalized with respect to the constant amplitude crack growth rate) as a function of crack length (normalized with respect to the crack length at which the overload occurred). Use both the Wheeler and the Willenborg models. How do the two results compare?

5.14. An "infinite" center-cracked panel is subjected to the uniform gross tensile stresses shown in the given block of load history. The panel contains an initial through thickness crack of length $2a = 0.4$ in. and is made of a material having the following properties:

$$K_c = 50 \text{ ksi}\sqrt{\text{in.}}$$

$$m = 3.0$$

$$C = 1.0 \times 10^{-10}$$

$$p = 2.5$$

(a) Predict the life of the panel in loading blocks using Wheeler's model to account for crack growth retardation effects. (*Hint:* Assume that the crack growth per block is small in comparison to the overall crack length.)

(b) What would be the predicted life (blocks to failure) if only damage caused by the overload cycles were considered (i.e., neglecting the damage caused by the smaller cycles)?

(c) How do these two life predictions compare?

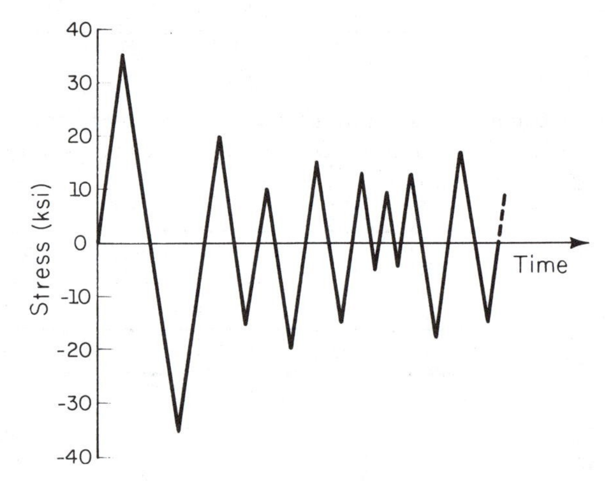

5.15. An "infinite" center-cracked panel is subjected to the uniform gross tensile stresses shown in the block loading history given in Problem 5.14. The panel contains an initial through thickness crack of length $2a = 0.4$ in. and is made of a material having the following properties:

$$K_c = 50 \text{ ksi} \sqrt{\text{in.}}$$

$$m = 3.0$$

$$C = 1.0 \times 10^{-10}$$

Predict the life of the panel in loading blocks using Barsom's model based on the root-mean-square stress intensity factor.

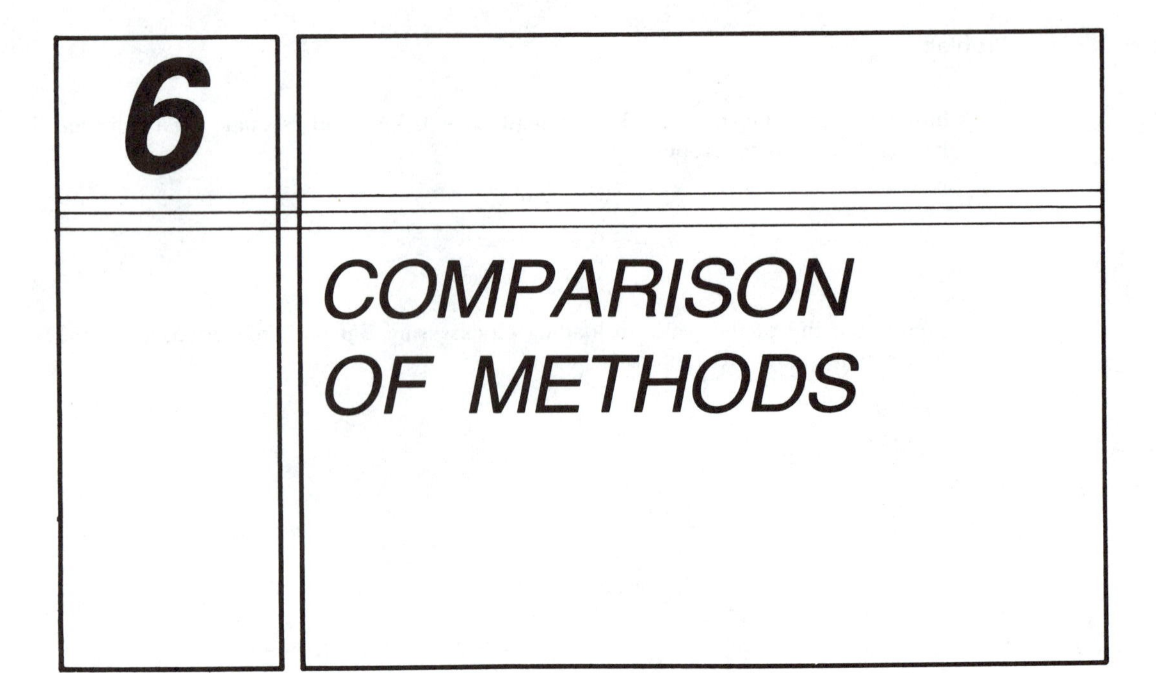

6.1 INTRODUCTION

In the previous five chapters, three different methods of fatigue analysis have been discussed. The purpose of this chapter is to make a comparison of these techniques and point out their relative strengths and limitations. This comparison is intended to be general in nature and is by no means meant to be definitive.

In the following discussion the stress–life, strain–life, and fracture mechanics approaches will be referred to as S–N, ϵ–N, and LEFM, respectively.

6.2 GENERAL POINTS FOR COMPARISON

Before discussing the individual approaches it is important to consider some general points related to making a comparison of the methods.

1. Are the methods to be used in the design cycle or to analyze an existing component?
2. What is the accuracy of the methods compared to input variables such as load history and material properties?
3. What are the relative economics?
4. What is the level of acceptance?
5. Uses in design versus research.

Any of the fatigue life estimation techniques can be used for either the initial sizing and design of a new component or for the analysis of an existing component. The latter case may involve a component that is experiencing in-service problems or a component for which service usage is altered. In either the case of design or analysis, the techniques are the same. The only difference is where and how they are used in the procedure to solve the problem at hand.

When any of the three methods are used in the design of new components, they all have the same objective, to correlate the behavior of small laboratory specimens to large structures or components. The initial design phase that sizes the component using baseline material data is incorporated into an overall design cycle which usually includes component testing. As the predictive capability of a method improves, it reduces the high cost of testing redesigned components. Currently, none of the methods are accurate enough to completely eliminate the need for component testing. Another important point is that the precision of any of the techniques is a function of how well the input variables are defined. The relative accuracy of the methods may be a minor consideration when compared to the unknowns involved in predicting service loading and environment. In most cases component testing and in-service simulation can be eliminated from the design cycle only at the expense of grossly overdesigning the component.

Economics of time and money is an important consideration when selecting an analysis technique. The S–N approach is the quickest and cheapest of the approaches (it is often referred to as a "back of the envelope" method), but the advantages of the other methods may far outweigh cost considerations. In recent years the development of less expensive computers and general analysis programs has greatly reduced the computational costs involved with any of the techniques. What is probably a more important economic consideration is that the computational costs may be quite small in relation to the costs involved in obtaining material fatigue constants and representative service loading histories. In addition, the consequences of catastrophic failure may dictate the use of more sophisticated analysis methods.

Very often a particular method is chosen on the basis of its level of acceptance. This acceptance is related to the amount of confidence the designer has in the method. The years since the introduction of the S–N, ϵ–N, and LEFM approaches are about 100, 30, and 20 years, respectively. Designers have had many more years of experience with the S–N approach and with this experience has come a greater level of confidence. The ϵ–N and LEFM methods may give better insights into the mechanisms of fatigue, but as of yet, some industries have not gained widespread confidence or experience in the use of the these approaches. This confidence will come only by correlating component design to service experience. In some industries, selection of an approach is specified by design code.

The level of acceptance may also be based on the apprehension felt for "new, analytical" techniques. The S–N approach is the technique usually taught at the undergraduate level in strength of materials, material science, and machine design courses. It only requires an understanding of elastic stress analysis while the other methods require increasing levels of technical background. The ϵ–N approach requires an understanding of elastic–plastic stress–strain properties, and the LEFM approach requires a basic background in fracture mechanics. Both of these topics are usually taught at the graduate level. This means that a method may be overlooked due only to a lack of understanding of more advanced strength of material concepts. The opposite can also be true: A method may be chosen simply because it is considered to be "high tech" when a technique requiring less analysis may be appropriate.

A distinction must be made between the use of these techniques in design and research. The S–N, ϵ–N, and LEFM approaches give increasing levels of understanding into the basic mechanisms of fatigue. The S–N approach is the most common design method but is of almost no interest to researchers. On the other hand, the LEFM approach is a current research topic, but as a design tool it is almost unknown outside the aerospace and nuclear industries.

Two problems must be considered: transfer of technology and design experience. Designers must be aware of and understand the available analytical tools. They must also gain experience in the application of these methods. Only through the understanding and application of these techniques can it be determined if they will provide better design tools.

There is no "general" fatigue analysis method that is best for all situations. Each technique has its own advantages and limitations, and a selection must be based on material, load history, service environment, component geometry, and consequences of component failure. In the following sections the relative strengths and limitations of each method are discussed and examples are given indicating where these methods are currently being applied.

6.3 STRESS–LIFE (S–N) APPROACH

6.3.1 Strengths of Method

1. The analysis and estimation of material constants necessary for this method are quite simple. This allows for quick "back of the envelope" calculations to get a reasonable estimate of life.
2. This method works well for designs involving long life, constant amplitude histories.
3. There are reams of data available for almost any variation in surface finish, load configuration, environment, and so on.

6.3.2 Weaknesses of Method

1. This method is completely empirical in nature and lacks the physical insights into the mechanisms of fatigue given by the other methods. Most of the empirical relationships come from tests of steels in the intermediate to long

life region. Care must be taken when extrapolating these relationships beyond the range of data used to determine them. When dealing with a new material, unique type of loading, and so on, a new set of tests must be run and empirical constants determined with curve fits.

2. The true stress–strain response of materials is ignored in favor of fictitious fully elastic strains. This has two principal implications:
   a. The plastic strains, which are critical at short lives, are ignored. This limits the S–N approach to long life applications.
   b. Since the true stress–strain relationship of the material at the root of notches is ignored, there is no way to model the mean residual stresses resulting from sequential loading effects. This means that the S–N approach has problems dealing with load histories that are not close to constant amplitude.
3. The S–N approach does not distinguish between initiation and propagation. Historically, this phenomenon was taken care of in one of the many empirical constants. This gives limited insights into the concept of damage.

6.3.3 Typical Applications

1. The S–N method can be used in almost any situation to get a rough estimate of life.
2. This method works very well in situations involving constant amplitude loading and long fatigue lives. The best examples of application of this method are in the design of various machine elements, such as power transmission shafts, valve springs, and gears.

The S–N approach may be used more intelligently when used in conjunction with the insights offered by the other methods.

6.4 STRAIN–LIFE (ϵ–N) APPROACH

6.4.1 Strengths of Method

Since this method takes into account the actual stress–strain response of the material, there are several implications.

1. Plastic strain, the mechanism that leads to crack initiation, is accurately modeled. This method can be used in high strain/low cycle situations.
2. This method can model the residual mean stresses resulting from the sequence effect in load histories. This allows for more accurate accounting of cumulative damage under variable amplitude loading.

3. The ϵ–N method can be more easily extrapolated to situations involving complicated geometries.
4. This method can be used in high temperature applications where fatigue–creep interaction is critical.
5. In situations where it is important, this method can incorporate transient material behavior.

6.4.2 Weaknesses of Method

1. This method involves a more complicated level of analysis. Some technique must be used to determine notch root strains, such as a Neuber analysis, a finite element analysis, or strain gage measurements. The life calculation involves numerical iterations which are best handled with computers.
2. The ϵ–N method only accounts for initiation life and cannot be used to predict propagation life.
3. There are several aspects of this method which are still very empirical in nature. Examples are mean stress effects and notch size effects (i.e., using K_f or K_t in the Neuber analysis).
4. The strain–life constants relate to the condition of the specimen tested. There is no defined way, other than additional testing, to account for differences in surface finish, plating, surface treatment, and so on.

6.4.3 Typical Applications

This method is used in:

1. Applications where plastic strains are significant. This may involve situations where load or stress levels are high, such as the root of a notch. It may also involve materials with very low yield points, such as low strength steels and some stainless steels.
2. High temperature applications, such as gas turbine engine components, where fatigue–creep interaction is important.
3. Applications involving variable amplitude load histories, where the load sequence effect on residual mean stress is important.
4. Smaller components, where initiation life is the primary concern.

The ϵ–N method can be used in conjunction with the LEFM approach to get a total initiation–propagation life. The ϵ–N method has no advantage over the S–N method when dealing with long life applications where the strains are primarily elastic and directly related to stress.

6.5 FRACTURE MECHANICS LEFM APPROACH

6.5.1 Strengths of Method

1. The LEFM approach is the only method that deals directly with the propagation of fatigue cracks. It also provides a method to characterize final failure due to fracture of the remaining cracked section.
2. Since crack length gives a physical measure of damage, crack growth rates can be incorporated with nondestructive inspection techniques to find the "safe life" of cracked components.
3. This method gives better insights into the actual mechanisms of fatigue that are important in research. It provides a method to deal with nonpropagating cracks and crack arrest behavior due to overloads.

6.5.2 Weaknesses of Method

1. This method has problems when used to deal with crack initiation. It is very difficult to estimate the initial crack size in situations where there is no obvious cracklike defect. In most cases the estimate of initial crack size has a major influence on the predicted fatigue life.
2. In certain situations the assumptions of linear elastic fracture mechanics are not valid and elastic–plastic fracture mechanics concepts must be used. This makes the analysis much more difficult. These situations include small crack growth in the plastic field near notches and crack growth at high loads.
3. This method requires an estimate of stress intensity factors which may be difficult to determine for complicated geometries.

6.5.3 Typical Applications

This method can be used:

1. To measure the crack growth from an assumed initial existing flaw. This analysis is used in conjunction with periodic in-service nondestructive testing (NDT) techniques to determine the safe life of damage-tolerant structures. In effect, the structure is designed to tolerate crack growth during a specified service period. This approach is used on large structures where propagation dominates fatigue life. Examples of applications are in the aerospace and nuclear reactor pressure vessel industries, where the consequences of failure are significant. This approach is tied to some type of maintenance procedure.
2. In situations where components have preexisting cracklike flaws. Examples are castings with large porosities and defects in welded components.
3. To determine the life of components with sharp notches, where only a small fraction of the life involves initiation.

The LEFM approach can also be used in conjunction with the ϵ–N approach to predict a total initiation–propagation life.

6.6 CONCLUSION

There is no general fatigue analysis method that is best for all design situations. Each technique has its own advantages and limitations and a selection must be based on material, load history, service environment, component and geometry, and consequences of component failure.

7

MULTIAXIAL FATIGUE

7.1 INTRODUCTION

In many applications, engineering components are subjected to complicated states of stress and strain. Components such as crank shafts, propeller shafts, and rear axles are often subjected to combined bending and torsion. Complex stress states—stress states in which the three principal stresses are nonproportional or whose directions change during a loading cycle—very often occur at geometric discontinuities such as notches or joint connections. Fatigue under these conditions, termed *multiaxial fatigue,* is an important design consideration for reliable operation and optimization of many engineering components.

In the past, a majority of fatigue research has been conducted under uniaxial loading conditions. To a lesser extent, work has been done in the area of torsional fatigue. Until the early 1970s, a limited amount of multiaxial fatigue research had been conducted. This was due, in part, to the difficulties involved in obtaining experimental multiaxial fatigue data. Recently, increased work has been conducted in this area, both experimentally and theoretically. This area continues to be a topic of concentrated research, with theories continuing to be developed and modified. Consequently, no general consensus has been reached on the "best" multiaxial fatigue theory. In light of this, the following chapter first presents a review of background material. Several multiaxial fatigue theories are then reviewed briefly, with emphasis placed on the concepts on which the theories are based.

7.2 BACKGROUND

As background for the following discussion on the development of multiaxial fatigue theories, an understanding of two areas is important. These are

1. The multiaxial stress/strain state at a point (aided by the use of Mohr's circle) and how this stress/strain changes with respect to time
2. The cracking behavior of a material under multiaxial fatigue loading

7.2.1 Stress State

Development of multiaxial fatigue analysis methods requires an understanding of the state of stress and strain in the component being analyzed. This in itself may sometimes be a complex task requiring use of strain gages to determine strains experimentally or use of finite element methods for analytical determination. However, a good understanding is necessary for identifying the parameters that correlate the stress/strain state to the damage developed in the material.

In multiaxial fatigue the directions of the principal stresses at a critical location often change during the loading cycle and are therefore a function of time. In addition, loading situations often arise where the three principal stresses are nonproportional. These difficulties complicate multiaxial fatigue analysis compared to the uniaxial loading condition. As an aid to understanding the stress/strain state as a function of time, a Mohr's circle representation of stress and strain at a point in time during the loading cycle is often used. The following example illustrates the foregoing concepts.

Figure 7.1 represents Mohr's circle for static tension and torsion loading.

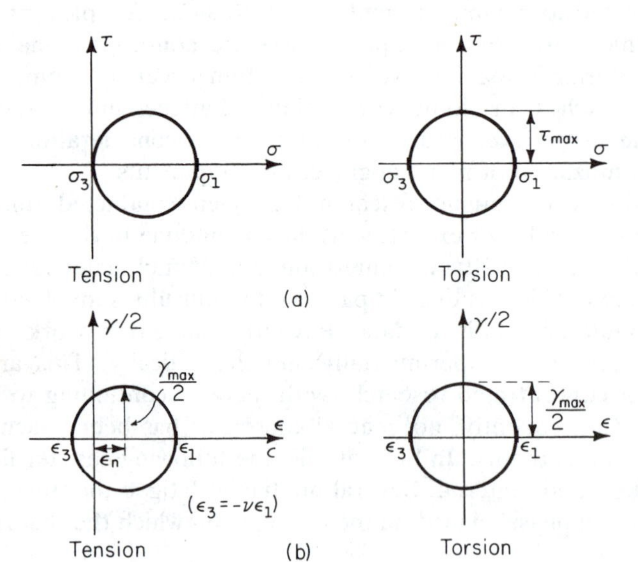

Figure 7.1 Mohr's circle representation of tension and torsion: (a) stress state; (b) strain state.

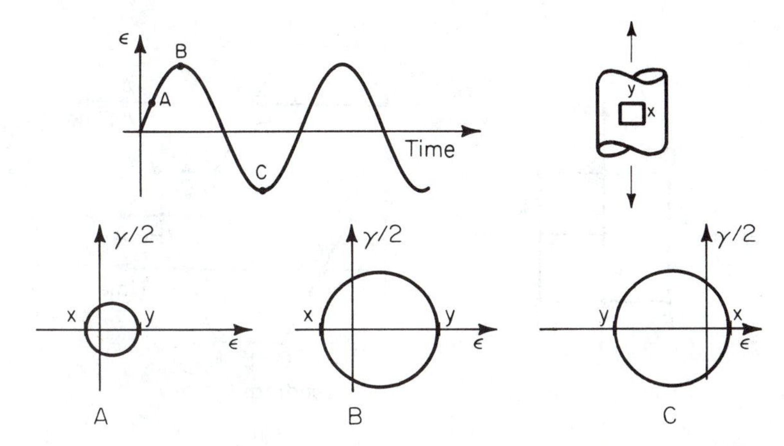

Figure 7.2 Uniaxial loading (constant strain amplitude): Mohr's circle representation of strain state at three points during loading cycle.

During cyclic loading, the size of Mohr's circle varies with time as shown for uniaxial loading in Fig. 7.2. In the uniaxial case, the direction of the principal stress and strain remains constant. (Point Y, corresponding to the Y plane as shown on the specimen, remains on the ϵ axis of Mohr's circle.) In contrast to the uniaxial case, Fig. 7.3 presents a nonproportional combined axial and torsional strain loading history. Figure 7.3a presents the variation of axial and torsional strain with respect to time as the loading completes one cycle. This is referred to as the *loading path.* Figure 7.3b presents Mohr's circle of strain for eight points in the loading cycle. Each circle represents the state of strain in the material at one point in time. [Although Mohr's circle and the loading path have similar axes (not identical), they represent different concepts.] In Fig. 7.3a, the ratio of shear strain, $\gamma/\sqrt{3}$, to normal strain, ϵ, changes during the cycle—resulting in nonproportional loading. In addition, the direction of the principal strains is no longer constant during the loading cycle. (Point Y no longer remains on the ϵ axis.) These features add to the complexity of a multiaxial fatigue analysis compared to the uniaxial case.

An important point that must be considered in the development and use of multiaxial fatigue theories is that extrapolation of concepts developed from uniaxial fatigue research must be used with caution. These may sometimes lead to erroneous results when used to evaluate fatigue under multiaxial loading conditions. An example of this is results of research conducted on combined loading—under both combined axial and torsional loading [1, 2] and combined bending and torsional loading [3–5]. Prior to this work, it has been assumed that for proportional loading, a phase difference between normal and shear stresses and strains could be neglected. It was believed that this was a conservative assumption. However, results of recent experimental tests have proven that combined out-of-phase loading, especially at high load levels, is more damaging

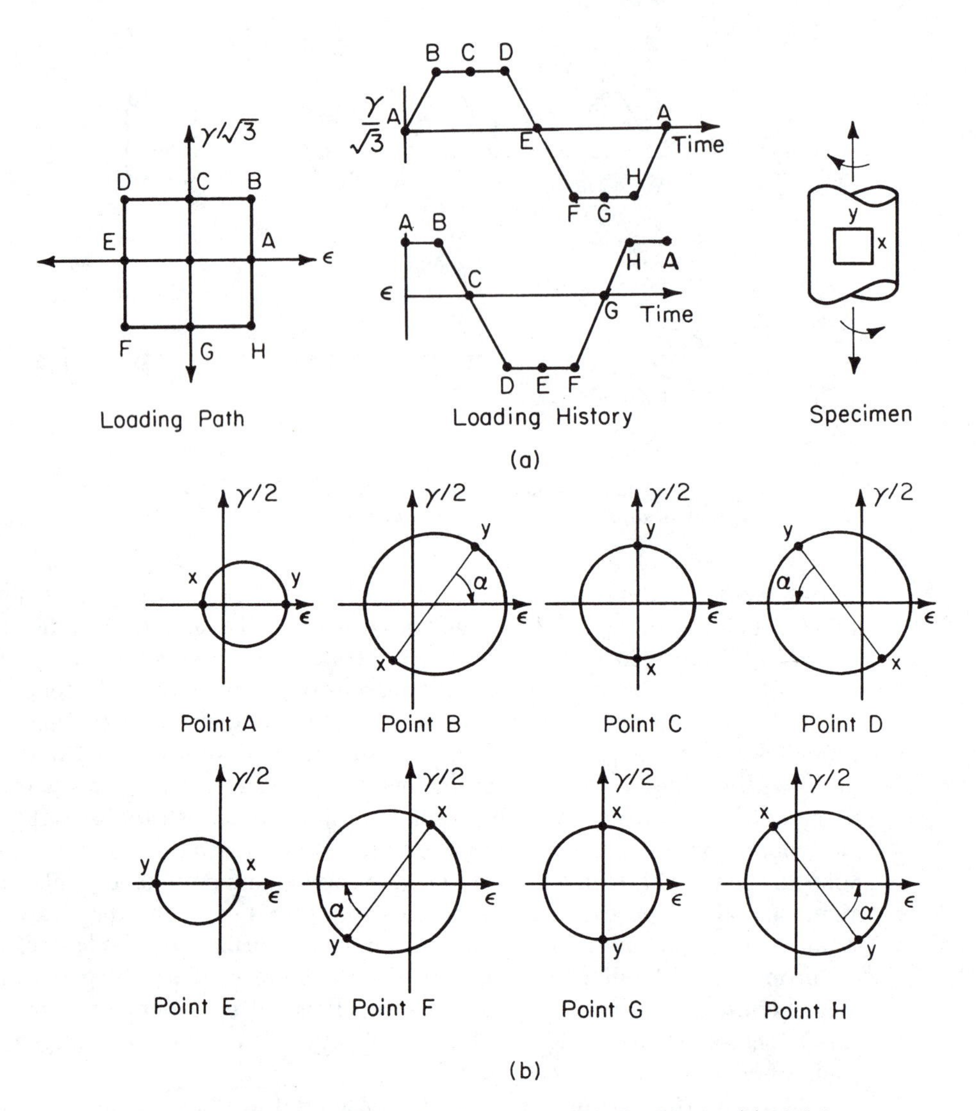

Figure 7.3 (a) Nonproportional loading path and loading history; (b) Mohr's circle representation of strain for eight points in loading cycle.

than in-phase loading [1, 2, 5] when tested at the same maximum shear strain range. Kanazawa et al. [1] found that phase angles of 90° gave the lowest fatigue lives, while in-phase cyclic strain resulted in the longest lives when tested at the same maximum shear strain range.

Note that in the past, when comparing proportional and nonproportional tests, a source of confusion has been what shear strain term is held constant. The following discussion attempts to clarify the difference between two cases.

When comparing nonproportional and proportional tests, either of the following two shear strain parameters may be constant:

1. The maximum applied shear strain
2. The maximum shear strain for the total cycle

In Fig. 7.4, a proportional loading path, as well as two 90° out-of-phase loading paths, are presented. In Fig. 7.4b, the 90° out-of-phase loading path has the same maximum *applied* shear strain as the proportional loading path shown in Fig. 7.4a. In this case, the maximum *applied* shear strain in the 90 out-of-phase test occurs at point C in the loading cycle. However, the 90 out-of-phase loading path shown in Fig. 7.4b has a smaller value of maximum shear strain for the total cycle than the proportional path. In other words, the maximum radius of Mohr's circle for the 90 out-of-phase loading shown in Fig. 7.4b is smaller than the maximum radius of Mohr's circle for the proportional path. To compare the proportional path and the 90 out-of-phase path on the basis of the same

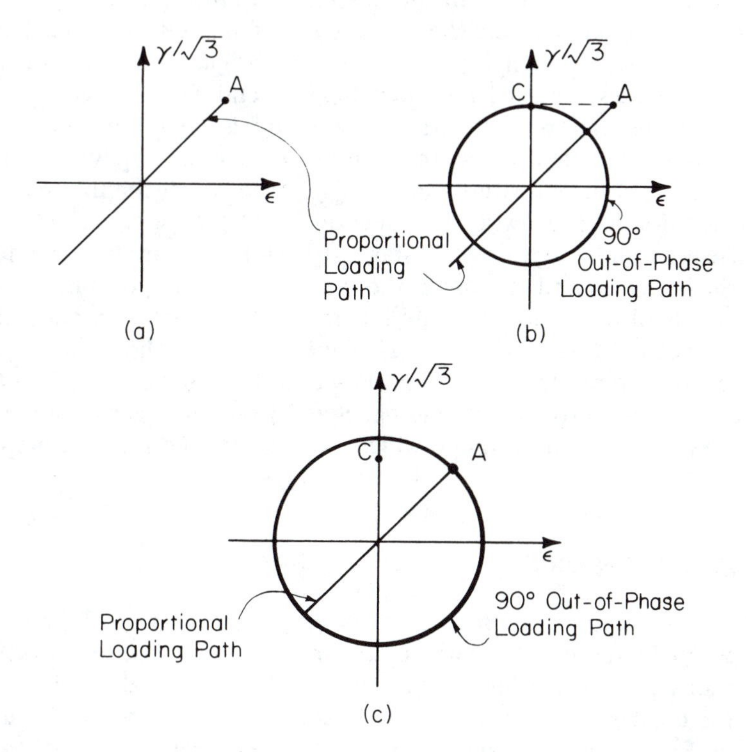

Figure 7.4 (a) Proportional loading path; (b) proportional loading path and 90° out-of-phase loading path with same maximum applied shear strain; (c) proportional loading path and 90° out-of-phase loading path with same maximum shear strain for total cycle.

maximum shear strain for the total cycle (maximum size of Mohr's circle for the two loading paths), the 90 out-of-phase loading path would be that shown in Fig. 7.4c. Notice that the maximum *applied* shear strain in the 90 out-of-phase loading path in Fig. 7.4c is now greater than the maximum *applied* shear strain in the proportional path.

7.2.2 Cracking Observations

As discussed in the introduction to this book a primary consideration in the development of fatigue analysis methods is the need to reconcile analytical approaches with physical observations. To this end, multiaxial theories must be consistent with physical observations. Consequently, in the following section we present a brief review of basic observations of fatigue cracking behavior.

Fatigue crack nucleation is a result of cyclic loading, which causes to-and-fro slip to occur. This slip results in the development of weakened bands of material, termed *slip bands,* from which cracks initiate. The formation of slip bands first occurs, for most materials, in those grains whose slip planes are most closely aligned with the plane of maximum shear strain. Slip occurs on these planes as a result of the macroscopic shear stress/strain applied to the component.

In 1961, Forsyth [6] observed that fatigue crack growth occurred in two distinct phases, which he termed Stage I and Stage II growth. Stage I growth includes nucleation and early growth and is spent on shear planes. Stage II growth is crack growth occurring on planes that are oriented perpendicular to the maximum principal stress range. (In fracture mechanics terms, this is often labeled Mode I growth; see Section 3.2.3.) It was later understood that in nucleation and early growth (Stage I) both the shear stresses and strains and the normal stresses and strains on the crack face are important. Normal stresses and strains tend to reduce the amount of Stage I growth. In general, the portion of life spent on Stage I or Stage II planes has been shown to depend on material type, loading mode, and strain amplitude [7]. However, it is interesting to note that even in out-of-phase loading, Stage I fatigue cracks initiate on the plane of maximum shear strain range which experiences the greatest amplitude of normal strain [1, 2].

7.3 MULTIAXIAL THEORIES

Early development of multiaxial fatigue theories were based on extensions of static yield theories to fatigue under combined stresses. In 1955, these were further extended by Sines, who developed a multiaxial theory that is very similar to the octahedral stress (von Mises) static yield theory but includes a hydrostatic term. From Sines' approach, several theories were developed. In the early 1970s, critical plane multiaxial fatigue theories were developed. These are theories based on the premise that failure occurs due to damage developed on a critical plane, and are based on cracking observations.

These approaches are reviewed below and the advantages and disadvantages of each are discussed.

7.3.1 Equivalent Stress/Strain Approaches

Early developments of multiaxial fatigue theories were based on extensions of static yield theories to fatigue under combined stresses. Three main static yield theories are listed below. They are given in terms of both stresses and strains. In regards to these, stress-based theories are usually limited to the high cycle fatigue (HCF) regime. In low cycle fatigue (LCF), the stresses and strains are no longer linearly related and the strain-based approaches are used in this regime. Examples of the application of these theories to fatigue problems are given.

NOTATION

$$\sigma_1 \geq \sigma_2 \geq \sigma_3 \qquad \epsilon_1 \geq \epsilon_2 \geq \epsilon_3$$

where σ_1, σ_2, σ_3 = principal stresses
ϵ_1, ϵ_2, ϵ_3 = principal strains

EFFECTIVE STRESS/STRAIN THEORIES

σ_e and ϵ_e = effective stress and strain values

(In a uniaxial test on a round bar, these are equal to the applied stress, σ_1, and strain, ϵ_1.)

Maximum principal stress

$$\sigma_1 = \sigma_e \tag{7.1}$$

Maximum principal strain

$$\epsilon_1 = \epsilon_e \tag{7.2}$$

Maximum shear stress (Tresca Criteria)

$$\left|\frac{\sigma_1 - \sigma_3}{2}\right| = \tau_e \tag{7.3}$$

Maximum shear strain (Tresca Criteria)

$$\left|\frac{\epsilon_1 - \epsilon_3}{2}\right| = \frac{\gamma_e}{2} = \frac{(1 + \nu)\epsilon_e}{2} \tag{7.4}$$

Distortion energy (octahedral or von Mises Criteria)

$$\frac{1}{\sqrt{2}}[(\sigma_1 - \sigma_2)^2 + (\sigma_2 - \sigma_3)^2 + (\sigma_3 - \sigma_1)^2]^{1/2} = \sigma_e \tag{7.5a}$$

$$\frac{1}{\sqrt{2}}[(\sigma_x - \sigma_y)^2 + (\sigma_y - \sigma_z)^2 + (\sigma_z - \sigma_x)^2 + 6(\tau_{xy}^2 + \tau_{xz}^2 + \tau_{yz}^2)]^{1/2} = \sigma_e \tag{7.5b}$$

$$\beta[(\epsilon_1 - \epsilon_2)^2 + (\epsilon_2 - \epsilon_3)^2 + (\epsilon_3 - \epsilon_1)^2]^{1/2} = \epsilon_e \tag{7.6a}$$

$$\beta\left\{(\epsilon_x - \epsilon_y)^2 + (\epsilon_y - \epsilon_z)^2 + (\epsilon_x - \epsilon_z)^2 + 6\left[\left(\frac{\gamma_{xy}}{2}\right)^2 + \left(\frac{\gamma_{xz}}{2}\right)^2 + \left(\frac{\gamma_{yz}}{2}\right)^2\right]\right\}^{1/2} = \epsilon_e \tag{7.6b}$$

$$\beta = \frac{1}{(1+\nu)\sqrt{2}} \qquad \text{for an arbitrary } \nu$$

For $\nu = \frac{1}{3}$: $\beta = \frac{3}{4\sqrt{2}}$

For $\nu = \frac{1}{2}$: $\beta = \frac{\sqrt{2}}{3}$

[*Note:* Poisson's ratio, ν, which is used in Hooke's law to relate stresses and strains, varies with load. For the fully elastic case, Poisson's ratio is generally considered to be equal to 0.3 ($\nu_e = 0.3$). For the fully plastic case, Poisson's ratio is usually assumed to be 0.5 ($\nu_p = 0.5$). For elastic–plastic cases, Poisson's ratio depends on the applied stress or strain. For a uniaxial case, an equivalent Poisson's ratio, ν^*, is sometimes determined. This is

$$\nu^* = \frac{\nu_e\epsilon_e + \nu_p\epsilon_p}{\epsilon_t} \tag{7.7}$$

where ν_e = elastic Poisson's ratio (usually taken to be $\frac{1}{3}$)
ν_p = plastic Poisson's ratio (usually taken to be $\frac{1}{2}$)
ϵ_e = elastic strain
ϵ_p = plastic strain
ϵ_t = total strain]

The following examples illustrate the use of equivalent stress/strain approaches. Using these theories, at a given effective stress/strain components are predicted to have the same fatigue life.

Example 7.1

A completely reversed uniaxial test with a stress amplitude of 20 ksi yields a fatigue life of 100,000 cycles,

$$\sigma_a = 20 \text{ ksi}$$

$$N_f = 100{,}000 \text{ cycles}$$

A torsion test conducted at the same equivalent stress is predicted to exhibit approximately the same fatigue life. Determine the value of the completely reversed shear stress that corresponds to a fatigue life of 100,000 cycles.

Solution The principal stresses σ_1 and σ_3, in a tension and torsion test (refer to Fig. 7.1 for convenience), are:

in tension,

$$\sigma_1 = 20 \text{ ksi}$$

and in torsion,

$$\sigma_1 = -\sigma_3 = \tau$$

Using the distortion energy criteria [also called the octahedral or von Mises criteria, Eq. (7.5a)], the equivalent stress, σ_e, for the tension case is simply equal to σ_1:

$$\sigma_e = \frac{1}{\sqrt{2}}[(\sigma_1 - \sigma_2)^2 + (\sigma_2 - \sigma_3)^2 + (\sigma_3 - \sigma_1)^2]^{1/2}$$

$$= \sigma_1$$

For the torsion case, the equivalent stress is equal to

$$\sigma_e = \frac{1}{\sqrt{2}}[(\sigma_1 - \sigma_2)^2 + (\sigma_2 - \sigma_3)^2 + (\sigma_3 - \sigma_1)^2]^{1/2}$$

$$= \frac{1}{\sqrt{2}}(\tau^2 + \tau^2 + 4\tau^2)^{1/2}$$

$$= \sqrt{3}\tau$$

Thus a torsion test with a completely reversed shear stress of

$$\tau = \frac{\sigma_e}{\sqrt{3}} = \frac{20}{\sqrt{3}}$$

$$= \underline{\underline{11.5 \text{ ksi}}}$$

is predicted to have a fatigue life of 100,000 cycles.

Example 7.2

Similarly, the same procedure may be used in terms of strain. For example, a completely reversed uniaxial strain controlled test results in a fatigue life of 10,000 cycles for a strain amplitude of 0.005,

$$\epsilon_a = 0.005$$

$$N_f = 10{,}000 \text{ cycles}$$

A torsion test run at the same equivalent strain would be predicted to have the same fatigue life. Determine the value of shear strain that would result in the same fatigue life.

Solution In tension

$$\epsilon_1 = 0.005$$

and in torsion

$$\epsilon_1 = -\epsilon_3 = \frac{\gamma}{2}$$

Assuming Poisson's ratio to be approximately $\frac{1}{2}$, from Eq. (7.7), in tension,

$$\epsilon_e = \epsilon_1$$

and in torsion, using Eq. (7.6a), we have

$$\epsilon_e = \frac{\sqrt{2}}{3}[(\epsilon_1 - \epsilon_2)^2 + (\epsilon_2 - \epsilon_3)^2 + (\epsilon_3 - \epsilon_1)^2]^{1/2}$$

$$= \frac{\sqrt{2}}{3}\left(\frac{\gamma^2}{4} + \frac{\gamma^2}{4} + \gamma^2\right)^{1/2}$$

$$= \frac{\gamma}{\sqrt{3}}$$

Thus at the same equivalent strain, in this case

$$\gamma = \sqrt{3}\,\epsilon$$

$$= \sqrt{3}\,(0.005)$$

$$= \underline{\underline{0.0087}}$$

The fatigue life obtained in the torsion test is predicted to be approximately equal to that obtained in the tension test, $N_f = 10{,}000$ cycles.

If equivalent stress/strain theories [Eqs. (7.1) through (7.6)] are assumed to be valid, they may be used to obtain the torsional strain-life curve in terms of uniaxial fatigue properties. Analogous to the derivation presented in Chapter 2, the expression for the strain-life curve for torsional loading may be developed. This is

$$\frac{\Delta\gamma}{2} = \frac{\tau_f'}{G}(2N_f)^b + \gamma_f'(2N_f)^c \tag{7.8}$$

In the following discussion, each of the three equivalent stress/strain theories will be used to express Eq. (7.8) in terms of uniaxial fatigue properties.

Using the von Mises theory [Eqs. (7.5) and (7.6)] and assuming that b and c remain nearly constant, this may be expressed as

$$\frac{\Delta\gamma}{2} = \frac{\sigma_f'}{\sqrt{3}\,G}(2N_f)^b + \sqrt{3}\,\epsilon_f'(2N_f)^c \tag{7.9}$$

Using the maximum principal stress/strain theory [Eqs. (7.1) and (7.2)] for stress:

$$\text{in tension:} \quad \sigma_e = \sigma_1$$

$$\text{in torsion:} \quad \sigma_e = \sigma_1 = \tau$$

and for strain:

$$\text{in tension:} \quad \epsilon_e = \epsilon_1$$

$$\text{in torsion:} \quad \epsilon_e = \epsilon_1 = \gamma/2$$

Thus Eq. (7.8) can be rewritten in terms of uniaxial properties using the maximum principal stress and strain criteria as

$$\frac{\Delta\gamma}{2} = \frac{\sigma_f'}{G}(2N_f)^b + 2\epsilon_f'(2N_f)^c$$

Using the maximum shear stress or strain (Tresca) criteria [Eqs. (7.3) and (7.4)] the stress state in tension is

$$\tau_e = \frac{\sigma_1}{2}$$

and in torsion

$$\tau_e = \tau$$

or

$$\tau_f' = \frac{\sigma_f'}{2}$$

For the strain state in tension

$$\frac{\gamma_e}{2} = \frac{\epsilon_1 - \epsilon_3}{2} = \frac{\epsilon_1 - (-\nu\epsilon_1)}{2} = \frac{(1+\nu)\epsilon_1}{2}$$

and in torsion

$$\frac{\gamma_e}{2} = \frac{\epsilon_1 - \epsilon_3}{2} = \frac{\gamma}{2}$$

Therefore,

$$\frac{(1+\nu)}{2}\epsilon_f' = \frac{\gamma_f'}{2}$$

$$\gamma_f' = (1+\nu)\epsilon_f'$$

Thus Eq. (7.8) can be rewritten in terms of uniaxial properties using the maximum shear (Tresca) criteria as

$$\frac{\Delta\gamma}{2} = \frac{\sigma_f'}{2G}(2N_f)^b + (1+\nu)\epsilon_f'(2N_f)^c$$

TABLE 7.1 Torsional Strain–Life Curve Coefficients Predicted by the Three Equivalent Stress/Strain Theories in Terms of Uniaxial Fatigue Constants

Torsional strain–life curve coefficients	τ'_f	γ'_f
Distortion energy Eqs. (7.5) and (7.6)	$\dfrac{\sigma'_f}{\sqrt{3}}$	$\sqrt{3}\,\epsilon'_f$
Maximum principal stress/strain, Eqs. (7.1) and (7.2)	σ'_f	$2\epsilon'_f$
Maximum shear stress/strain, Eqs. (7.3) and (7.4)	$\dfrac{\sigma'_f}{2}$	$1.5\epsilon'_f$

Substituting in the plastic Poisson's ratio value, $\nu = 0.5$, yields

$$\frac{\Delta\gamma}{2} = \frac{\sigma'_f}{2G}(2N_f)^b + (1.5)\epsilon'_f(2N_f)^c$$

In the development above, the three equivalent stress/strain theories used to obtain the torsional strain–life curve in terms of uniaxial fatigue properties predict different results for the torsional strain–life coefficients. This is summarized in Table 7.1.

Some general points regarding the effective stress/strain approaches are:

1. Of all the equivalent stress/strain approaches, the von Mises (distortion energy, octahedral) criteria [Eqs. (7.5) and (7.6)] have been shown to have the highest degree of acceptance among researchers [8], with both conservative and nonconservative results reported.
2. In the Tresca criteria, the value of σ_2 or ϵ_2 does not affect the equivalent stress or strain value, respectively, while the von Mises effective stress or strain varies with σ_2 or ϵ_2.
3. A discrepancy (often approximately a factor of 2 in strain) generally results when using these equivalent stress/strain criteria to correlate results of axial and torsional tests. For the same plastic strain, torsional loading results in longer lives.
4. Both the Tresca (maximum shear) [Eqs. (7.3) and (7.4)] and the von Mises (octahedral or distortion energy) [Eqs. (7.5) and (7.6)] criteria do not vary with the application of a hydrostatic stress or strain ($\sigma_1 = \sigma_2 = \sigma_3$ or $\epsilon_1 = \epsilon_2 = \epsilon_3$). However, fatigue failure has been shown to be sensitive to the application of a hydrostatic stress (pressure) or strain.
5. Effective stress/strain methods do not account for the fact that fatigue failure is observed to occur on specifically oriented planes relative to the principal stress/strain axes. Rather, these approaches "average" the stresses/strains to obtain a failure criterion with no regard to direction of crack growth.

6. Equivalent strain approaches do not account for mean stresses. An attempt to rationalize this is based on the argument that at high loads (LCF regime), where equivalent strain approaches are used, mean stresses tend to relax out (see Section 2.6). In the HCF regime, it has been suggested that mean stresses may be accounted for in the equivalent stress approach by constructing an effective Haigh (modified Goodman) diagram. In this method, the effective alternating stress, σ_a^e, is plotted against the effective mean, σ_m^e, where

$$\sigma_m^e = \frac{1}{\sqrt{2}}[(\sigma_{1m} - \sigma_{2m})^2 + (\sigma_{2m} - \sigma_{3m})^2 + (\sigma_{3m} - \sigma_{1m})^2]^{1/2} \tag{7.10}$$

$$\sigma_a^e = \frac{1}{\sqrt{2}}[(\sigma_{1a} - \sigma_{2a})^2 + (\sigma_{2a} - \sigma_{3a})^2 + (\sigma_{3a} - \sigma_{1a})^2]^{1/2} \tag{7.11}$$

(Refer to Section 1.3 for a more complete description and explanation of the use of the Haigh diagram.)

7. Significant complications arise in this approach when trying to deal with out-of-phase loading. Approaches such as those described in the following sections are preferred.

Despite some of the fundamental weaknesses, equivalent stress/strain approaches are often used. They are easy to implement and may be useful in obtaining a "first approximation." In addition, they relate multiaxial conditions to uniaxial cases for which large quantities of fatigue data are available.

7.3.2 Sines' Model and Similar Approaches

Sines' Model. In 1955, Sines [9] reviewed the results of experiments on the effect of different combinations of tensile, compressive, and torsional static (mean) and alternating stresses on fatigue life. He reported that the alternation of shear stresses seemed to cause fatigue failure. Because of this, the influence of mean static stresses on the planes of maximum shear alternation was studied.

Sines observed that a torsional mean stress did not affect the fatigue life of a specimen subjected to alternating torsion or bending stresses. He further reported that a tensile mean stress reduced the fatigue life of a component subjected to cyclic torsional loading. Finally, he observed the familiar results that a tensile mean stress decreases the fatigue life and a compressive mean increases the fatigue life of a specimen subjected to cyclic uniaxial (tension–compression) loading.

From this study he developed the relationship

$$\frac{1}{3}[(P_1 - P_2)^2 + (P_2 - P_3)^2 + (P_1 - P_3)^2]^{1/2} + \alpha(S_x + S_y + S_z) \leq A \tag{7.12}$$

where P_1, P_2, P_3 = amplitudes of the alternating principal stresses
S_x, S_y, S_z = orthogonal (any coordinate system) static (mean) stresses
A = material constant proportional to reversed fatigue strength
α = material constant which gives variation of the permissible range of stress with static stress

(A and α are material properties for a given life level.)

The first term on the left-hand side of Eq. (7.12) is the octahedral shear stress, τ_{oct}. Sines suggested that τ_{oct} averages the effect of shear stresses on many differently oriented slip planes. In addition, a hydrostatic stress term is included in this model by the second term on the left-hand side of the equation.

In Sines' equation, A and α may easily be determined. For example, in a completely reversed uniaxial test, Eq. (7.12) is

$$\frac{\sqrt{2}}{3}P_1 = A \qquad (P_2 = P_3 = S_x = S_y = S_z = 0)$$

Letting $P_1 = f_1$ gives us

$$A = \frac{\sqrt{2}}{3}f_1$$

where f_1 is the amplitude of reversed axial stress that would cause failure at desired cyclic life. For 0 to σ_{max} loading, Eq. (7.12) becomes

$$S'_x = P'_1 \qquad (P'_2 = P'_3 = S'_y = S'_z = 0)$$

$$\frac{\sqrt{2}}{3}P'_1 = A - \alpha P'_1$$

Letting $P'_1 = f'_1$ yields

$$\alpha = \frac{A}{P'_1} - \frac{\sqrt{2}}{3} = \frac{\sqrt{2}}{3}\frac{f_1}{f'_1} - \frac{\sqrt{2}}{3}$$

where f'_1 is the amplitude of fluctuating stress that would cause final failure at the same life as f_1. Thus A and α are described in terms of stress amplitudes, f_1 and f'_1.

Equation (7.12) may be represented graphically by concentric ellipses as shown in Fig. 7.5. The size of the ellipses depends on the sum of the static normal stresses. The larger (more positive) the sum, representing larger tensile static stresses, the smaller the ellipse. The permissible combinations of alternating stress amplitude may then be chosen for a given value of static stress. The ellipse corresponding to the sum of the stresses is selected. Thus any combination of cyclic stresses within the area of the ellipse is safe.

A disadvantage of this approach is that it is limited to applications in which the principal axes of the alternating stress components are fixed in the body. To

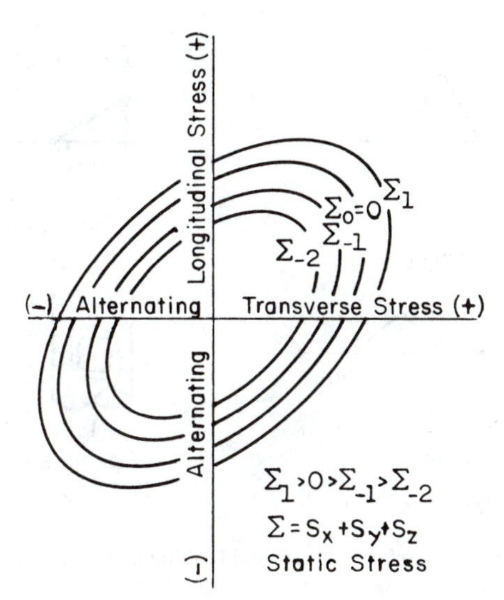

Figure 7.5 Sines' criteria plotted as concentric ellipses. (From Ref. 10.)

overcome this problem, Fuchs [11] suggested a modification to Sines' theory which is described in the following section.

Maximum range of shear stress criterion. Fuchs suggested a modification to Sines' theory which eliminates the restriction that the principal axes of the body must remain fixed. In this modification, termed the *maximum range of shear stress criterion,* the first term in Eq. (7.12) is modified, giving

$$\frac{1}{6}[(\Delta S_{11} - \Delta S_{22})^2 + (\Delta S_{22} - \Delta S_{33})^2 + (\Delta S_{33} - \Delta S_{11})^2 + 6(\Delta S_{12}^2 + \Delta S_{23}^2 + \Delta S_{31}^2)]^{1/2} + \alpha(S_x + S_y + S_z) = A \quad (7.13)$$

where the term in brackets is maximized with respect to time and where

$$\Delta S_{ij} = \sigma_{ij}(t_1) - \sigma_{ij}(t_2) = \text{stress differences}$$

$$\sigma_{ij}(t_1) = \text{stress component at time } t_1$$

$$\sigma_{ij}(t_2) = \text{stress component at time } t_2$$

As an example using Eq. (7.13), Fig. 7.6 presents the load history for the two components of stress (the remaining four components of stress are equal to zero). The first term in the previous expression is then calculated as follows:

$$\begin{aligned}\Delta S_{11} &= \sigma_{11}(t_1) - \sigma_{11}(t_2)\\ &= (5 - 0)\text{ ksi}\\ &= 5\text{ ksi}\end{aligned}$$

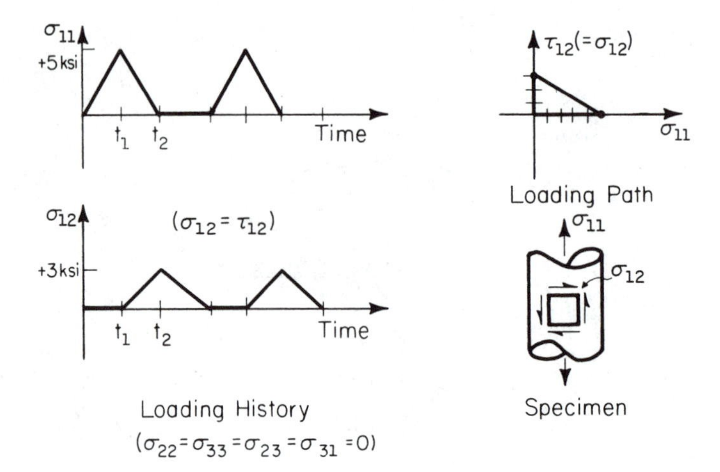

Figure 7.6 Nonproportional axial and torsional loading history and loading path.

Similarly,

$$\Delta S_{12} = \sigma_{12}(t_1) - \sigma_{12}(t_2)$$
$$= (0 - 3) \text{ ksi}$$
$$= -3 \text{ ksi}$$

Thus the first term in Eq. (7.13) is

$$\frac{1}{6}[(5-0)^2 + (0)^2 + (0-5)^2 + 6(-3)^2]^{1/2}$$

or 1.70 ksi.

Equivalent strain range criterion. In the LCF regime, an equation similar to Eq. (7.13) has been developed in terms of strain and is used in an ASME code procedure [12]. This criterion requires the calculation of an equivalent strain range, $\Delta\epsilon_{eq}$.

$$\Delta\epsilon_{eq} = \text{value of } \left\{ \frac{\sqrt{2}}{3}[(\Delta\epsilon_{11} - \Delta\epsilon_{22})^2 + (\Delta\epsilon_{22} - \Delta\epsilon_{33})^2 + (\Delta\epsilon_{33} - \Delta\epsilon_{11})^2 + 6(\Delta\epsilon_{12}^2 + \Delta\epsilon_{23}^2 + \Delta\epsilon_{31}^2)]^{1/2} \right\} \quad \text{maximized with respect to time} \tag{7.14}$$

where $\Delta\epsilon_{ij} = \epsilon_{ij}(t_1) - \epsilon_{ij}(t_2)$ are strain differences
$\epsilon_{ij}(t_1)$ = components of the strain tensor at time t_1
$\epsilon_{ij}(t_2)$ = components of the strain tensor at time t_2

For proportional loading this term is proportional to the octahedral shear strain. In general, though, it is not equal to the octahedral shear strain (distortional energy) theory.

A shortcoming of this theory is that it does not predict a dependence of the fatigue life on hydrostatic stress which contradicts experimental evidence. Another problem with this theory and the maximum range of shear stress criterion [Eqs. (7.13) and (7.14)] is that they are dependent only on the stresses/strains at two instants in time, t_1 and t_2. The theories state that fatigue life is path independent. In other words, it does not matter how these

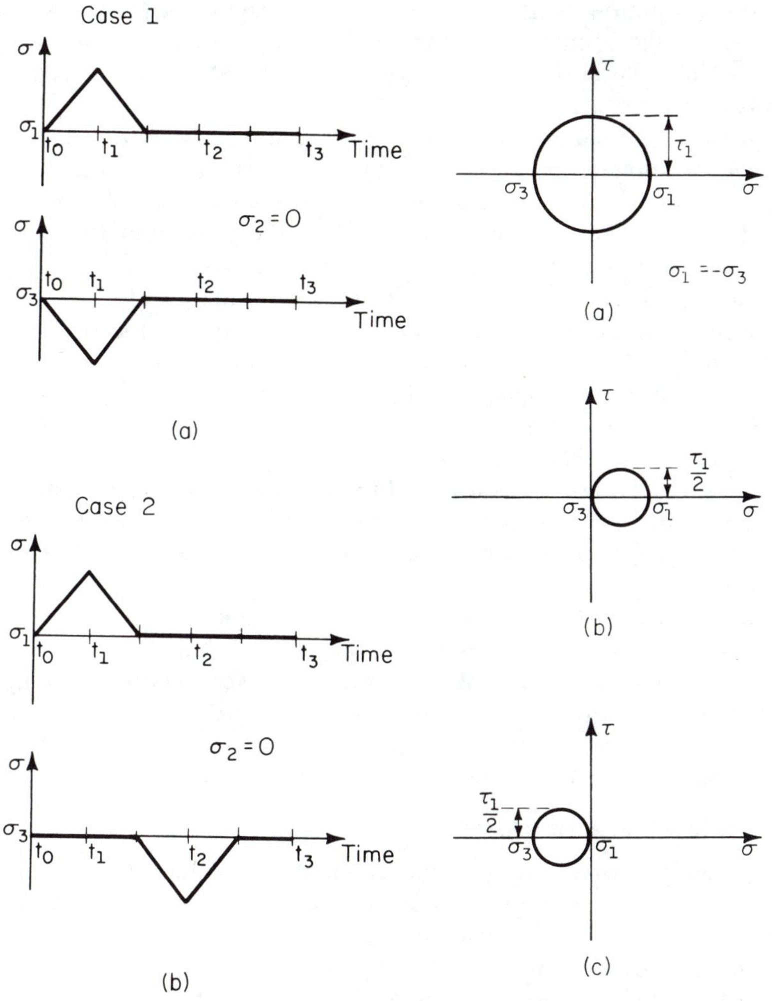

Figure 7.7 (a) Loading history for case 1; (b) loading history for case 2.

Figure 7.8 (a) Mohr's circle for case 1; (b) Mohr's circle for first half of case 2; (c) Mohr's circle for second half of case 2.

stresses/strains were reached from one another. This predicted path independence has been experimentally shown to be incorrect. The damage developed in the material has been shown to be path dependent and not just dependent on the maximum stress/strain difference at two specified points in the loading cycle. Additionally, the following case presents an example of a problem that may develop when using the type of methods presented in Eqs. (7.13) and (7.14).

Figure 7.7 presents stress–time plots for σ_1 and σ_3 for the two load cases. The first term of Eq. (7.13) is equal in both cases between times t_1 and t_2. In case 1 (Fig. 7.7a) the loads σ_1 and σ_3 are applied simultaneously, so that the Mohr's circle representation of stress at that time is as shown in Fig. 7.8a. (*Note:* This is equivalent to the Mohr's circle for a torsion test.) In case 2 the Mohr's circle is shown in Fig. 7.8b at time t_1, and in Fig. 7.8c the Mohr's circle representation of stress is shown for time t_2.

In case 1 the shear stress is τ_1 at time t_1 and between the time t_0 and t_3 only one cycle is applied. In case 2 during the time period between t_0 and t_3, the component experiences two loading cycles—one loading cycle of $+\sigma_1$ and another loading cycle of $-\sigma_3$. In this case, though, the maximum shear stress is $\tau_1/2$. Thus, although case 2 experiences two loading cycles for every time period (time between t_0 and t_3), case 1 results in a shorter fatigue life. This is due to the applied shear stress, which, in case 1, is twice as large as in case 2. Consequently, different fatigue lives result even though the maximum shear stress criterion [Eq. (7.13)] predicts equal fatigue lives for the two cases. Subtleties such as these are lost when using the methods given by Eqs. (7.13) and (7.14).

An additional difficulty with these general types of approaches [Eqs. (7.12), (7.13), and (7.14)] are that they tend to "average" the effect of shear stresses on many differently oriented slip planes. In contrast, as discussed in Section 7.2.2, it has been experimentally observed that fatigue cracks develop on planes of maximum shear.

A significant advantage of the approaches above is that they may easily be implemented. Garud [13] states that Eq. (7.12) may be "good enough" for proportional loading in the HCF regime, while for complex loading Eq. (7.13) may be used. In the LCF regime, Eq. (7.14) may be used. He qualifies this recommendation in regard to Eqs. (7.13) and (7.14) with the statement: "Experimental evidence suggests that some caution may have to be exercised."

7.3.3 Critical Plane Approaches

As discussed previously, fatigue cracks initiate on planes of maximum shear. The normal stress and strain on these planes have also been shown to have an influence on fatigue crack behavior and consequently fatigue life.

The importance of cracking behavior was recognized by early fatigue researchers such as Sines [10], Gough [14], and Findley [15]. In the development of Eq. (7.12), Sines recognized that "the alternation of shear stress seems to cause fatigue cracks." Because of this, he studied the effect of static stresses acting on the maximum shear planes. Findley also recognized the significance of the maximum shear plane and normal stress acting on it. He developed a theory based on the assumption that cracks would nucleate on the planes with a critical combination of shear and normal stresses.

In 1973, Brown and Miller [16] developed a multiaxial fatigue theory which they stated was based on a physical interpretation of the mechanisms of fatigue crack growth. This critical plane approach considered the maximum shear strain plane and the tensile or normal strain acting on it.

Equation (7.15) presents a form of their theory developed by Kandil et al. [17]. This is

$$\frac{\Delta\gamma}{2} + \Delta\epsilon_n = C \tag{7.15}$$

In this equation, the first term represents the shear strain amplitude on the maximum shear strain plane. The second term represents the tensile strain normal to this plane. C is a material constant. This theory predicts that equivalent fatigue lives result for equivalent values of the material constant, C.

An example of the physical significance of this equation is explained by considering the two cases shown in Fig. 7.9. In the first case (Fig. 7.9a) a shear strain is applied to the crack face. Due to the friction between the crack faces, the shear load experienced by the crack tip is reduced as some of the applied shear load is carried by the crack face material. In the second case, however (Fig. 7.9b), the strain normal to the plane of the crack causes the crack faces to separate. Consequently, friction between the crack faces is eliminated and the material at the crack tip now experiences all of the applied shear load. This is more damaging than the first case and results in shorter fatigue lives.

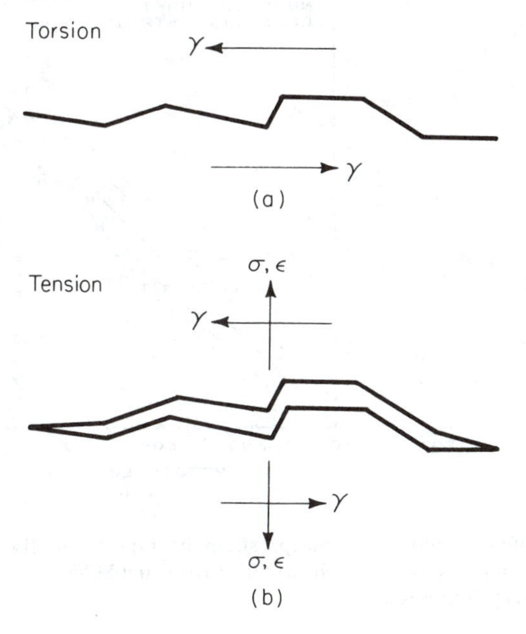

Figure 7.9 (a) Shear load applied at crack faces; (b) shear and tensile load applied at crack faces.

From Brown and Miller's initial work [16], modifications and additions have been made to develop new critical plane multiaxial fatigue theories [2, 17, 18] that have had varying success for different materials and loading conditions. Further research is required in this area before successful multiaxial fatigue life predictions can be made for a wide variety of loading modes and materials. The major advantage of critical plane approaches and the significant motivating factor in the development of these approaches is that they relate predicted fatigue life to experimentally observed cracking behavior.

7.4 SAE NOTCHED SHAFT PROGRAM

In an effort to evaluate the predictive capabilities of several multiaxial fatigue theories, a series of combined bending and torsion tests were conducted using the SAE notched shaft shown in Fig. 7.10 (see Ref. 19 for details). These were made

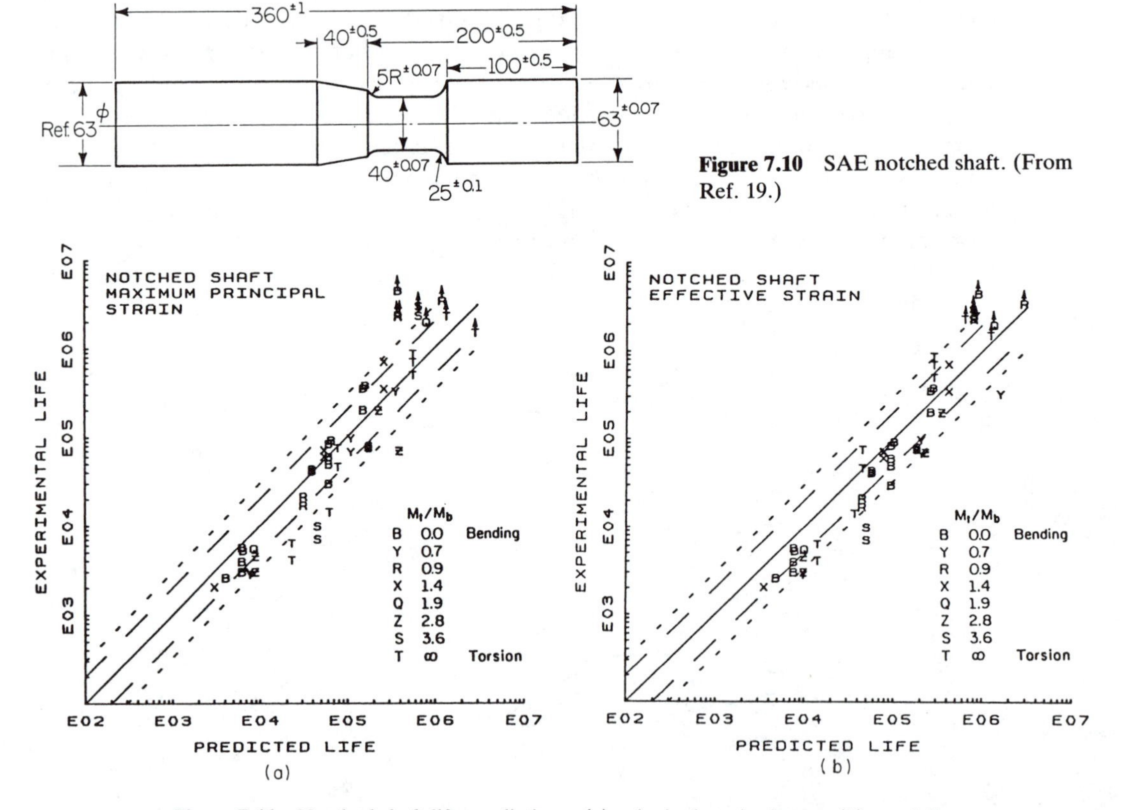

Figure 7.10 SAE notched shaft. (From Ref. 19.)

Figure 7.11 Notched shaft life predictions: (a) principal strain theory; (b) von Mises effective strain theory; (c) maximum shear strain theory; (d) Brown and Miller theories; (e) Lohr and Ellison theory [17]. (From Ref. 19.)

from SAE 1045 steel. The results of comparisons made between experimental data and results of predictions made using five multiaxial fatigue theories are presented in Fig. 7.11. As shown, all five theories do an adequate job of predicting the fatigue life for constant amplitude, completely reversed, in-phase, combined bending and torsion.

For more complicated loading situations such as out-of-phase, nonproportional, and variable amplitude loading, it is not apparent how these multiaxial theories will perform. Further research is required to answer this question.

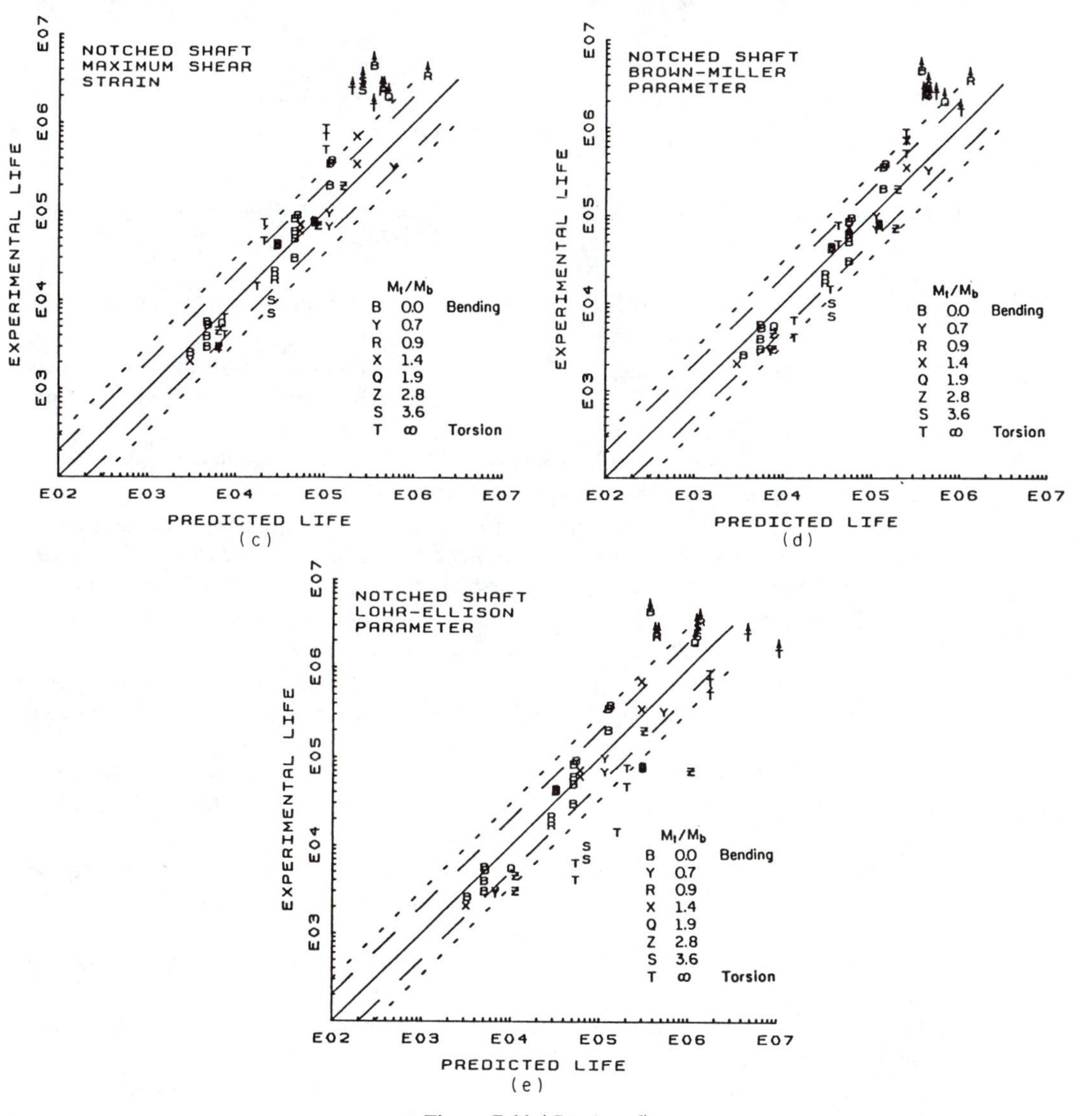

Figure 7.11 (*Continued*)

7.5 CONCLUSION

Multiaxial fatigue continues to be a topic of concentrated research. It has been shown that fatigue life estimates for in-phase loading are reasonably good, with no one theory being superior for a wide range of materials and loading situations. The degree of difficulty in analyzing components subjected to nonproportional, out-of-phase multiaxial states of stress and strain is significantly greater than that of a component subjected to in-phase loading. However, predictive multiaxial fatigue theories are needed for the safe, cost-efficient design of many engineering components.

REFERENCES

1. Kanazawa, M. W. Brown, and K. J. Miller, "Low Cycle Fatigue under Out-of-Phase Loading Conditions," *J. Eng. Mater. Tech., Trans. ASME,* Vol. H99, No. 3, 1977, pp. 222–228.
2. D. F. Socie, P. Kurath, and J. Koch, "A Multiaxial LCF Parameter," *2nd Int. Symp. Biaxial Multiaxial Fatigue,* 1985.
3. F. Gruibisic and A. Simbürger, "Fatigue under Combined Out-of-Phase Multiaxial Stresses," *Intl. Conf. Fatigue Testing and Design,* Society of Environmental Engineers, London, 1976, pp. 27.1–27.8.
4. R. E. Little, "A Note on the Shear Stress Criterion for Fatigue Failure under Combined Stress," *Aeronaut. Q.,* Vol. 20, Feb. 1969, pp. 57–60.
5. J. W. Fash, N. J. Hurd, C. T. Hua, and D. F. Socie, "Damage Development during Multiaxial Fatigue of Notched and Unnotched Specimens," in *Low Cycle Fatigue,* ASTM STP 942, American Society for Testing and Materials, Philadelphia, 1988.
6. P. J. E. Forsyth, "A Two Stage Process of Fatigue Crack Growth," *Proc. Symp. Crack Propagation,* Cranfield, England, 1961, pp. 76–94.
7. J. A. Bannantine and D. F. Socie, "Observations of Cracking Behavior in Tension and Torsion Low Cycle Fatigue," in *Low Cycle Fatigue,* ASTM STP 942, American Society for Testing and Materials, Philadelphia, 1988.
8. E. Krempl, "The Influence of State of Stress on Low-Cycle Fatigue of Structural Materials: A Literature Survey and Interpretive Report," in ASTM STP 549, American Society for Testing and Materials, Philadelphia, 1974.
9. G. Sines, "Failure of Materials under Combined Repeated Stresses Superimposed with Static Stresses," Tech. Note 3495, National Advisory Council for Aeronautics, Washington, D.C., 1955.
10. G. Sines, "Behavior of Metals under Complex Static and Alternating Stresses," in *Metal Fatigue,* G. Sines and J. L. Waisman (eds.), McGraw-Hill, New York, 1959.
11. H. O. Fuchs, "Fatigue Research with Discriminating Specimens," *Fatigue Eng. Mater. Struct.,* Vol. 2, No. 2, 1979, pp. 207–215.
12. American Society of Mechanical Engineers, "Cases of the ASME Boiler and Pressure Vessel Code," Case N-47-12 (1592-2), ASME, New York, 1977.

13. Y. S. Garud, "Multiaxial Fatigue: A Survey of the State of the Art," *J. Test. Eval.*, Vol. 9, No. 3, 1981, pp. 165–178.
14. H. J. Gough, "Crystalline Structure in Relation to Failure of Metals—Especially by Fatigue," Edgar Marburg Lecture, *Am. Soc. Test. Mater. Proc.*, Vol. 33, Part 2, 1933, pp. 3–114.
15. W. N. Findley, "A Theory for the Effect of Mean Stress on Fatigue of Metals under Combined Torsion and Axial Load or Bending," *Trans. ASME, J. Eng. Ind.*, Vol. B81, Nov. 1959, pp. 301–306.
16. M. W. Brown and K. J. Miller, "A Theory for Fatigue Failure under Multiaxial Stress–Strain Conditions," *Proc. Inst. Mech. Eng.*, Vol. 187, No. 65, 1973, pp. 745–755.
17. F. A. Kandil, M. W. Brown, and K. J. Miller, *Biaxial Low-Cycle Fatigue Fracture of 316 Stainless Steel of Elevated Temperatures,* Book 280, The Metals Society, London, 1982, pp. 203–210.
18. R. D. Lohr and E. G. Ellison, "A Simple Theory for Low Cycle Multiaxial Fatigue," *Fatigue Eng. Mater. Struct.*, Vol. 3, 1980, pp. 1–17.
19. J. W. Fash, "An Evaluation of Damage Development during Multiaxial Fatigue of Smooth and Notched Specimens," Materials Engineering Report No. 123, University of Illinois at Urbana–Champaign, 1985.
20. H. O. Fuchs and R. I. Stephens, *Metal Fatigue in Engineering,* Wiley-Interscience, New York, 1980.

PROBLEMS

SECTION 7.2

7.1. Given the loading history shown below, sketch the loading path in strain–strain space ($\gamma/\sqrt{3}$ on the vertical axis and ϵ on the horizontal axis). Draw a Mohr's circle

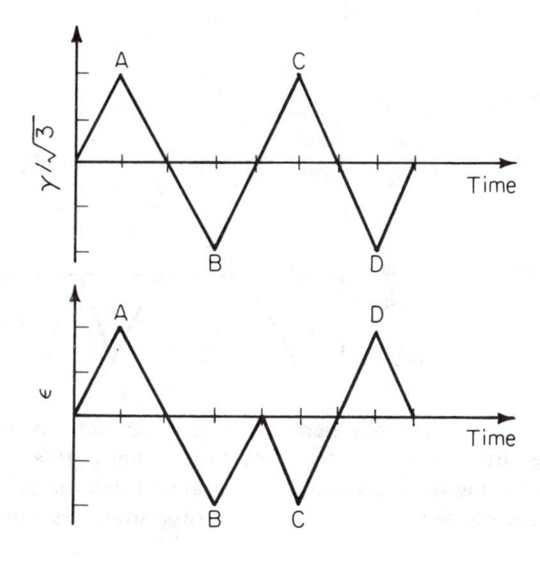

representation of strain for each point A, B, C, and D. (*Note:* This represents the strain state at four points in time.) Determine the maximum shear strain in the cycle. At which point or points does this occur? Is this proportional or nonproportional loading? For a 90° out-of-phase test (the torsional strain, $\gamma/\sqrt{3}$, and the axial strain, ϵ, vary in time by 90°), sketch the 90° out-of-phase loading path both on strain versus time plots and in strain–strain space. Compare to the loading path drawn below. (Assume plane strain conditions.)

7.2 A thin-walled tube, with ends capped, is subjected to a fluctuating internal pressure and an axial load. The loading histories are shown below. The applied axial load, P, is proportional to the pressure, p, such that

$$P = -3r^2p\pi$$

where r = radius of the tube. Assume that the cross-sectional area of the tube can be approximated as

$$A = 2tr\pi$$

where t = wall thickness. Sketch Mohr's circle for this loading at peak load. From inspection of Mohr's circle, to what state of stress does this loading correspond? If only torsional loading was available, sketch the load history (applied torque) as a function of time that would be required to obtain the same state of stress as the combined loading history shown below.

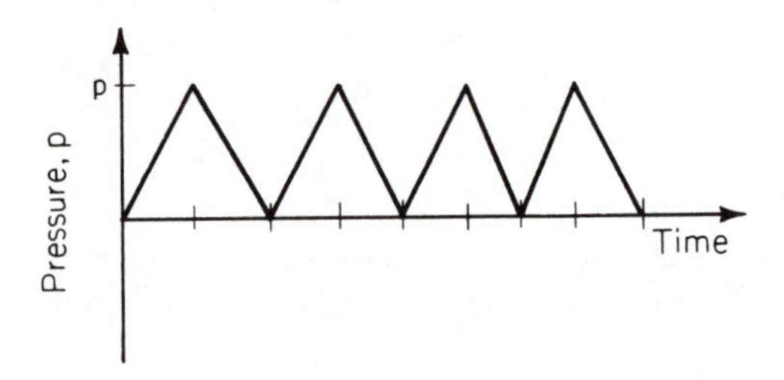

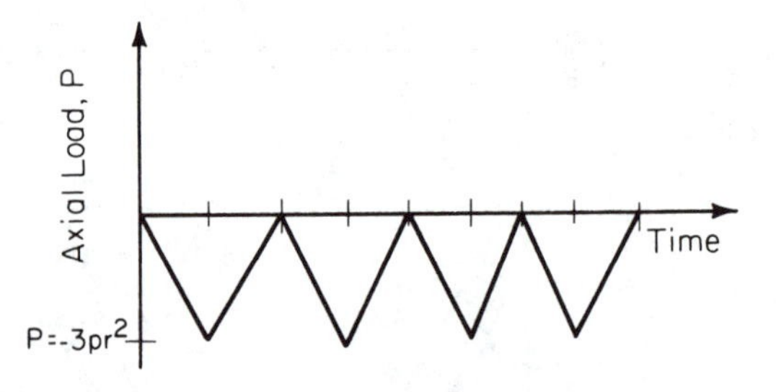

7.3. Below are differential elements of material with loading as shown. Sketch Mohr's circles for stress and strain for the loadings. If a material fails on the plane(s) experiencing the maximum shear strain, sketch the failure planes for each. (Assume plane stress conditions and elastic stress–strain response.)

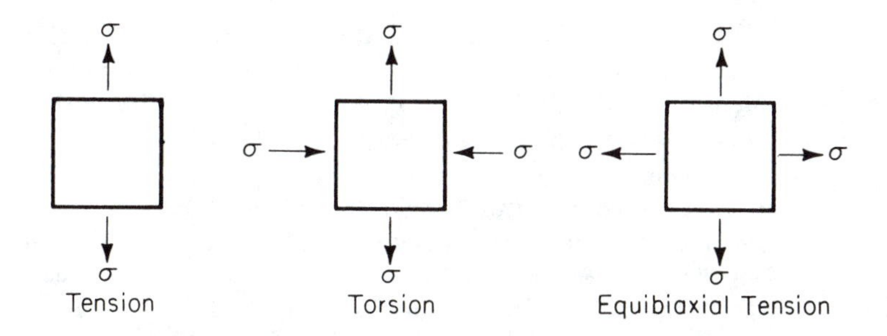

SECTION 7.3

7.4. Given the uniaxial material properties

$$\sigma_f' = 120 \text{ ksi} \qquad b = -0.1$$
$$E = 30{,}000 \text{ ksi} \qquad c = -0.5$$
$$\epsilon_f' = 1$$

determine the life to failure, in cycles, for a shaft subjected to a completely reversed torsional loading of $\Delta\gamma/2 = 0.42\%$ (0.0042). Use von Mises equivalent stress/strain theory.

7.5. A gear-shaft assembly made from 4340 steel is loaded as shown below. Using Sines' criteria [Eq. (7.12)], determine the maximum load, P, that may be applied to the gear for a fatigue life of 10,000 cycles. (*Hint:* Remember that S_x, S_y, and S_z are static normal stresses.) Use Fig. 1.8. If a static stress of 100 ksi in the radial direction was somehow added to the shaft near the bearing, determine the maximum load, P, for a fatigue life of 10,000 cycles.

Could Sines' criteria be used if a zero to maximum ($R = 0$) alternating axial force was applied to the end of the shaft such that the axial force reaches a maximum value every time the shaft completes a half revolution? Discuss your answer.

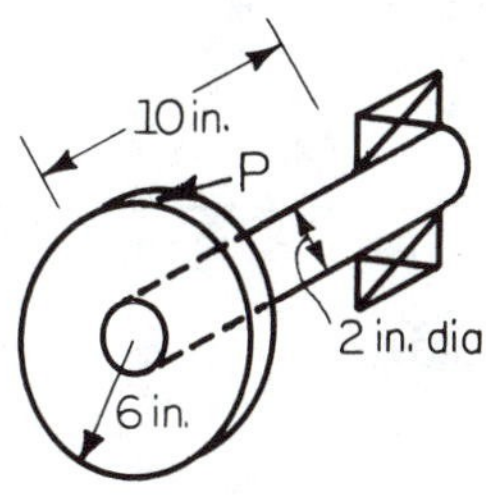

7.6. Another form of Sines' criteria presented in Ref. 20 is simply to reduce the alternating stresses to an equivalent alternating stress using an equation such as Eq. (7.5), and reduce the mean stress to an equivalent mean stress by summing the three orthogonal static normal stresses. These equivalent stress expressions are

$$\sigma_{ae} = \frac{1}{\sqrt{2}}[(P_1 - P_2)^2 + (P_2 - P_3)^2 + (P_1 - P_3)^2]^{1/2}$$

$$\sigma_{me} = S_x + S_y + S_z$$

These values are then used with a master diagram (Fig. 1.8) or a Haigh diagram (Fig. 1.7) combined with a mean stress equation such as the Goodman relationship [Eq. (1.9)]. Solve Problem 7.5 using this method with the master diagram and compare your solution to that obtained in Problem 7.5.

7.7. Another equivalent stress approach is described in Section 7.3.1. In this method, both the alternating stresses and the mean stresses are reduced to equivalent alternating and mean stresses [see Eqs. (7.10) and (7.11)]. A master diagram or a Haigh diagram used with a Goodman relationship is then used with these values similar to the method described in Problem 7.6. Solve Problem 7.5 using this method and compare your solution to that obtained in Problems 7.6. Are your answers similar? Would you always expect this to be the case? Discuss.

7.8. Using the information given in Problem 7.5, rework the gear problem with new static stresses. Let $S_x = S_y = S_z = 50$ ksi. Use the methods presented in Problems 7.6 and 7.7 and compare the solutions. Given the fact that hydrostatic pressure ($S_x = S_y = S_z$) affects the fatigue life, which theory would you choose for a situation such as this? Discuss your answer.

7.9. A capped thin-walled tube is subjected to internal pressure, axial loading, and torsional loading. The ratio of internal pressure to the stress due to the axial load is

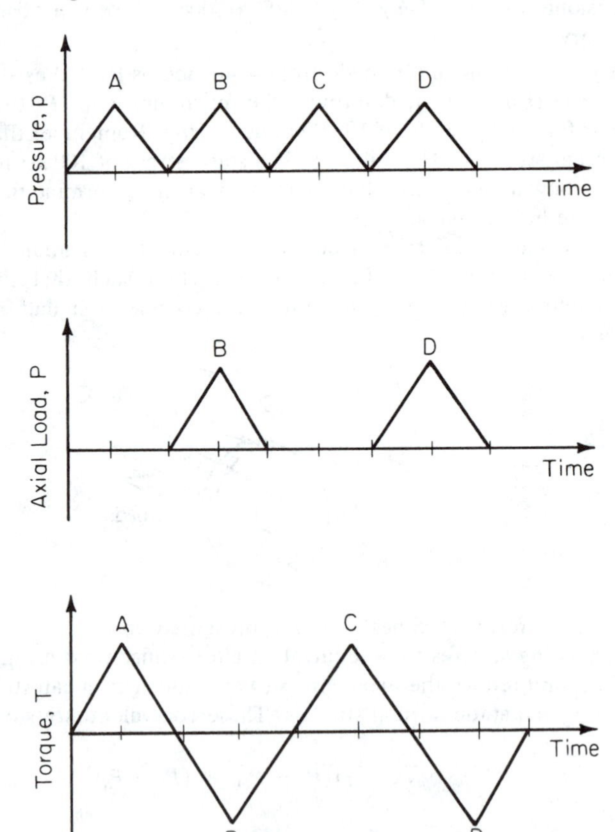

1:2, and the ratio of internal pressure to the shear stress due to the torque is 2:1. Given the loading histories shown below (load histories not to scale) and using Eq. (7.13) and the material properties in Problem 7.5 ($A = 35.35$ and $\alpha = 0.073$), determine the maximum pressure, axial stress, and shear stress for the part to last a minimum 10,000 cycles. (Let the cross-sectional area of the tube be approximated by $A = 2tr$. The ratio of the tube thickness to the radius is 0.1.)

INDEX

FIGURE CREDITS

The following are reprinted with permission:

Figures 1.13, 4.2, 4.4, 4.10, Table 1.3:
R. C. Juvinall, *Engineering Considerations of Stress, Strain and Strength* (New York: McGraw-Hill, 1967). Reprinted by permission.

Figure 1.8: U.S. Department of Defense, MIL-HBDK-5.

Figure 1.14: R. C. Johnson, *Machine Design Magazine* (Ohio: Penton Publishing, Inc., 1973), p. 108. Reprinted from Machine Design, May 3, 1973. Copyright 1973 by Penton Publishing Inc., Cleveland, Ohio. Reprinted with permission.

Figure 1.17, Table 1.4: C. C. Osgood, *Fatigue Design,* 2nd ed. (Oxford: Pergamon Press, 1982), pp. 126, 471, Table 4.8. Reprinted with permission from Pergamon Books Ltd.

Figures 1.18, 1.19, 1.22: J. O. Almen and P. H. Black, *Residual Stresses and Fatigue in Metals* (New York: McGraw-Hill, 1963), pp. 31, 141, 149.

Figures 1.20, 1.26, Table 1.6: P. G. Forrest, *Fatigue of Metals* (Oxford: Pergamon Press, 1962), pp. 184, 212–13, 222–23. Reprinted with permission from Pergamon Books Ltd.

Figure 1.21: L. P. Tarasov and H. J. Grover, "Effects of Grinding and Other Finishing Processes on the Fatigue Strength of Hardened Steel," *American Society for Testing and Materials,* Proceedings (American Society for Testing and Materials, 1950), Vol. 50, p. 668. Copyright ASTM. Reprinted with permission.

Figure 1.23: J. C. Straub, "Shot-Peening," *Metals Engineering Design,* 2nd ed., O. J. Horger, Ed. (New York: McGraw-Hill, 1965), p. 258. Reprinted with permission from McGraw-Hill, Inc.

Figure 1.24: O. J. Horger, "Mechanical and Metallurgical Advantages of Shot Peening", *The Iron Age* (Pennsylvania: Fairchild Publications, 1945). Reprinted with permission from Iron Age.

Figure 2.9: R. W. Landgraf, J. Morrow, and T. Endo, "Determination of the Cyclic Stress-Strain Curve," *J. of Materials* (American Society for Testing and Materials, 1969) 4, No. 1, p. 176. Copyright ASTM. Reprinted with permission.

Figure 2.10: J. Morrow, "Cyclic Plastic Strain Energy and Fatigue of Metals," *Internal Friction, Damping, and Cyclic Plasticity* (American Society for Testing and Materials, 1965) ASTM STP 37B, p. 45. Copyright ASTM. Reprinted with permission.

Figure 2.12: J. F. Martin, "Cyclic Stress-Strain Behavior and Fatigue Resistance of Two Structural Steels," FCP Report (University of Illinois at Urbana-Champaign, 1973), No. 9. Reprinted with permission from Dr. J. F. Martin.

Figure 2.13: D. Weinacht, "Fatigue Behavior of Gray Cast Iron Under Torsional Loads," Materials Engineering Report (University of Illinois at Urbana-Champaign, 1986) No. 126, p. 54. Reprinted with permission from D. Weinacht.

Figures 2.18, 2.19: R. W. Landgraf, "The Resistance of Metals to Cylic Deformation," *Achievement of High Fatigue Resistance in Metals and Alloys* (American Society for Testing and Materials, 1970) STP 467, pp. 3–36. Copyright ASTM. Reprinted with permission.

Figures 3.10, 3.14, E3.2: Stanley T. Rolfe and John M. Barsom, *Fracture and Fatigue Control in Structures: Applications of Fracture Mechanics* © 1977, pp. 225, 233, p. 41, Fig. 2.10. Adapted by permission of Prentice Hall, Englewood Cliffs, NJ.

Figure 3.15: A. Yuen and others, "Correlations between Fracture Surface Appearance and Fracture Mechanics Parameters for Stage II Fatigue Crack Propagation in Ti-6Al-4V," *Metallurgical Transactions,* © 1974, Vol. 5, p. 1834. Reprinted with permission from the Metallurgical Society of American Inst. of Mining, Metallurgical and Petroleum Eng'rs Inc.

Figure 3.16: T. W. Crooker and D. J. Krause, "The Influence of Stress Ratio and Stress Level on Fatigue Crack Growth Rates in 140 KSi 4S Steel" Report on NRL Progress, Naval Research

FIGURE CREDITS (cont'd)

Laboratory, Washington, DC, 1972, pp. 33–35. Reprinted with permission from the Naval Research Laboratory.

Figure 3.17: R. J. Bucci, Ph.d. Dissertation, Lehigh University, 1970. Reprinted with permission from Dr. R. J. Bucci.

Figure 4.21: N. E. Dowling, "Fatigue at Notches and the Local Strain and Fracture Mechanics Approaches," *Fracture Mechanics,* C. W. Smith (ed) (American Society for Testing and Materials, 1979) STP 677, pp. 247–273. Copyright ASTM. Reprinted with permission.

Figure 4.24: R. A. Smith and K. J. Miller, "Fatigue Cracks at Notches," *Int. J. of Mech. Sciences* (Oxford: Pergamon Press Inc., 1977) 19, pp. 11–122. Reprinted with permission from Pergamon Books Ltd.

Figure 5.3: J. H. Crews, Jr., "Crack Initiation at Stress Concentrations as Influenced by Prior Local Plasticity," *Achievement of High Fatigue Resistance in Metals and Alloys* (American Society for Testing and Materials, 1970), STP 467, p. 37. Copyright ASTM. Reprinted with permission.

Figure 5.13: E. F. J. von Euw, R. W. Hertzberg, and R. Roberts, (American Society for Testing and Materials, 1972), STP 513, p. 230. Copyright ASTM. Reprinted with permission.

Figures 5.16, 5.17, 5.18: D. V. Nelson, "Review of Fatigue Crack Growth Prediction Under Irregular Loading," Society for Experimental Stress Analysis, 1975. *Spring Meeting,* Experimental Mechanics, 1977, 17, pp. 41–49. Adapted with permission from Society for Experimental Mechanics, Inc.

Figures 5.23, 5.24, 5.25, 5.26, 5.27: R. M. Wetzel (ed), *Fatigue under Complex Loading: Analyses and Experiments* (Society of Automotive Engineers, 1977), pp. 4, 57. Reprinted with permission © 1977 from Society of Automotive Engineers, Inc.

Figures 5.28, 7.9: D. F. Socie, "Estimating Fatigue Crack Initiation and Propagation Lives in Notched Plates under Variable Loading Histories," T. & A.M. Report (University of Illinois at Urbana-Champaign, 1977), No. 417. Reprinted with permission from Dr. D. F. Socie.

Figure 7.5: G. Sines, "Behavior of Metals under Complex Static and Alternating Stresses," *Metal Fatigue,* G. Sines and J. L. Waisman, eds. (New York: McGraw-Hill, 1959), p. 161. Reprinted with permission from Professor G. Sines.

Figures 7.10, 7.11, Prob 4.20: J. W. Fash, "An Evaluation of Damage Development During Multiaxial Fatigue of Smooth and Notched Specimens," Materials Engineering Report (University of Illinois at Urbana-Champaign, 1985), No. 123, p. 108, Fig. 22. Reprinted with permission from Dr. J. Fash.

Table Problem 2.28a: T. Fugger, Jr., "Service Load Histories Analyzed by the Local Strain Approach," Materials Engineering Report (University of Illinois, 1985), No. 120, p. 20, table 3. Reprinted with permission from Mr. T. Fugger, Jr.

Table Problem 2.28c: R. Landgraf, "Cyclic Deformation and Fatigue Behavior of Hardened Steels, T. & A.M. Report, University of Illinois at Urbana-Champaign, 1968, No. 320, pp. 34, 40, Tables 2, 3. Reprinted with permission from Dr. R. Landgraf.

Table Problem 2.29: T. Endo and J. Morrow, "Cyclic Stress-Strain and Fatigue Behavior of Representative Aircraft Metals," *J. Mater.,* (American Society for Testing and Materials, 1969), Vol. 4, No. 1, p. 167, Table 3. Copyright ASTM. Reprinted with permission.

Table Problem 2.35: J. Koch, 'Proportional and Non-Proportional Biaxial Fatigue of Incone 718," Materials Engineering Report (University of Illinois at Urbana-Champaign, 1985), No. 121, pp. 232–4, table 2 & table 3. Reprinted with permission from Mr. J. Koch.

Figure Problems 3.8, 3.15: E. Caulfield, "Evaluation of Fracture Mechanics Parameters for A27 Cast Steel," FCP Report, (University of Illinois at Urbana-Champaign, 1977), No. 28, pp. 20, 22, Figs. 2, 4. Reprinted with permission from Dr. E. Caulfield.

FIGURE CREDITS (cont'd)

Table Problems 2.28b, 4.26, 4.27, 4.32, 4.38c: D. F. Socie, N. E. Dowling and P. Kurath, "Fatigue Life Estimation of Notched Members," *Fracture Mechanics,* R. J. Sanford, ed. (American Society for Testing and Materials, 1984), STP 833, pp. 284–299, Fig. 1, 2, 3, 4, 5, 6, 7. Copyright ASTM. Reprinted with permission.

Table Problem 4.29: J. Morrow and S. J. Stadnick, "Techniques for Smooth Specimen Simulation of the Fatigue Behavior of Notched Members," *Testing for Prediction of Material Performance in Structures and Components* (American Society for Testing and Materials, 1972), STP 515, pp. 229–252, Fig. 5, Tables 4, 5. Copyright ASTM. Reprinted with permission.

Table Problem 4.38: H. Sehitoglu, "Fatigue of Low Carbon Steels as Influenced by Repeated Strain Aging," FCP Report (University of Illinois at Urbana-Champaign, 1981), No. 40, Tables V, VI, VIII, IX. Reprinted with permission from Dr. H. Sehitoglu.

Table Problem 4.38: P. Kurath, "Investigation into a Non-arbitrary Fatigue Crack Size Concept," T.&A.M. Report (University of Illinois at Urbana-Champaign, 1978), No. 429, p. 15, 1978. Reprinted with permission from Dr. P. Kurath.

Program pp. 195–96: S. D. Downing and D. F. Socie, "Simplified Rainflow Counting Algorithms," International Journal of Fatigue (Guildford, United Kingdom: Butterworth Scientific Ltd., 1982), Vol. 4, No. 1, pp. 31–40. Reprinted with permission from International Journal of Fatigue.